T0134404

Springer Series in Optical Sciences

Volume 222

Founded by

H. K. V. Lotsch

Editor-in-Chief

William T. Rhodes, Georgia Institute of Technology, Atlanta, GA, USA

Series Editors

Ali Adibi, School of Electrical and Computer Engineering, Georgia Institute of Technology, Atlanta, GA, USA
Toshimitsu Asakura, Toyohira-ku, Hokkai-Gakuen University, Sapporo, Hokkaido, Japan
Theodor W. Hänsch, Max Planck Institute of Quantum, Garching, Bayern, Germany
Ferenc Krausz, Garching, Bayern, Germany
Barry R. Masters, Cambridge, MA, USA
Herbert Venghaus, Fraunhofer Institute for Telecommunications, Berlin, Germany
Horst Weber, Berlin, Berlin, Germany
Harald Weinfurter, München, Germany
Katsumi Midorikawa, Laser Tech Lab, RIKEN Advanced Sci Inst, Saitama, Japan

Springer Series in Optical Sciences is led by Editor-in-Chief William T. Rhodes, Georgia Institute of Technology, USA, and provides an expanding selection of research monographs in all major areas of optics:

– lasers and quantum optics
– ultrafast phenomena
– optical spectroscopy techniques
– optoelectronics
– information optics
– applied laser technology
– industrial applications and
– other topics of contemporary interest.

With this broad coverage of topics the series is useful to research scientists and engineers who need up-to-date reference books.

More information about this series at http://www.springer.com/series/624

Paulo Ribeiro · David L. Andrews ·
Maria Raposo

Editors

Optics, Photonics and Laser Technology 2017

Springer

Editors
Paulo Ribeiro
Faculdade de Ciências e Tecnologia
Universidade Nova de Lisboa
Caparica, Portugal

David L. Andrews
School of Chemistry
University of East Anglia
Norwich, UK

Maria Raposo
Faculdade de Ciências e Tecnologia
Universidade Nova de Lisboa
Caparica, Portugal

ISSN 0342-4111   ISSN 1556-1534   (electronic)
Springer Series in Optical Sciences
ISBN 978-3-030-12694-0   ISBN 978-3-030-12692-6   (eBook)
https://doi.org/10.1007/978-3-030-12692-6

Library of Congress Control Number: 2019930725

© Springer Nature Switzerland AG 2019
This work is subject to copyright. All rights are reserved by the Publisher, whether the whole or part of the material is concerned, specifically the rights of translation, reprinting, reuse of illustrations, recitation, broadcasting, reproduction on microfilms or in any other physical way, and transmission or information storage and retrieval, electronic adaptation, computer software, or by similar or dissimilar methodology now known or hereafter developed.
The use of general descriptive names, registered names, trademarks, service marks, etc. in this publication does not imply, even in the absence of a specific statement, that such names are exempt from the relevant protective laws and regulations and therefore free for general use.
The publisher, the authors and the editors are safe to assume that the advice and information in this book are believed to be true and accurate at the date of publication. Neither the publisher nor the authors or the editors give a warranty, expressed or implied, with respect to the material contained herein or for any errors or omissions that may have been made. The publisher remains neutral with regard to jurisdictional claims in published maps and institutional affiliations.

This Springer imprint is published by the registered company Springer Nature Switzerland AG
The registered company address is: Gewerbestrasse 11, 6330 Cham, Switzerland

Organization

Conference Co-chairs

David L. Andrews University of East Anglia, UK
Paulo A. Ribeiro CEFITEC/FCT/UNL, Portugal

Programme Chair

Maria Raposo CEFITEC, FCT/UNL, Portugal

Programme Committee

Jean–Luc Adam   Université de Rennes 1—CNRS, France
Marcio Alencar   Universidade Federal de Sergipe, Brazil
Augusto A. Iribarren Alfonso   Instituto de Ciencia y Tecnología de Materiales,
                                  Universidad de La Habana, Cuba
Tatiana Alieva   Universidad Complutense de Madrid, Spain
David Andrews   University of East Anglia, UK
Maria Flores Arias   Universidade de Santiago de Compostela, Spain
Andrea Armani   University of Southern California, USA
Jhon Fredy Martinez Avila   Universidade Federal de Sergipe, Brazil
Benfeng Bai   Tsinghua University, China
Francesco Baldini   Istituto di Fisica Applicata "Nello Carrara", Italy
Almut Beige   University of Leeds, UK
Antonella Bogoni   Scuola Superiore Sant'Anna-TeCIP, Italy
Svetlana V. Boriskina   Massachusetts Institute of Technology, USA

Remy Braive	Laboratory for Photonics and Nanostructures/CNRS, France
Giovanni Breglio	Università degli Studi di Napoli Federico II, Italy
Neil Broderick	University of Auckland, New Zealand
Tom Brown	University of St Andrews, UK
Alexandre Cabral	Laboratory of Optics, Lasers and Systems (LOLS), Faculdade de Ciências da Universidade de Lisboa, Portugal
Giampaolo Campana	Università di Bologna, Italy
John Canning	University of Technology, Sydney, Australia
Maurizio Casalino	IMM, Italian National Research Council (CNR), Italy
Eric Cassan	Centre de Nanosciences et de Nanotechnologie (UMR 9001), Université Paris-Sud, Orsay, France
Christophe Caucheteur	University of Mons, Belgium
Calvin C. K. Chan	The Chinese University of Hong Kong, Hong Kong
Zhao Changming	Beijing Institute of Technology, China
Sima Chaotan	Huazhong University of Science and Technology, China
Xianfeng Chen	Shanghai Jiao Tong University, China
Dominique Chiaroni	Alcatel-Lucent Bell Labs, France
João Coelho	Faculdade de Ciências da Universidade de Lisboa, Portugal
Lorenzo Colace	University of Rome "Roma Tre", Italy
Elisabetta Collini	University of Padova, Italy
Florinda Costa	Departamento de Física Universidade de Aveiro, Portugal
Manuel Filipe Costa	Universidade do Minho, Portugal
Judith Dawes	Macquarie University, Australia
Jean-Marc Delavaux	Cybel, LLC, USA
Arif Demir	Kocaeli University, Turkey
Maria Dinescu	National Institute for Lasers, Plasma and Radiation Physics, Romania
Ivan Divliansky	University of Central Florida—College of Optics and Photonics—CREOL, USA
John F. Donegan	Trinity College, Ireland
Jonathan Doylend	Intel Corporation, USA
Abdulhakem Youssef Elezzabi	University of Alberta, Canada
Andrea Di Falco	St Andrews University, UK
Sergio Fantini	Tufts University, USA
Robert Ferguson	National Physical Laboratory, UK

Maurizio Ferrari	National Research Council of Italy (CNR), Institute for Photonics and Nanotechnologies (IFN), Italy
Quirina Ferreira	Instituto de Telecomunicações/IST, Portugal
Henryk Fiedorowicz	Institute of Optoelectronics, Military University of Technology, Poland
José Figueiredo	Universidade de Lisboa, Portugal
Luca Fiorani	Researcher at ENEA—Professor at "Lumsa", "Roma Tre" and "Tor Vergata" Universities, Italy
Orlando Frazão	INESC Porto, Portugal
Ivana Gasulla	Polytechnic University of Valencia, Spain
Sanka Gateva	Institute of Electronics, Bulgarian Academy of Sciences, Bulgaria
Marco Gianinetto	Politecnico di Milano, Italy
John Girkin	Department of Physics, Biophysical Sciences Institute, Durham University, UK
Guillaume Gomard	Light Technology Institute, Germany
Yandong Gong	Institute for Infocomm Research, Singapore
Norbert Grote	Fraunhofer Heinrich Hertz Institute, Germany
Mircea D. Guina	Tampere University of Technology, Finland
David J. Hagan	University of Central Florida, USA
Rosa Ana Perez Herrera	Universidad Publica de Navarra, Spain
Daniel Hill	Universitat de Valencia, Spain
Charles Hirlimann	Institut de Physique et Chimie des Matériaux de Strasbourg, France
Werner Hofmann	Center of Nanophotonics at Technische Universität Berlin, Germany
Weisheng Hu	Shanghai Jiao Tong University, China
Silvia Soria Huguet	IFAC- N. Carrara Institute of Applied Physics/C.N.R., Italy
Nicolae Hurduc	Gheorghe Asachi Technical University of Iasi, Romania
Baldemar Ibarra-Escamilla	Instituto Nacional de Astrofísica, Optica y Electrónica, Mexico
M. Nazrul Islam	Farmingdale State University of New York, USA
Leszek R. Jaroszewicz	Military University of Technology, Poland
Phil Jones	Optical Tweezers Group, Department of Physics and Astronomy, University College London, UK
José Joatan Rodrigues Jr.	Universidade Federal de Sergipe, Brazil
Yoshiaki Kanamori	Tohoku University, Japan
Fumihiko Kannari	Keio University, Japan
Christian Karnutsch	Karlsruhe University of Applied Sciences, Faculty of Electrical Engineering and Information Technology, Germany

Henryk Kasprzak	Wroclaw University of Science and Technology, Poland
Gerd Keiser	Photonics Comm. Solutions Inc., USA
Qian Kemao	School of Computer Engineering, Singapore
Chulhong Kim	POSTECH, Korea, Republic of
Young-Jin Kim	Nanyang Technological University, Singapore
Stefan Kirstein	Humboldt-Universitaet zu Berlin, Germany
Stanislaw Klosowicz	Military University of Technology, Faculty of Advanced Technologies and Chemistry Warsaw, Poland
Tomasz Kozacki	Warsaw University of Technology (WUT), Institute of Micromechanics and Photonics (IMiF), Institute of Microelectronics and Optoelectronics (IMiO), Poland
Leonid Krivitsky	Data Storage Institute, Singapore
Jesper Lægsgaard	Technical University of Denmark, Denmark
Vasudevan Lakshminarayanan	University of Waterloo, Canada
Emma Lazzeri	Scuola Superiore Sant'Anna-TeCIP, Italy
Wei Lee	National Chiao Tung University, Taiwan
Henri J. Lezec	National Institute of Standards and Technology, United States, USA
Dawei Liang	FCT/UNL, Portugal
João Carlos Lima	REQUIMTE-LAQV, Departamento de Química, Faculdade de Ciências e Tecnologia, Universidade NOVA de Lisboa, Portugal
Milton S. F. Lima	Instituto de Estudos Avancados, Brazil
Paulo Limão-Vieira	CEFITEC/FCT/UNL, Portugal
Yu-Lung Lo	National Cheng Kung University, Taiwan
Boris A. Malomed	Tel Aviv University, Israel
Edwin A. Marengo	Northeastern University, USA
Manuel Marques	FCUP, Portugal
Francisco Medina	University of Seville, Spain
Robert Minasian	The University of Sydney, Australia
Silvia Mittler	University of Western Ontario, Canada
Paulo Monteiro	Instituto de Telecomunicações, Portugal
Phyllis R. Nelson	California State Polytechnic University Pomona, USA
André Nicolet	Institut Fresnel, AIX Marseillle Université, CNRS, France
Luca Palmieri	University of Padua, Italy
Krassimir Panajotov	Free University of Brussels, Belgium
Jisha Chandroth Pannian	Centro de Física da Universidade do Porto, Portugal
Matteo G. A. Paris	Università degli Studi di Milano, Italy

Seoung-Hwan Park	Catholic University of Daegu, Korea, Republic of
YongKeun Park	Department of Physics, KAIST, Korea, Republic of
Lorenzo Pavesi	University of Trento, Italy
Anna Peacock	University of Southampton, UK
Luís Pereira	Faculdade de Ciências e Tecnologia, Universidade Nova de Lisboa, Portugal
Romain Peretti	Swiss Federal Institute of Technology in Zurich (ETHZ), Switzerland
Francisco Pérez-Ocón	Universidad de Granada, Spain
Klaus Petermann	Technische Universität Berlin, Germany
Wiktor Piecek	Military University of Technology, Poland
Angela Piegari	ENEA, Italy
Alexander Popov	Institute of Nuclear Physics, Moscow State University, Russian Federation
Sandro Rao	University "Mediterranea" of Reggio Calabria, Italy
Maria Raposo	CEFITEC, FCT/UNL, Portugal
Ioannis Raptis	Nat. Center for Scientific Research Demokritos, Greece
João Rebola	Instituto de Telecomunicações, ISCTE-IUL, Portugal
Paulo A. Ribeiro	CEFITEC/FCT/UNL, Portugal
Murilo Romero	University of Sao Paulo, Brazil
Luis Roso	Centro de Laseres Pulsados, CLPU, Spain
Luigi L. Rovati	Università di Modena e Reggio Emilia, Italy
Svilen Sabchevski	Institute of Electronics, Bulgarian Academy of Sciences, Bulgaria
Manuel Pereira dos Santos	Centro de Física e Investigação Tecnológica, Faculdade de Ciências e Tecnologia, Universidade Nova de Lisboa, Portugal
Susana Sério	CEFITEC, Faculdade de Ciências e Tecnologia, Portugal
Gaurav Sharma	University of Rochester, USA
Josmary Silva	Universidade Federal do Mato Grosso-Campus Universitário do Araguaia, Brazil
Ronaldo Silva	Federal University of Sergipe, Brazil
Susana Silva	INESC TEC, Portugal
Stefan Sinzinger	Technische Universität Ilmenau, Germany
Javier Solis	Laser Processing Group—LPG, Spain
Viorica Stancalie	National Institute for Laser, Plasma and Radiation Physics, Romania
Dimitar Stoyanov	Institute of Electronics, Bulgarian Academy of Sciences, Bulgaria

Kate Sugden	Aston University, UK
Slawomir Sujecki	University of Nottingham, UK
Takenobu Suzuki	Toyota Technological Institute, Japan
João Manuel R. S. Tavares	FEUP—Faculdade de Engenharia da Universidade do Porto, Portugal
R. K. Thareja	Indian Institute of Technology (IIT) Kanpur, India
José Roberto Tozoni	Universidade Federal de Uberlandia, Brazil
Cosimo Trono	Istituto di Fisica Applicata "Nello Carrara", CNR, Italy
Valery V. Tuchin	Saratov State University, Russian Federation
Moshe Tur	Tel Aviv University, Israel
M. Selim Ünlü	Boston University, USA
Ignacio del Villar	Universidad Pública de Navarra, Spain
Zinan Wang	University of Electronic Science and Technology, China
Lech Wosinski	KTH—Royal Institute of Technology, Sweden
Kaikai Xu	University of Electronic Science and Technology of China (UESTC), China
Lianxiang Yang	Oakland University, Rochester, USA
Kiyotoshi Yasumoto	Kyushu University, Japan
Anna Zawadzka	Institute of Physics, Faculty of Physics, Astronomy and Informatics Nicolaus Copernicus University, Poland
Xin-Liang Zhang	Huazhong University of Science and Technology, China
Chao Zhou	Lehigh University, USA

Invited Speakers

José Luís Santos	Universidade do Porto, Portugal
Jean–Luc Adam	Université de Rennes 1—CNRS, France
David J. Richardson	University of Southampton, UK

Preface

This book brings together review chapters on a selected range of themes in the science of light—developed from, and inspired by, papers presented at the 5th International Conference on Photonics, Optics and Laser Technology (PHOTOPTICS 2017), held in Porto, Portugal, from 27 February to 1 March 2017. These papers have been selected by the chairs from over a hundred paper submissions, originating from thirty countries, using criteria-based assessment and comments from the programme committee members, feedback from the session chairs, and also the programme chairs' global view of all contributions included in the technical programme. Accordingly, this volume includes papers on both theoretical and practical aspects, across the fields of Optics, Photonics and Laser Technology, all of which contribute to a developed understanding of current research trends. The book contents comprise twelve chapters, including imaging and microscopy—techniques and apparatus, with contributions including the synchronization of active fibre mode-locked lasers for stimulated emission depletion microscopy, feature detection, resolution enhancement in visually impaired images through physics-based computational imaging; laser device technology, addressing high-power continuous-wave Er-doped fibre lasers and high-power SOA-based tunable fibre compound-ring lasers and finally, the field of photonic materials, including devices, methodologies and techniques, the suppression of zeroth-order diffraction in phase-only spatial light modulators via destructive interference, higher-order diffraction suppression by means of two-dimensional quasi-periodic gratings, two-dimensional connected hole gratings for high-order diffraction suppression, liquid crystal polarization-independent tunable spectral filters, bistable scattering fluorescent materials based on polymeric matrixes, advances in fs-laser micromachining for the development of optofluidic devices, microfiber knot resonator sensors and charge transfer features in enhanced pinned photodiodes with a collection gate.

We thank all of the authors for their contributions and also the reviewers who have helped ensure the quality of this publication.

Norwich, UK   David L. Andrews
Caparica, Portugal   Paulo Ribeiro
Caparica, Portugal   Maria Raposo

Contents

Contributors

D. Alexandre Department of Physics, University of Trás-os-Montes e Alto Douro, Vila Real, Portugal;
CAP - Centre for Applied Photonics, INESC TEC, Porto, Portugal

Vítor A. Amorim Department of Physics and Astronomy, Faculty of Sciences, University of Porto, Porto, Portugal;
CAP - Centre for Applied Photonics, INESC TEC, Porto, Portugal

Mohammad Asghari Department of Electrical and Computer Engineering, University of California Los Angeles, Los Angeles, CA, USA;
Department of Electrical Engineering and Computer Science, Loyola Marymount University, Los Angeles, CA, USA

Simeon Bikorimana The City College of New York, City University of New York, New York, USA

Roger Dorsinville The City College of New York, City University of New York, New York, USA

O. Frazão Department of Physics and Astronomy, Faculty of Sciences and INESC TEC, University of Porto, Porto, Portugal

Nan Gao Key Laboratory of Microelectronic Devices and Integrated Technology, Institute of Microelectronics, Chinese Academy of Sciences, Beijing, People's Republic of China

L. Girgenrath ELMOS Semiconductor AG, Dortmund, Germany;
Faculty of Engineering, EIT, EBS, University of Duisburg-Essen, Duisburg, Germany

A. D. Gomes Department of Physics and Astronomy, Faculty of Sciences and INESC TEC, University of Porto, Porto, Portugal

M. Hofmann ELMOS Semiconductor AG, Dortmund, Germany

Abdullah Hossain The City College of New York, City University of New York, New York, USA

Wynn Dunn Gil D. Improso National Institute of Physics, University of the Philippines, Quezon City, Philippines

Bahram Jalali Department of Electrical and Computer Engineering, Department of Bioengineering, Department of Surgery, David Geffen School of Medicine, California NanoSystems Institute, University of California Los Angeles, Los Angeles, CA, USA

Leonid V. Kotov College of Optical Sciences, University of Arizona, Tucson, AZ, USA;
Fiber Optics Research Center of the Russian Academy of Sciences, Moscow, Russia

Shree Krishnamoorthy Indian Institute of Technology, Chennai, India

R. Kühnhold ELMOS Semiconductor AG, Dortmund, Germany

Hailiang Li Key Laboratory of Microelectronic Devices and Integrated Technology, Institute of Microelectronics, Chinese Academy of Sciences, Beijing, People's Republic of China

Mikhail E. Likhachev Fiber Optics Research Center of the Russian Academy of Sciences, Moscow, Russia

Ziwei Liu Key Laboratory of Microelectronic Devices and Integrated Technology, Institute of Microelectronics, Chinese Academy of Sciences, Beijing, People's Republic of China

Ata Mahjoubfar Department of Electrical and Computer Engineering, California NanoSystems Institute, University of California Los Angeles, Los Angeles, CA, USA

João M. Maia Department of Physics and Astronomy, Faculty of Sciences, University of Porto, Porto, Portugal;
CAP - Centre for Applied Photonics, INESC TEC, Porto, Portugal

P. V. S. Marques Department of Physics and Astronomy, Faculty of Sciences, University of Porto, Porto, Portugal;
CAP - Centre for Applied Photonics, INESC TEC, Porto, Portugal

Jiebin Niu Key Laboratory of Microelectronic Devices and Integrated Technology, Institute of Microelectronics, Chinese Academy of Sciences, Beijing, China

Anil Prabhakar Indian Institute of Technology, Chennai, India

Tanchao Pu Key Laboratory of Microelectronic Devices and Integrated Technology, Institute of Microelectronics, Chinese Academy of Sciences, Beijing, People's Republic of China

Mitsunori Saito Department of Electronics and Informatics, Ryukoku University, Seta, Otsu, Japan

Caesar A. Saloma National Institute of Physics, University of the Philippines, Quezon City, Philippines

Lina Shi Key Laboratory of Microelectronic Devices and Integrated Technology, Institute of Microelectronics, Chinese Academy of Sciences, Beijing, People's Republic of China

Madhuri Suthar Department of Electrical and Computer Engineering, University of California Los Angeles, Los Angeles, CA, USA

Giovanni A. Tapang National Institute of Physics, University of the Philippines, Quezon City, Philippines

S. Thiruthakkathevan Valarkathir Creatronics Private Limited, Chennai, India

Muhammad A. Ummy New York City College of Technology, City University of New York, New York, USA

H. Vogt Faculty of Engineering, EIT, EBS, University of Duisburg-Essen, Duisburg, Germany;
Fraunhofer Institute for Microelectronic Circuits and Systems, Duisburg, Germany

Changqing Xie Key Laboratory of Microelectronic Devices and Integrated Technology, Institute of Microelectronics, Chinese Academy of Sciences, Beijing, People's Republic of China

Chapter 1
Suppression of Zeroth-Order Diffraction in Phase-Only Spatial Light Modulator

Wynn Dunn Gil D. Improso, Giovanni A. Tapang and Caesar A. Saloma

Abstract A correction beam is created using a spatial light modulator (SLM) to suppress the zeroth-order diffraction (ZOD) that is produced by the unmodulated light coming from the dead areas of the said SLM. The correction beam is designed to interfere destructively with the undesirable ZOD that degrades the overall quality of the propagated SLM signal. Two possible techniques are developed and tested for correction-beam generation: aperture division and field addition. With a properly-calibrated SLM, ZOD suppression is demonstrated numerically and experimentally at sufficiently high area factor (AF) values where suitable matching is achieved between the correction beam and the ZOD profiles to result in a 39% reduction of the ZOD intensity via angular aperture division, 32% reduction via annular aperture division, and 24% reduction via vertical aperture division. At low AF values however, meaningful ZOD suppression is not obtained. With the field addition method, a ZOD reduction as high as 99% is gained numerically which was not realized experimentally using an SLM with a fill factor of 0.81 due to limitations posed by an iterative phase-recovery algorithm (ghost image) as well as unwanted signal contributions from the SLM anti-reflection coating, SLM surface variations, optical misalignment and aberrations.

1.1 Introduction

Manipulating the amplitude and phase of light had been extensively studied to achieve desired complex light distribution in different practical applications [1–6]. Controlling the amplitude or the phase, or both at the same time can be done as necessary.

W. D. G. D. Improso (✉) · G. A. Tapang · C. A. Saloma
National Institute of Physics, University of the Philippines, Diliman,
Quezon City, Philippines
e-mail: wimproso@nip.upd.edu.ph

G. A. Tapang
e-mail: gtapang@nip.upd.edu.ph

C. A. Saloma
e-mail: csaloma@nip.upd.edu.ph

© Springer Nature Switzerland AG 2019
P. Ribeiro et al. (eds.), *Optics, Photonics and Laser Technology 2017*,
Springer Series in Optical Sciences 222,
https://doi.org/10.1007/978-3-030-12692-6_1

Phase modulation has become more popular than amplitude modulation due to its higher efficiency, avoiding light loss due to spatial filtering [7]. Phase modulation is done using lenses, prisms, and in the past decade, the spatial light modulator (SLM).

The SLM is a device that allows control over the phase of the incident light via a computer generated hologram (CGH) in its input giving it pixel by pixel control. Because of this versatility, the SLM is used in many different applications such as optical trapping [8, 9], microfabrication [5, 10], microscopy [11, 12] and astronomy [13].

Each SLM pixel has an active and inactive area. The ratio of the active area to the whole area of a pixel in an SLM is called the fill factor. Due to the electronic addressing of the SLM, the fill factor is typically less than one, which results to an inactive area that does not modulate incident light [14]. In a Fourier reconstruction set-up, where a lens is used to reconstruct the desired light configuration, this unmodulated light gets focused, and manifests as a high intensity spot called the zeroth order diffraction (ZOD).

The ZOD disrupts the complex light distribution due to its localized high intensity. It is usually removed in applications by shifting the desired light pattern away from the optical axis or placing a physical beam block [2]. Both of these techniques limit the functional area and decreases the diffraction efficiency. Another way is to create a correction beam together within the desired area [3]. The correction beam interferes with the ZOD, thereby lessening the intensity of the focused spot. This technique becomes slow in cases where multiple variable targets are to be shown in succession since recalculation of the target with the correction beam is necessary.

In this paper, we suppress the unwanted ZOD intensity with a correction beam that is generated via the SLM without the introduction of a physical block or grating. The final phase input to the SLM is calculated using the aperture division method or the field addition method as discussed by Hilario et al. [6]. We calculate the hologram input to SLM that contains the phase information needed for constructing the correction beam and the desired target. The field addition and the aperture division method are described and evaluated in the next Section.

1.2 The Spatial Light Modulator

The SLM is a device that allows versatile and dynamic light manipulation due to its pixel by pixel control capability and fast refresh rate [9]. The SLM can be used for amplitude modulation, phase modulation or both simultaneously. Newer SLMs can also control the polarization of incident light [15] thus allowing for a high degree of freedom control on the incident light. When using the SLM for phase modulation, a hologram is programmed to display on the SLM through a computer. This changes the phase of the incident light without changing the intensity. The corresponding desired pattern is then reconstructed using a Fourier lens.

In this work, we use a programmable phase modulator (PPM, Hamamatsu X8267) as our SLM. The structure is shown in Fig. 1.1 [16]. The PPM is an electrically

Fig. 1.1 Structure of the
SLM [16]

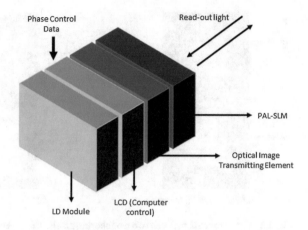

addressed phase modulator, which uses an optical image transmitting element to couple an optically-addressed PAL-SLM (Parallel Aligned Nematic Liquid Crystal Spatial Light Modulator) with an electrically addressed intensity modulator. From the figure, the electrically addressed intensity modulator is an LCD controlled via an external computer. The LCD is coupled to the PAL-SLM using a fiber optic plate (FOP) as the optical image transmitting element to remove the diffraction noise from the pixel structure of the LCD. The photoconductive layer of the PAL-SLM will then be modulated with electric fields by the image from the FOP. This in turn changes the optical path of the incident light thereby modulating the phase in a pixel by pixel manner. Ideally, the SLM will be able to impose the exact phase to the incident light and all of the incident light is modulated. However, limitations in the SLM exists that affect the phase input itself, and limits the area of the incident light that can be controlled. We will discuss some of the limitations in the next section.

1.2.1   Limitations of the SLM

Spatial Phase Variation. Many factors affect the phase that is input to the SLM. An example is the distortion of the phase caused by variation in ambient temperature [17]. Due to the thermal expansion of the SLM, the phase response maybe different for different ambient temperature and thus it becomes necessary to compensate for the change in phase response. Another is the effect of the imperfect flatness of the surface of the SLM. This results to a different phase response per pixel since there is variation in the thickness of the SLM [18]. These kinds of spatial variation in phase response is static, which means these have the same effect for different holograms. Therefore by calibrating the SLM, the effect of these variations in phase can be compensated [19].

Pixel Crosstalk. Phase retardation of the SLM comes from the reaction of the photoconductive layer of the PAL-SLM to the image brought by the FOP from the

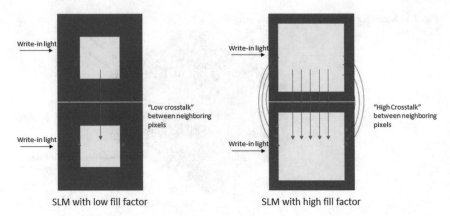

Fig. 1.2 Pixel crosstalk between two neighboring pixels. The closer the working areas, the stronger the crosstalk

LCD. Thus, if the working areas of each pixel are in close proximity, gradual voltage changes (known as fringing fields) occur across the border of neighboring pixels [20–22]. This effect is called pixel crosstalk, and studies have shown that it low-pass filters the desired phase pattern. Pixel crosstalk changes the effective phase imposed to the incident light and is modelled as a convolution of the ideal phase with a point-spread-function given by the SLM [20, 21].

Figure 1.2 shows the crosstalk in a schematic between two pixels of an SLM. If the working areas of two neighboring pixels are near to each other, the effect of fringing fields is greater than if the working areas are farther. In a multiple spot reconstruction, pixel crosstalk causes non-uniformity of the intensities of the spots. The effect of pixel crosstalk varies from hologram to hologram since it depends on the fields between adjacent pixels which is different for each hologram. Pixel crosstalk is describe as a non-linear dynamic phase response [19, 22].

Persson et al. [20] solved this problem by modifying the phase retrieval algorithm, specifically the GS algorithm, used to calculate the hologram that is programmed to the SLM. The calculated hologram compensates for the effect of the low-pass filtering thus producing the desired reconstruction. In their modified algorithm, the hologram is oversampled to a higher resolution before being convolved with the point-spread function that represents the pixel crosstalk. The field is then propagated. After which, the output field is undersampled to the original matrix size. The amplitude is then replaced by a weighted sum between the desired amplitude and the obtained amplitude before being back propagated to the SLM plane. They tested the algorithm using multiple spots usually used in optical trapping. Using this modified GS algorithm resulted in the increased uniformity of the diffraction spots and increased diffraction efficiency.

Fill Factor. In order to lessen the effect of the pixel crosstalk, there must be a separation between the working area of one pixel to another. The outcome of this separation is that the fill factor (F), or the ratio of the working area of a pixel to the whole pixel area is not equal to one [3, 14, 23], which results to areas that do not modulate light. As F decreases, the effect of pixel crosstalk also decreases, but the area that does not modulate light increases.

It is assumed that F is the same for all pixels over the PAL-SLM area. If the length of one side of a pixel is D and the length of one side of the working area is d, then $F = \frac{d^2}{D^2}$. These non-modulating areas result to unmodulated light that, when propagated using a lens, results to the zeroth order diffraction (ZOD). The ZOD will be discussed in the next section.

1.2.2 The Zeroth Order Diffraction

The non-unity F results to areas that do not modulate the incident light. The consequence of this unmodulated light is a highly localized bright spot in the reconstruction known as the zeroth order diffraction (ZOD) pattern [14]. The ZOD can be described by calculating the propagation of the nonmodulating areas of the SLM. We start with the transmission function of the SLM with $F < 1$, given by [14]:

$$t(\eta, \chi) = a(\eta, \chi) \left\{ \text{rect}\left(\frac{\eta}{d}, \frac{\chi}{d}\right) \otimes q(\eta, \chi) + \left[\text{rect}\left(\frac{\eta}{D}, \frac{\chi}{D}\right) - \text{rect}\left(\frac{\eta}{d}, \frac{\chi}{d}\right)\right] \otimes p(\eta, \chi) \right\}$$
(1.1)

where $\otimes$ denote the convolution operation, η and χ are the field coordinates, rect is the rectangular function, $a(\eta, \chi)$ is the aperture function, and $q(\eta, \chi)$ and $p(\eta, \chi)$ are given by:

$$q(\eta, \chi) = \sum_{m,n=0}^{M-1} \delta(\eta - mD, \chi - nD) \exp(i\phi_{mn})$$
(1.2)

$$p(\eta, \chi) = \sum_{m,n=0}^{M-1} \delta(\eta - mD, \chi - nD) \exp(i\phi_c)$$
(1.3)

ϕ_{mn} describes the input hologram to reconstruct the desired target while ϕ_c describes the phase imposed by the nonmodulating areas of the SLM. In this work, ϕ_c is assumed to be the same for all non-working areas and that, all working areas are strictly square-shaped [14].

Equation (1.1) can then be rewritten in terms of fields incident to the SLM:

$$U_{\text{SLM}}(\eta, \chi) = U_w(\eta, \chi) + U_{nw}(\eta, \chi)$$
(1.4)

where U_w is the field due to the working areas of the SLM and U_{nw} is the field due to the non-working areas. In relation to the transmission function, we have:

$$U_w(\eta, \chi) = a(\eta, \chi) \times \left(\text{rect}\left(\frac{\eta}{d}, \frac{\chi}{d}\right) \otimes q(\eta, \chi) \right)$$
(1.5)

$$U_{nw}(\eta, \chi) = a(\eta, \chi) \times \left\{ \left[\text{rect}\left(\frac{\eta}{D}, \frac{\chi}{D}\right) - \text{rect}\left(\frac{\eta}{d}, \frac{\chi}{d}\right) \right] \otimes p(\eta, \chi) \right\} \quad (1.6)$$

Propagating (1.4) forward using Fourier transform, we have:

$$\mathscr{F}\{U_{\text{SLM}}(\eta, \chi)\} = \mathscr{U}_{\text{recon}}(x, y) = \mathscr{U}_{\text{target}}(x, y) + \mathscr{U}_{\text{ZOD}}(x, y) \quad (1.7)$$

where (x, y) are the field coordinates in the Fourier reconstruction, $\mathscr{U}_{\text{target}}$ and $\mathscr{U}_{\text{ZOD}}$ are the field of the desired target and ZOD, respectively, with the following correspondence:

$$\mathscr{F}\{U_w(x, y)\} = \mathscr{U}_{\text{target}} \quad (1.8)$$

$$\mathscr{F}\{U_{nw}(x, y)\} = \mathscr{U}_{\text{ZOD}} \quad (1.9)$$

inserting (1.5)–(1.8) and (1.6)–(1.9), we obtain the expressions for both the desired reconstruction and the ZOD, given by:

$$\mathscr{U}_{\text{target}} = \mathscr{A}(x, y) \otimes \{d^2 \text{sinc}(xd, yd) Q(x, y)\} \quad (1.10)$$

$$\mathscr{U}_{\text{ZOD}} = \mathscr{A}(x, y) \otimes \{(\text{sinc}(xD, yD) - F\text{sinc}(xd, yd))D^2 P(x, y) \quad (1.11)$$

where:

$$\text{sinc}(\zeta_x, \zeta_y) = \frac{\sin(\pi \zeta_x)}{\pi \zeta_x} \frac{\sin(\pi \zeta_y)}{\pi \zeta_y} \quad (1.12)$$

and $P(x, y)$ and $Q(x, y)$ are Fourier transforms of (1.2) and (1.3), respectively. $\mathscr{U}_{\text{ZOD}}$ is the result of propagating the nonmodulating areas of the SLM and describes the ZOD. It distorts the reconstruction in the low spatial frequencies and as such reduces the functionality of the reconstructed pattern [3]. It also introduces unnecessary illumination that will saturate the camera due to its high intensity. In applications involving microscopy, the ZOD may introduce heating in the sample. The ZOD also affects the diffraction efficiency of the SLM [14].

One common way to remove the ZOD is by placing a physical beam block in an intermediate plane [2]. The physical beam block will fully block the ZOD, and remove it from the final reconstruction. However, a non-accessible area in the final reconstruction arises, since any part of the reconstruction that is near the ZOD will be blocked by the physical beam block. Thus reconstructing holographic traps near the center will entail moving the whole set-up physically so that the non-accessible area will also move.

Ronzitti et al. [24] characterized the SLM by using binary gratings and checkerboards with different modulation depths as holograms. By comparing it with the theoretical output, they solved for the correcting function in the calculation of the hologram, and were able to reduce the power of the ZOD by about 90%. However, the main cause of the ZOD for their SLM is the pixel crosstalk due to high fill factor of their SLM.

Liang et al. [25] proposed phase compression method to lessen the ZOD intensity. Phase compression is applied to the hologram, making recalculation of the hologram unnecessary. However, the signal to noise ratio of the resulting reconstruction decreased.

One way to remove the ZOD is by introducing a correction beam in the location of the ZOD. Daria and Palima [14] constructed the correction beam by deriving the field at $(x, y) = (0, 0)$, given by:

$$\mathcal{U}_{\text{recon}}(0, 0) = d^2 Q(0, 0) + (1 - F)D^2 \exp(i\phi_c) \tag{1.13}$$

The field given by (1.13) is then combined with the desired target. A phase retrieval algorithm is then used to obtain the phase ϕ_{mn} that will reconstruct both the desired target and the correction beam, with a constraint that the phase of the correction beam must be equal to $\phi_c + \pi$. Using this, they were able to show that the ZOD can be totally removed.

However, the main limitation with this method is that, if the desired target is changed to a new one, then the phase must be recalculated so that the correction beam is combined with the new desired target. This takes longer computation time. Normally, this is not an issue, but in applications where the change in the desired target depends on an external factor then the recalculation of the phase greatly hinders the experiment. An example is in optical trapping. If the traps are moving and the configuration depends on the experimental particles to be trapped, then recalculation of the phase to construct both of the traps and the correction beam makes it hard to control the traps.

In this work, the objective is to create a suppression method that calculates the field for the correction beam separate from the calculation of the desired target so that if there is a change in the desired target, then there is no recalculation needed for the correction beam. It is assumed that only the non-modulating areas of the SLM contributes to the ZOD. Methods to independently calculate the desired target and correction beam is discussed in the next section.

1.3 Suppression of the Zeroth Order Diffraction

In order to suppress the ZOD, we construct a correction beam in the location of the ZOD, independent of the construction of the desired target. Essentially, the field from the SLM given by (1.7) becomes:

$$\mathcal{U}_{\text{recon}}(x, y) = \mathcal{U}_{\text{target}}(x, y) + \mathcal{U}_{\text{corr}}(x, y) + \mathcal{U}_{\text{ZOD}}(x, y) \tag{1.14}$$

where $\mathcal{U}_{\text{corr}}$ is the correction beam that will be used to destructively interfere with the ZOD by adding the correct phase to this field.

Thus, (1.8) becomes:

$$\mathcal{F}\{U_w(\eta, \chi)\} = \mathcal{U}_{\text{target}}(x, y) + \mathcal{U}_{\text{corr}}(x, y) \tag{1.15}$$

Obtaining U_w so that both the desired target and correction beam employs the use of a phase retrieval algorithm. Here, we discuss the construction of the correction beam.

1.3.1 The ZOD Suppression Method

Creating the Correction Beam. We suppress the ZOD by inducing a destructive interference between the ZOD and a correction beam. This correction beam will be created either using aperture division, or by addition of fields, as discussed by Hilario et al. [6]. Using the GS algorithm, the phase that is needed to construct the correction beam is calculated, given by $\phi_{corr}(\eta, \chi)$. We then add a constant phase ϕ_{shift} to the whole field until total destructive interference occurs.

The inputs to the GS algorithm are the amplitude of the source and the amplitude of the ZOD. The amplitude of the source is a part of the aperture, as is the case in aperture division, or the whole aperture, as the case in addition of fields. Both methods of constructing the correction beam needs the ZOD amplitude as target to the phase retrieval algorithm.

In order to have the ZOD amplitude, we obtain the light distribution of the non-modulating areas, $U_{nw}(\eta, \chi)$. This is from oversampling the light distribution that is incident to the SLM. The distribution is then obtained by separating the non-modulating areas, $U_{nw}(\eta, \chi)$, from the modulating areas, $U_w(\eta, \chi)$. Due to over-sampling, the size of the matrix becomes 15360×15360 from 768×768. Fourier transform is then performed to $U_{nw}(\eta, \chi)$. The ZOD is then obtained from the middle 768×768 of the reconstruction. From this, we obtain the ZOD amplitude from an SLM with $F = 0.81$. The SLM to be used in the experiment, Hamamatsu PPM X8267, has $F = 0.8$. In our method, we also create a desired target. In order to properly describe the effectiveness of suppressing the ZOD, the desired target does not have intensity in $(x, y) = (0, 0)$ and surrounding area. This desired target represents the application that will be done using the SLM, whether it be optical trapping, lithography or others. In our case, we are focused on suppressing the ZOD beam, therefore we need to observe the behaviour of the intensity of the ZOD itself. To do this, the field due to the desired application must be separated from the location of the ZOD so that only the ZOD and the correction beam will interact. In this case, reconstructing the target will suffice. We use GS algorithm to obtain the phase needed to construct the desired target, $\phi_{target}(\eta, \chi)$.

After calculating ϕ_{corr} and ϕ_{target}, we combine the fields in order to obtain the phase input to the SLM, ϕ_{SLM}. From this, we can obtain the field of the incident light to the SLM, given by:

$$U_{SLM}(\eta, \chi) = A_{SLM}(\eta, \chi)e^{i\phi_{SLM}(\eta, \chi)} \qquad (1.16)$$

We use two methods in constructing the correction beam. First is the aperture division, and the second is addition of fields, to be discussed below.

(a) (b) (c)

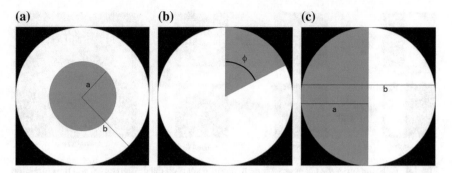

Fig. 1.3 Three aperture division to be employed in creating the correction beam. **a** Annular aperture division. **b** Angular aperture division. **c** Vertical aperture division

Aperture Division. In this method, we divide the aperture into two parts, $A_{\text{corr}}(\eta, \chi)$ and $A_{\text{target}}(\eta, \chi)$. $A_{\text{corr}}(\eta, \chi)$ is used as source to calculate for $\phi_{\text{corr}}(\eta, \chi)$ while $A_{\text{target}}(\eta, \chi)$ is used to calculate for $\phi_{\text{target}}(\eta, \chi)$. The aperture division is shown in Fig. 1.3. The relationship between A_{corr} and A_{target} is given by following:

$$A_{\text{SLM}}(\eta, \chi) = A_{\text{corr}}(\eta, \chi) + A_{\text{target}}(\eta, \chi) \qquad (1.17)$$

where $A_{\text{SLM}}(\eta, \chi)$ is the aperture function of the SLM. Since A_{corr} and A_{target} are independent spatially, their intensities are simply added together.

We divide the aperture three ways: (1) annular aperture division, where the aperture is divided into an inner circle and an annulus; (2) angular aperture division, where the aperture is divided in an angular fashion; and (3) vertical aperture division, where the aperture is divided vertically.

In the three aperture divisions, a factor that determines the total area used in constructing the correction beam is applied. This factor is called the area factor (AF), given by:

$$AF = \frac{\text{area used to create correction beam}}{\text{whole aperture area}} \qquad (1.18)$$

In each aperture divisions, we calculate $\phi_{\text{corr}}(\eta, \chi)$ using $A_{\text{corr}}(\eta, \chi)$ as source input and ZOD as target input to the GS algorithm. We then calculate $\phi_{\text{target}}(\eta, \chi)$ using $A_{\text{target}}(\eta, \chi)$ as source input and the desired target as input to the GS algorithm. This is to simulate the application where the SLM is to be used. Another reason is to move the area used to construct the desired target away from the location where the ZOD and correction beam are located so that when measuring the change in the total intensity of the ZOD, it is only the ZOD and the correction beam that contributes to the total intensity. In order to induce destructive interference, we add a spatially constant phase shift ϕ_{shift} to ϕ_{corr} Since $A_{\text{corr}}(\eta, \chi)$ and $A_{\text{target}}(\eta, \chi)$ are spatially independent, the final phase input to the SLM is given by:

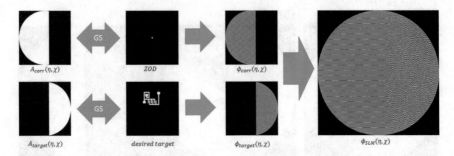

Fig. 1.4 Calculation of the necessary phase to reconstruct both the desired target and the correction beam for the aperture division

$$\phi_{\text{SLM}}(\eta, \chi) = \phi_{\text{target}}(\eta, \chi) + \phi_{\text{corr}}(\eta, \chi) + \phi_{\text{shift}} \qquad (1.19)$$

We add ϕ_{shift} to ϕ_{SLM} in order to induce destructive interference between correction beam and the ZOD. From (1.1), the non modulating areas give phase shift ϕ_c to the incident light. However, this ϕ_c is not known. Thus, ϕ_{shift} is added so that the correct phase shift is obtained and destructive interference is induced.

For this method, we obtain the optimal AF which controls the total energy and the information used to create the correction beam. The optimal AF gives the maximum energy to the desired target and creates a correction beam that is most similar to the ZOD in terms of total energy and profile. Therefore, we scan the AF from 0 to 1 in order to obtain the optimal value. The procedure in obtaining ϕ_{SLM} is shown in Fig. 1.4.

Addition of Fields. The second method in creating the correction beam and the desired target is the addition of fields, as discussed by Hilario et al. [6, 26, 27]. ϕ_{target} and ϕ_{corr} are calculated separately. The source input for the GS algorithm for each phase is the whole aperture. We have:

$$A_{\text{target}}(\eta, \chi) = A_{\text{corr}}(\eta, \chi) = A_{\text{SLM}}(\eta, \chi) \qquad (1.20)$$

where $A_{\text{SLM}}(\eta, \chi)$ is the aperture function of the SLM. The amplitudes and phase give us the fields needed to reconstruct the correction beam and target separately:

$$U_{\text{target}}(\eta, \chi) = A_{\text{SLM}}(\eta, \chi) e^{i\phi_{\text{target}}(\eta, \chi)} \qquad (1.21)$$

$$U_{\text{corr}}(\eta, \chi) = A_{\text{SLM}}(\eta, \chi) e^{i\phi_{\text{corr}}(\eta, \chi)} \qquad (1.22)$$

where U_{target} is the field of the SLM when only the target is to be reconstructed, and U_{corr} is the field when only the correction beam is to be reconstructed. If the fields U_{target} and U_{corr} are propagated independently, we have:

$$\mathscr{F}\{U_{\text{target}}(\eta, \chi)\} = \mathscr{U}_{\text{target}}(x, y) \qquad (1.23)$$

$$\mathcal{F}\{U_{\text{corr}}(\eta, \chi)\} = \mathcal{U}_{\text{corr}}(x, y) \tag{1.24}$$

The phase to construct both target and correction beam is given by calculating:

$$\phi_{\text{inp}}(\eta, \chi) = \arg(c_{\text{corr}} U_{\text{corr}}(\eta, \chi) + c_{\text{target}} U_{\text{target}}(\eta, \chi)) \tag{1.25}$$

c_{corr} and c_{target} are the constants used in order to control the total energy that goes to the corresponding reconstructions of each field. The field that hits the SLM is then given by:

$$U_{\text{SLM}}(\eta, \chi) = A_{\text{SLM}}(\eta, \chi) e^{(i\phi_{\text{inp}}(\eta,\chi) + \phi_{\text{shift}})} \tag{1.26}$$

If we calculate the amplitude and phase of the added field (inside the parenthesis in (1.25)), we obtain:

$$\begin{aligned} U_{\text{sum}}(\eta, \chi) &= c_{\text{corr}} U_{\text{corr}}(\eta, \chi) + c_{\text{target}} U_{\text{target}}(\eta, \chi) \\ &= c_{\text{corr}} A_{\text{SLM}}(\eta, \chi) e^{i\phi_{\text{corr}}(\eta,\chi)} + c_{\text{target}} A_{\text{SLM}}(\eta, \chi) e^{i\phi_{\text{target}}(\eta,\chi)} \end{aligned} \tag{1.27}$$

$U_{\text{sum}}(\eta, \chi)$ has the same phase profile as U_{SLM} since:

$$\begin{aligned} \phi_{\text{SLM}}(\eta, \chi) &= \arg(c_{\text{corr}} U_{\text{corr}}(\eta, \chi) + c_{\text{target}} U_{\text{target}}(\eta, \chi)) + \phi_{\text{shift}} \\ &= \arg(U_{\text{sum}}) + \phi_{\text{shift}} \end{aligned}$$

and ϕ_{shift} is constant for the whole field. However, the amplitude of U_{sum} is different. We have:

$$\begin{aligned} |U_{\text{SLM}}(\eta, \chi)| &= |U_{\text{corr}} + U_{\text{target}}| \\ &= A_{\text{SLM}}(\eta, \chi) \times \\ &\sqrt{(c_{\text{corr}}^2 + c_{\text{target}}^2 + 2c_{\text{corr}} c_{\text{target}} \cos(\phi_{\text{corr}}(\eta, \chi) - \phi_{\text{target}}(\eta, \chi))} \end{aligned}$$

If we set

$$A_{\text{cross}}(\eta, \chi) = \sqrt{(c_{\text{corr}}^2 + c_{\text{target}}^2 + 2c_{\text{corr}} c_{\text{target}} \cos(\phi_{\text{corr}}(\eta, \chi) - \phi_{\text{target}}(\eta, \chi))},$$

then we have the following:

$$U_{\text{SLM}}(\eta, \chi) = \frac{U_{\text{sum}}(\eta, \chi)}{A_{\text{cross}}(\eta, \chi)} \tag{1.28}$$

If U_{SLM} is propagated forward using Fourier transform, we obtain the following:

$$\mathcal{F}\{U_{\text{SLM}}(\eta, \chi)\} = \mathcal{F}\left\{\frac{1}{A_{\text{cross}}(\eta, \chi)}\right\} * \mathcal{F}\{U_{\text{sum}}(\eta, \chi)\} \tag{1.29}$$

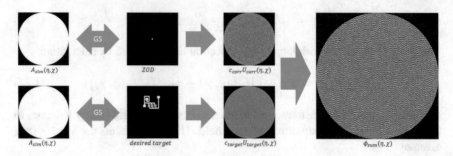

Fig. 1.5 Calculation of the phase required to reconstruct both the desired target and correction beam in field addition. First the fields for reconstructing the desired target and correction beam are calculated separately, with the amplitude being the aperture function of the SLM. Then the fields are added. The phase of the output field is then ϕ_{SLM}

invoking the convolution theorem of the Fourier transform. Then finally, the reconstruction is given by:

$$\mathscr{F}\{U_{SLM}(\eta, \chi)\} = \mathscr{F}\left\{\frac{1}{A_{cross}(\eta, \chi)}\right\} * \{\mathscr{U}_{corr}(x, y) + \mathscr{U}_{target}(x, y)\} \qquad (1.30)$$

The cross term in the amplitude of U_{SLM}, $A_{cross}(\eta, \chi)$, is not constant and thus, we cannot analytically isolate the total intensities of $\mathscr{U}_{corr}(x, y)$ and $\mathscr{U}_{target}(x, y)$ without explicit knowledge of $\phi_{corr}(\eta, \chi)$ and $\phi_{target}(\eta, \chi)$. Because of this, the exact relationship between c_{corr} and c_{target} cannot be determined analytically. In this case, we set $c_{corr} + c_{target} = 1$.

The cross term in (1.30) results to the presence of ghost orders [24] due to the cross terms between the added field. The ghost orders are undesired reconstructions that make a symmetry with the reconstructed target. When constructing a correction beam using the field addition method, the location of its ghost order is at the same spot, which can affect the suppression of the ZOD. Since the ghost order has the conjugate phase of the correction beam, it lessens the total intensity of the correction beam itself.

The procedure in calculating for ϕ_{SLM} is summarized in Fig. 1.5.

1.3.2 Suppression of the ZOD

The constant phase shift ϕ_{shift} is varied from 0 to 3π. For the aperture division method, AF is changed from 0 to 1 to achieve the maximum AF with maximum suppression while still reconstructing the desired target. For the field addition method, c_{corr} is changed from 0 to 1.

To measure the effect of the method, we calculate the relative intensity R, given by the following:

$$R = \frac{I_{\text{method}} - I_{\text{ZOD}}}{I_{\text{ZOD}}} \times 100\% \qquad (1.31)$$

where I_{ZOD} is the total intensity of the ZOD before the suppression method is applied. This is when U_{mod} is used entirely to create the desired pattern. I_{method} is the total intensity after application of the suppression method.

If R is greater than zero, this means the intensity of the ZOD increased, which indicates that either constructive interference occurred between the correction beam and the ZOD, or the total energy of the correction beam overshoots that of the ZOD, that even if full destructive interference was induced, there was still enough energy from the correction beam to create another ZOD. If R is equal to zero, this means nothing changed. If R is less than zero, the total intensity of the ZOD decreased. The ideal result is that I_{method} equals zero, which means R is equal to -100%.

1.4 Experiment

1.4.1 Suppression of the ZOD Experiment

We input the hologram which reconstructs the desired target without the correction beam to the SLM. From this, we add or remove NDF until the image captured by camera is unsaturated (Fig. 1.6). We capture this image and extract the total intensity of the ZOD by summing the intensities of the pixels around the ZOD. This total intensity is then our I_{ZOD}. We then input holograms that reconstruct both the desired target and the phase shifted correction beam. For each hologram, we capture the image without changing the NDF and obtain the total intensity of the ZOD. This is then I_{method}.

Aperture Division. Sample holograms for different AF with different ϕ_{shift} is shown in Fig. 1.7 for angular, annular and vertical aperture division. These holograms are input to the SLM. In the experiment, the number of ϕ_{shift} added is 128 ranging from 0 to 3π for 32 values of AF from 0 to 1. Images are captured for each hologram. Three trials are done per AF. For the capture images, we obtain I_{method} by summing the total intensity of a 15×15 box around the ZOD. This size is the smallest box to capture the whole ZOD while avoiding the intensity due to ambient light capture by the camera. We plot R versus ϕ_{shift} in Fig. 1.8.

Figure 1.8 shows that for the three aperture divisions, the minimum relative intensity is found when ϕ_{shift} is equal to 0 or 2π. At $\phi_{\text{shift}} = 0$, R is negative. But when a small ϕ_{shift} is added, R shoots up to a positive value, at around 20–30%. Then as ϕ_{shift} increases, R steadily decreases until ϕ_{shift} is equal to 2π, where it reaches a small value of R, comparable to when ϕ_{shift} is equal to 0. However, when it increases again, a sudden jump is again observed. This can be attributed to the fact that, when saving the holograms, it is ensured that the phase range is from 0 to 2π, which means that the hologram when $\phi_{\text{shift}} = 2\pi + \theta$ is the same as the hologram when $\phi_{\text{shift}} = \theta$.

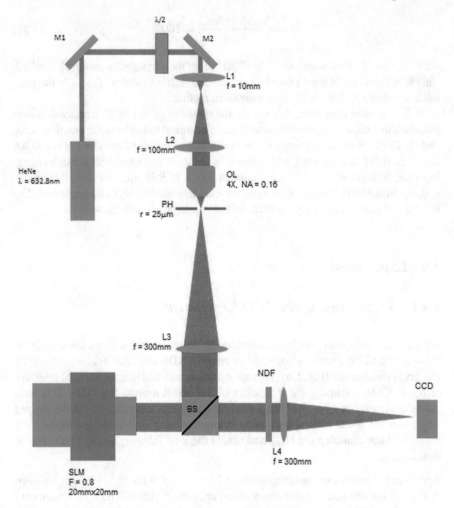

Fig. 1.6 Optical set-up used to verify the suppression of the ZOD

We shown in Fig. 1.9 a sample reconstruction of the desired target with the ZOD. This is to show that in each aperture division, we are still constructing a desired target together with the correction beam. We also show sample intensities of the ZOD for annular, angular and vertical aperture division when R is minimum for an AF. The intensities are visually lower as AF becomes higher. We plot the minimum R for each AF versus AF in Fig. 1.10.

Maximum suppression versus AF is plotted in Fig. 1.10 for the three different aperture divisions. For the angular aperture division, -39% of the ZOD is suppressed at $AF = 0.94$. For the annular aperture division, -32% of the ZOD intensity is suppressed at $AF = 0.97$. Finally, for the vertical aperture division, -24% of ZOD is suppressed at $AF = 0.88$. For the three aperture divisions, there is no clear trend

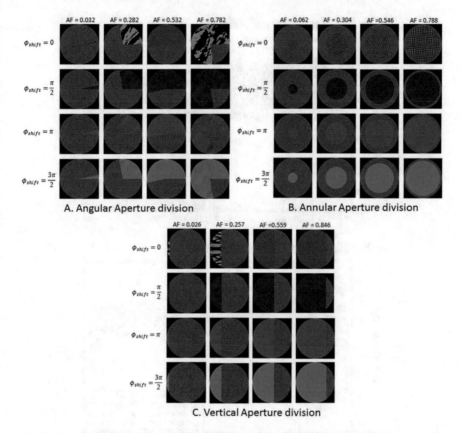

Fig. 1.7 Sample ϕ_{SLM} for different AF and different ϕ_c for the angular aperture division (top left), annular aperture division (top right) and vertical aperture division (bottom)

between the maximum suppression and AF. However, it is observed that as higher suppression occurs at high AF for the three aperture divisions.

Field Addition Method. We plot R versus phase shift in Fig. 1.11 for different c_{corr}. Similar to the Fig. 1.8 in the aperture division, maximum suppression is at phase shift equal to 0 or 2π. For higher values of c_{corr}, the discontinuity is evident to phase shift below 2π and after. For lower values of c_{corr}, the plot approaches a sinusoid-like behavior.

We show the ZOD when maximum suppression occurs per c_{corr} in Fig. 1.12a [27]. It is observed that the change in the ZOD intensity is gradual as c_{corr} increases. This is attributed to the fact that c_{corr} only changes the total energy that goes into reconstructing the correction beam, as opposed to the aperture division where AF dictates both the energy and the information used to construct the correction beam.

We plot the maximum suppression versus c_{corr} in Fig. 1.12b. The minimum R is steady for low values of c_{corr} (up to c_{corr} equal to 0.3). Then its value decreases until

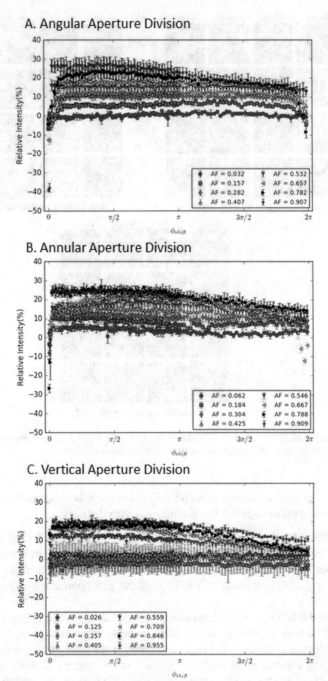

Fig. 1.8 Relative intensity versus ϕ_{shift} for different AF. **a** Angular aperture division, **b** Annular aperture division and **c** vertical aperture division

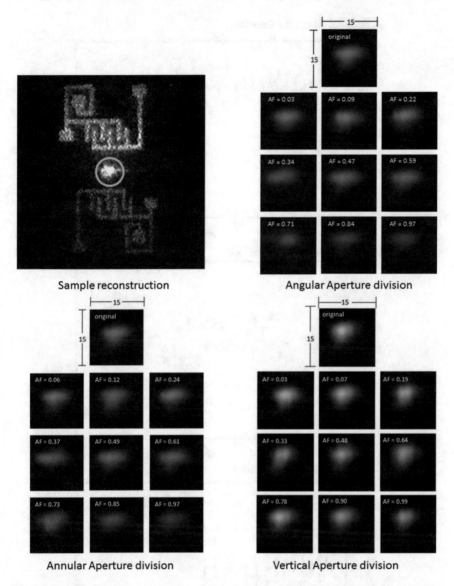

Fig. 1.9 Top left: Sample reconstruction of desired target and ZOD. Image is saturated so that the desired target is observed; Top right, bottom row: ZOD with minimum R for each AF for annular, angular and vertical aperture division

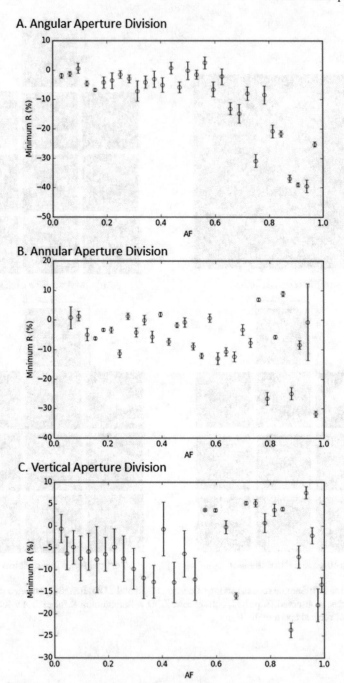

Fig. 1.10 Minimum R versus *AF* for annular, angular and vertical aperture divisions

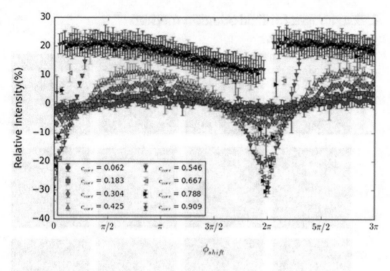

Fig. 1.11 Relative intensity versus ϕ_{shift} for different c_{corr}

c_{corr} is equal to 0.82, where ZOD is suppressed up to -32% of its original value. It does not change significantly for c_{corr} greater than 0.82.

1.4.2   Discussion of Results

To separate the contribution of the correction beam from the two methods mentioned above, we calculate the effect of the correction beam numerically. We simulate the suppression method with the assumption that only dead areas contribute to the ZOD. Moreover, the SLM used in the simulation has no limitation other than its F value being less than one. We also assume that the phase input to the SLM is imposed without error.

The simulation of the SLM with $F < 1$ is done by oversampling the hologram where every pixel is sampled at higher resolution. In our case, each pixel is represented by 400 sampled points to form a 20×20 image. To simulate the non-working areas, the outer pixels of the 20×20 image are assigned non-modulating zero phase shift values.

The oversampled holograms with F constraint are then used as phase with an oversampled aperture size. Hence the fields that are originally represented as 768×768 pixel image sizes now become 15360×15360 pixel images. These oversampled fields are then propagated via Fourier transform and a reconstruction is obtained at higher frequencies.

We choose the middle reconstruction, in the zeroth order, and take the total intensity of the ZOD in a 40×40 square area that is used so that the ZOD intensity is obtained without the desired target while avoiding ambient noise (see Fig. 1.13).

A. ZOD images for field addition method

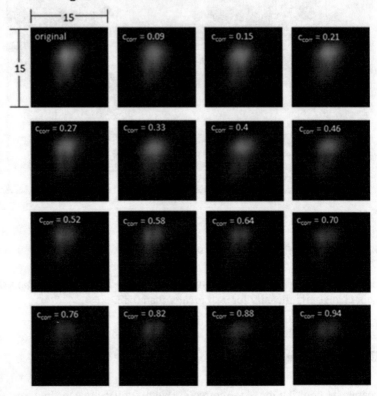

B. Minimum R for field addition method

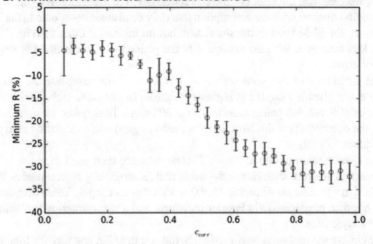

Fig. 1.12 Top image: ZOD with minimum R for each c_{corr}. Bottom image: Maximum suppression versus c_{corr} for the field addition method

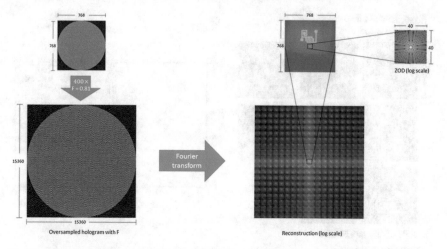

Fig. 1.13 Representation of SLM operation. Each hologram is oversampled and F is imposed. The resulting hologram is used as phase and propagated using Fourier transform. Reconstruction by Fourier transformation is shown in log scale. Middle reconstruction is then obtained from the result and the total intensity of the ZOD is given by the sum of all the intensities inside the 40×40 area

1.4.3 Result of Numerical Simulation

The introduction of the correction beam suppresses the ZOD intensity by destructive interference. We first evaluate the profile of the correction beam by comparing it with the ZOD profile using the Linfoots criteria of merit [28]: Fidelity (F), Correlation quality (Q) and Structural content (C), which measure the overall similarity of two signals, alignment of peaks and the relative sharpness of peak profiles, respectively. If the ZOD and correction beam profiles are identical then: $F = C = Q = 1$. In general, $2Q - C = F$.

We find the dependence of R with ϕ_{shift} and locate the ZOD when R is minimum for different AF and c_{corr} values when using the aperture division and field addition method, respectively. The minimum R values are then plotted as a function of AF and c_{corr}.

Aperture Division. Figure 1.14 presents the ZOD and correction beam profiles for different AF when only the modulating areas are propagated via Fourier transform. Each image is normalized to its maximum thereby observing only the profile and the effect of AF while not observing the the total energy to the correction beam.

We also show the corresponding cross-sections of the ZOD profiles. At low AF values, the correction beam and the ZOD profiles do not match with each other because only a part of the aperture is utilized to construct the correction beam thereby limiting its bandwidth. On the other hand, the ZOD is produced by the contributions of all the dead areas present in the entire SLM. Because the correction-beam bandwidth increases with AF a matching of the correction beam and ZOD profiles is eventually

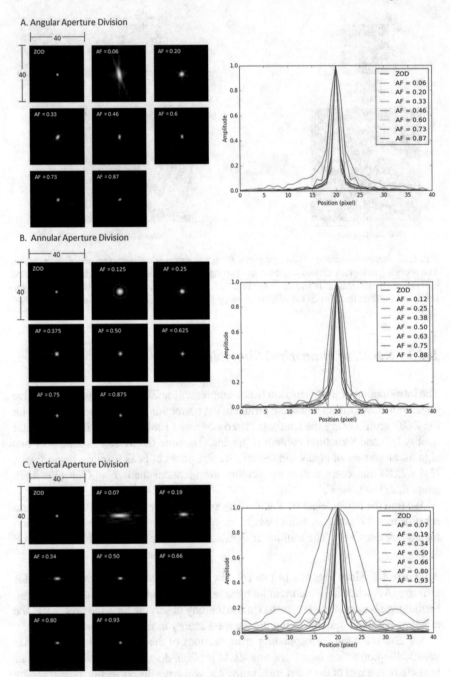

Fig. 1.14 Left column: The ZOD and correction beam intensities for angular aperture division (top), annular aperture division (middle) and vertical aperture division. Right column: Cross section profile of each beam and the ZOD

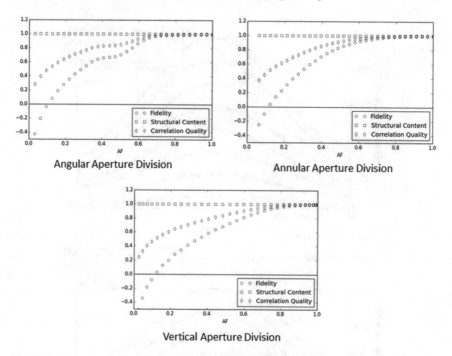

Fig. 1.15 Linfoot's criteria of merit plot for the three aperture divisions

achieved at sufficiently high AF values. We note that the profile matching is reached without taking into account the total energy that is used to construct the correction beam since it is the AF that dictates how much energy is used.

Figure 1.15 plots the values of F, C and Q as a function of AF. With the angular and annular aperture division, both F and C reach the value of unity when AF equals 0.8 while Q remains equal to one all throughout. For the vertical aperture division, $F = C = Q = 1$ only when AF reaches 0.9. Note that the total intensity of the correction beam for a given AF value and that of the ZOD are held equal because we are only concerned with the beam profiles and not with the total energy in the correction beam.

Figure 1.16 plots the values of R versus ϕ_{shift} for the angular, annular and vertical aperture division, respectively. A sinusoidal trend is observed for R as a function of phase shift since the interference intensity depends on the individual intensities of the interfering beams and the cosine of the phase difference between them. Thus the ZOD intensity after suppression would depend on the ZOD and the correction beam intensities as well as on the value of the phase shift ϕ_{shift}. The minimum R is found when ϕ_{shift} becomes equal to 0 or 2π since a phase shift of π is produced when light is reflected. A π-phase difference already exists between the correction beam and the ZOD even when no phase shift is inputted into the SLM itself. Figure 1.17 shows the original ZOD and the ZOD intensities at different AF values when R is minimum for three different methods of aperture division. Instead of suppressing the

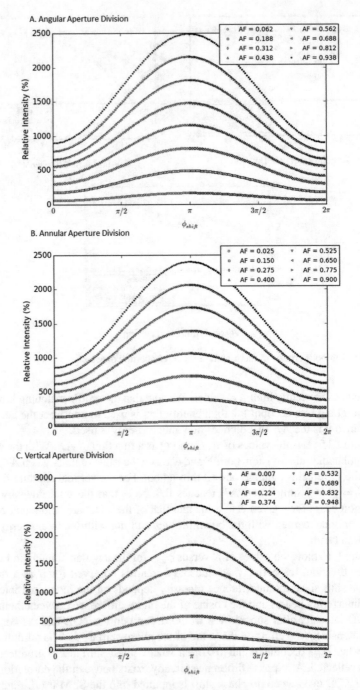

Fig. 1.16 R versus ϕ_{shift} for sample values of AF for the angular, annular and vertical aperture division

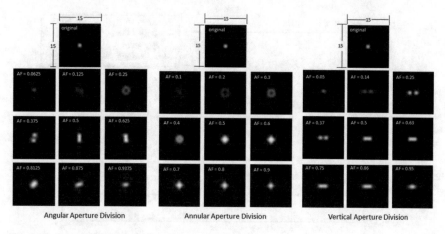

Fig. 1.17 Sample ZOD intensities when R is minimum

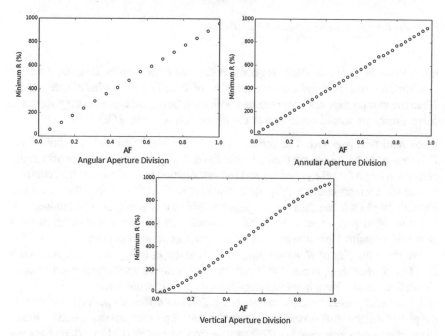

Fig. 1.18 Minimum R versus AF for the angular, annular and vertical aperture division

ZOD, the correction beam only alters the profile and increases the total intensity of the ZOD due to a significant mismatch between the ZOD and the correction beam profiles that occurs at low AF and high correction beam energy at high AF values.

Figure 1.18 plots maximum suppression (minimum R) versus AF. It can be seen that maximum suppression per AF increases as AF increases—at low AF values, limited information is employed to construct the correction beam and the profiles of the correction beam and the ZOD do not match. As the AF increases, more

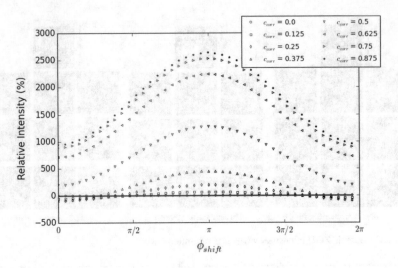

Fig. 1.19 R versus ϕ_{shift} for different values of c_{corr}

information becomes available for constructing the correction beam at the expense of additional beam energy that exceeds that of the ZOD (beam imbalance). Even when the two profiles are now matched with each other, the excess energy from the correction beam would contribute to the generation of a new ZOD.

Field Addition Method. The profile of the correction beam matches that of the ZOD using the field addition method since the entire aperture is used to construct the correction beam for all c_{corr} values and full information is available for the construction of the correction beam [27]. As a result, $F = C = Q = 1$ for all c_{corr} values. Figure 1.19 plots R as a function of ϕ_{shift} for different c_{corr} values. The minimum R is found when ϕ_{shift} is equal to 0 and 2π, similar to the aperture division method simulation results. Here however, negative values of R are observed.

We show the plot of R versus ϕ_{shift} for different c_{corr} in Fig. 1.19. The minimum R is found when ϕ_{shift} is equal to 0 and 2π, similar to the aperture division method simulation results. Here however, negative values of R is observed.

Figure 1.20 presents sample images of the ZOD for different c_{corr} values as well as the plot of minimum R versus c_{corr}. The minimum R plot reveals that the ZOD intensity is reduced to near zero i.e. 99% suppression, at $c_{\text{corr}} = 0.3125$, which happens when the correction beam and ZOD profiles are totally matched. The only remaining problem is in balancing the energies of the two beams. We determine the total energy of the correction beam including its ghost order by changing c_{corr}. The possible suppression that is attainable numerically higher than that achieved experimentally.

Figure 1.10 shows experimental plots for the minimum R versus AF for the three methods of aperture division, which are unlike their corresponding simulation plots at high AF. The difference is explained as follows: at low AF values, the correction beam profile is wider than that of the ZOD (see Fig. 1.14) while at high AF values, the profiles become narrower and more similar making the beams more sensitive

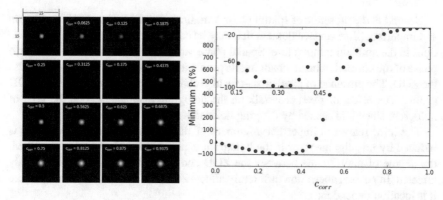

Fig. 1.20 Left: The ZOD and the correction beam for different values of c_{corr}. Middle: Cross section profile of each beam and the ZOD. Right: Linfoot's criteria of merit versus c_{corr}

to relative misalignment. Slight misalignments are sufficient to significantly reduce the effectivity of two-beam interference in suppressing the ZOD. Figure 1.18 shows that numerical results do not predict any suppression with the intensity of the ZOD increasing linearly with AF because at low AF values, correction beam and the ZOD profiles do no match does not match while at high AF, the total energies of correction beam and the ZOD are unequal. Instead of full destructive interference, a ZOD is created by the excess energy from the correction beam.

For the field addition method, the experiment yields negative values for minimum R at all c_{corr}. The minimum R-value decreases with increasing c_{corr} until $c_{corr} = 0.82$, (see Fig. 1.11). The numerical results reveal a minimum R-value that decreases with increasing c_{corr} until $c_{corr} = 0.3125$ where minimum R achieves its lowest value of $R = -99\%$ (see Fig. 1.20). For $c_{corr} > 0.3125$, minimum R increases with c_{corr}. At low c_{corr} values, the simulation and the experiment results are in agreement and they result from low energy of the correction beam that limits its effectivity to suppress the ZOD. However the two results start to deviate from each other as c_{corr} increases due in part to the presence of a ghost image that strengthens with the correction beam energy. The undesirable ghost image which reduces the R value, is an artifact that arises from the iterative nature of the GS algorithm [29] which could not distinguish between the mirror/ghost image from the primary image. The ghost or twin image may be removed using various methods [24, 30].

Three reasons may be cited to cause the difference between the numerical and the experimental results. First, we have simply assumed that only the non-modulating areas of the SLM contribute to the ZOD in the construction of the correction beam in the numerical simulations. In practice other factors might affect the total intensity and profile of the ZOD. For example, minute imperfections in the anti-reflection coating of the SLM could result in a fraction of the incident beam being left unmodulated [24]. The presence of random phase fluctuations [31] can also alter the ZOD intensity profile.

Second is the presence of spatial phase variations caused by uneven illumination and imperfect SLM surface flatness that may be caused by manufacturing and variations in the ambient temperature. Spatial phase variations contribute to the effective phase of the correction beam changing its profile and lessening its ability to suppress the ZOD. The presence of pixel cross-talk also changes the input phase and the ZOD profile. The effect of pixel crosstalk on the hologram also depends on the type of hologram since it is caused by fringing fields that exist between pixels.

The third reason is imperfect alignment of the optical system. If aberration is induced by misalignment, or if the hologram is not correctly inputted to the SLM, the degree of interference between the ZOD and the correction beam is seriously affected. In general aberration differently affects ZOD and the correction beam since it is location dependent.

For the correction beam to completely suppress the ZOD, the above mentioned issues must be addressed satisfactorily. The experimental ZOD profile can be determined accurately and then used to match with the profile of the correction beam. By properly calibrating the SLM, the effect of the spatial phase variation can be taken into account and the correction beam profile may be designed to match to that of the ZOD. Accurate alignment of the different elements comprising the optical set-up can reduce significantly the degree of aberration present. Satisfying the aforementioned conditions will lead to full ZOD suppression.

1.5 Summary and Conclusions

We have suppressed the unwanted ZOD by inducing a destructive interference between the ZOD and a correction beam. Two methods were tested to create the correction beam—aperture division and addition of fields [6]. We have calculated the fields necessary to create the desired target and correction beam separately—the input source to the GS algorithm serving as the aperture amplitude of the SLM. The final phase input to the SLM is obtained by calculating the phase of the sum of the two fields. The energy that is given to the correction beam is controlled using multiplicative constants c_{corr} and c_{target}.

We have observed a discontinuity in the dependence of R with ϕ_{shift}, which could be attributed to the algorithm used to generate the hologram input to the SLM. The phase maps were wrapped from 0 to 2π, which greatly limits the number gray levels that are possible for the hologram. In the experiments, we were able to suppress the ZOD up to: 39% at $AF = 0.94$ with the angular aperture division method, 32% at $AF = 0.97$ with the annular aperture division method and 24% at $AF = 0.88$ with the vertical aperture division method. At lower AF values, the correction beam and the ZOD profiles do not match thus limiting the effectivity of the correction beam to suppress the ZOD. At higher AF, the correction beam becomes narrower and more sensitive to misalignment. For the field addition method, we have been able to suppress the ZOD up to 32% at $c_{corr} = 0.82$.

For the field addition method, simulations show that ZOD suppression is possible up to 99% of its original total intensity since the whole aperture is used to construct the correction beam allowing maximum similarity between the ZOD and correction beam. Suitable choice of the multiplicative constants allowed us to match the total energy of the correction beam to that of the ZOD and making full destructive interference between the two profiles nearly possible.

To fully suppress the ZOD and improve the quality of reconstruction using the SLM, we have cited a number of experimental issues that needed to be addressed by calibrating the SLM in order to neutralize the effect of spatial phase variations in the SLM surface. By obtaining a more accurate ZOD profile, the profile of the correction beam can be matched with the ZOD profile. The degree of ZOD suppression is highly sensitive to aberrations that may be caused by minute misalignments among the different optical elements in the set-up and has to be seriously taken into consideration.

Acknowledgements This work was partly funded by the UP System Emerging Interdisciplinary Research Program (OVPA-EIDR-C2-B-02-612-07) and the UP System Enhanced Creative Work and Research Grant (ECWRG 2014-11). This work was also supported by the Versatile Instrumentation System for Science Education and Research, and the PCIEERD DOST STAMP (Standards and Testing Automated Modular Platform) Project.

References

1. R. Eriksen, V. Daria, J. Gluckstad, Fully dynamic multiple-beam optical tweezers. Opt. Express **10**, 597–602 (2002)
2. M. Polin, K. Ladavac, S.H. Lee, Y. Roichman, D.G. Grier, Optimized holographic optical traps. Opt. Express **13**, 5831–5845 (2005)
3. D. Palima, V. Daria, Effect of spurious diffraction orders in arbitrary multifoci patterns produced via phase-only holograms. Appl. Opt. **45**, 6689–6693 (2006)
4. V. Nikolenko, B.O. Watson, R. Araya, A. Woodruff, D. Peterka, R. Yuste, Slm microscopy: scanless two-photon imaging and photostimulation using spatial light modulators. Front. Neural Circuits **2**, 5 (2008)
5. N.J. Jenness, R.T. Hill, A. Hucknall, A. Chilkoti, R.L. Clark, A versatile diffractive maskless lithography for single-shot and serial microfabrication. Opt. Express **18**, 11754–11762 (2010)
6. P.L. Hilario, M.J. Villangca, G. Tapang, Independent light fields generated using a phase-only spatial light modulator. Opt. Lett. **39**, 2036–2039 (2014)
7. L. Zhu, J. Wang, Arbitrary manipulation of spatial amplitude and phase using phase-only spatial light modulators. Sci. Rep. **4**, 7441 (2014)
8. E.R. Dufresne, G.C. Spalding, M.T. Dearing, S.A. Sheets, D.G. Grier, Computer-generated holographic optical tweezers arrays. Rev. Sci. Instrum. **72**, 1810–1816 (2001)
9. H. Melville, G.F. Milne, G.C. Spalding, W. Sibbett, K. Dholakia, D. McGloin, Optical trapping of three-dimensional structures using dynamic holograms. Opt. Express **11**, 3562–3567 (2003)
10. M. Farsari, S. Huang, P. Birch, F. Claret-Tournier, R. Young, D. Budgett, C. Bradfield, C. Chatwin, Microfabrication by use of a spatial light modulator in the ultraviolet: experimental results. Opt. Lett. **24**, 549–550 (1999)
11. Y. Shao, W. Qin, H. Liu, J. Qu, X. Peng, H. Niu, B.Z. Gao, Addressable multiregional and multifocal multiphoton microscopy based on a spatial light modulator. J. Biomed. Opt. **17**, 0305051–0305053 (2012)

12. F.O. Fahrbach, V. Gurchenkov, K. Alessandri, P. Nassoy, A. Rohrbach, Light-sheet microscopy in thick media using scanned bessel beams and two-photon fluorescence excitation. Opt. Express **21**, 13824–13839 (2013)

13. M.A. Alagao, M.A. Go, M. Soriano, G.A. Tapang, Improving the point spread function of an aberrated 7-mirror segmented reflecting telescope, in *4th International Conference on Optics, Photonics and Laser Technology, Institute for Systems and Technologies of Information, Control and Communication* (2015)

14. D. Palima, V. Daria, Holographic projection of arbitrary light patterns with a suppressed zero-order beam. Appl. Opt. **46**, 4197–4201 (2007)

15. I. Moreno, J.A. Davis, T.M. Hernandez, D.M. Cottrell, D. Sand, Complete polarization control of light from a liquid crystal spatial light modulator. Opt. Express **20**, 364–376 (2012)

16. Hamamatsu: SLM module, programmable phase modulator (2003)

17. Y. Takiguchi, T. Otsu, T. Inoue, H. Toyoda, Self-distortion compensation of spatial light modulator under temperature-varying conditions. Opt. Express **22**, 16087–16098 (2014)

18. T. Inoue, H. Tanaka, N. Fukuchi, M. Takumi, N. Matsumoto, T. Hara, N. Yoshida, Y. Igasaki, Y. Kobayashi, Lcos spatial light modulator controlled by 12-bit signals for optical phase-only modulation, in *Integrated Optoelectronic Devices 2007* (International Society for Optics and Photonics, 2007), pp. 64870Y–64870Y

19. S. Reichelt, Spatially resolved phase-response calibration of liquid-crystal-based spatial light modulators. Appl. Opt. **52**, 2610–2618 (2013)

20. M. Persson, D. Engström, M. Goksör, Reducing the effect of pixel crosstalk in phase only spatial light modulators. Opt. Express **20**, 22334–22343 (2012)

21. E. Hällstig, J. Stigwall, T. Martin, L. Sjöqvist, M. Lindgren, Fringing fields in a liquid crystal spatial light modulator for beam steering. J. Mod. Opt. **51**, 1233–1247 (2004)

22. L. Yang, J. Xia, C. Chang, X. Zhang, Z. Yang, J. Chen, Nonlinear dynamic phase response calibration by digital holographic microscopy. Appl. Opt. **54**, 7799–7806 (2015)

23. V. Arrizon, E. Carreon, M. Testorf, Implementation of fourier array illuminators using pixelated slm: efficiency limitations. Opt. Commun. **160**, 207–213 (1999)

24. E. Ronzitti, M. Guillon, V. de Sars, V. Emiliani, LCOS nematic SLM characterization and modeling for diffraction efficiency optimization, zero and ghost orders suppression. Opt. Express **20**, 17843–17855 (2012)

25. J. Liang, Z. Cao, M.F. Becker, Phase compression technique to suppress the zero-order diffraction from a pixelated spatial light modulator (SLM), in *Frontiers in Optics* (Optical Society of America, 2010), p. FThBB6

26. W.D.G.D. Improso, P.L.A.C. Hilario, G.A. Tapang, Zero order diffraction suppression in a phase-only spatial light modulator via the gs algorithm, in *Frontiers in Optics* (Optical Society of America, 2014), p. FTu4C–3

27. W.D.G.D. Improso, G.A. Tapang, C.A. Saloma, Suppression of zeroth-order diffraction in phase-only spatial light modulator via destructive interference with a correction beam, in *5th International Conference on Photonics, Optics, and Laser Technology, Institute for Systems and Technologies of Information, Control and Communication* (2017), pp. 208–214

28. G. Tapang, C. Saloma, Behavior of the point-spread function in photon-limited confocal microscopy. Appl. Opt. **41**, 1534–1540 (2002)

29. J.R. Fienup, Phase retrieval algorithms: a comparison. Appl. Opt. **21**, 2758–2769 (1982)

30. C. Gaur, K. Khare, Sparsity assisted phase retrieval of complex valued objects, in *SPIE Photonics Europe* (International Society for Optics and Photonics, 2016), p. 98960G

31. A. Lizana, I. Moreno, A. Márquez, C. Iemmi, E. Fernández, J. Campos, M. Yzuel, Time fluctuations of the phase modulation in a liquid crystal on silicon display: characterization and effects in diffractive optics. Opt. Express **16**, 16711–16722 (2008)

Chapter 2
Towards High-Order Diffraction Suppression Using Two-Dimensional Quasi-Periodic Gratings

Changqing Xie, Lina Shi, Hailiang Li, Ziwei Liu, Tanchao Pu and Nan Gao

Abstract Two-dimensional (2D) diffraction gratings are playing an increasingly important role in the optics community due to their promising dispersion properties in two perpendicular directions. However, conventional 2D diffraction gratings often suffer from wavelength overlapping caused by high-order diffractions, and producing diffraction gratings with nanometer feature size still remains a challenge. In recent years, 2D quasi-periodic diffraction gratings have emerged that seek to suppress high-order diffractions, and to be compatibility with silicon planar process. This chapter reviews the optical properties of 2D quasi-periodic gratings comprised of quasi-triangle array of holes, and details the effects of hole shape and location distribution on the high-order diffraction suppression. It is also discuss the feasibility of various nanofabrication techniques for high volume manufacturing 2D quasi-periodic gratings at the nanoscale.

2.1 A Quick Tour of Diffraction Gratings

Diffraction gratings composed of periodic structures are simple and fundamental optical elements that separate incident light into its constituent wavelength components. The dispersive feature of diffraction gratings makes them attractive for fundamental studies and photonics applications in spectroscopy, microscopy, and interferometry [1–4]. Actually, since Joseph von Fraunhofer laid the foundation of diffraction gratings about 200 years ago, diffraction grating has been the most successful single optical device, and has found promising applications in widely diverse areas of physics, chemistry, biology, and engineering. For more details, one can see [5]. Among them, some related work on absorption and emission line spectroscopy investigations has been awarded Nobel Prizes [6].

C. Xie · L. Shi (✉) · H. Li · Z. Liu · T. Pu · N. Gao
Key Laboratory of Microelectronic Devices and Integrated Technology, Institute of
Microelectronics, Chinese Academy of Sciences, Beijing 100029, People's Republic of China
e-mail: shilina@ime.ac.cn

© Springer Nature Switzerland AG 2019
P. Ribeiro et al. (eds.), *Optics, Photonics and Laser Technology 2017*,
Springer Series in Optical Sciences 222,
https://doi.org/10.1007/978-3-030-12692-6_2

2.1.1 Grating Equation

Various diffraction gratings with different configurations have been developed, they can be divided into two broad categories, reflection and transmission types [6, 7]. The former utilizes diffracted light on the same side of the grating normal, and can be further classified as plane and concave types. The latter utilizes diffracted light cross over the grating normal. The angle of diffraction measured from the grating normal can be quantitatively described by a simple expression, the well-known grating equation [6]

$$\sin \theta_m = (m\lambda - d \sin \theta_i)/d, \qquad (2.1)$$

where d represents the grating period, λ is the incident light wavelength, θ_i denotes the angle between the incident light direction and the normal to the grating surface, and m is an integer known as the order of the diffracted light. Compared to the reflection gratings, transmission gratings are much simpler to use in monochromator and spectrometer. This is because they have higher figure error tolerances, smaller weight and size, wider bandwidth and easy-to-fabricate grating profiles, and greatly simplify the alignment at the expense of lower diffraction efficiency [8]. For most practical applications, only the generated +1st or −1st diffraction order is needed to realize the unique light dispersion. The resolving power R of a planar diffraction grating is the ability to separate and distinguish adjacent spectral lines in the constituent spectrum as a function of wavelength. It can be expressed as [6]

$$R = \lambda/\Delta\lambda = mN, \qquad (2.2)$$

where N is the total number of grooves illuminated on the grating surface.

2.1.2 Overlapping of Diffraction Orders

Compared with a prism, one major disadvantage of planar diffraction gratings is that they suffer from higher-order diffraction contamination and limited free-spectral range due to the diffractive nature of planar periodic structures. When the bandwidth of the incident light is large enough, the grating equation can be fulfilled by an infinite set of wavelength values, i.e., light with wavelength λ in order m is always diffracted at the same angle as light with wavelength $\lambda/2$ in order $2m$, and as light of wavelength $\lambda/3$ in order $3m$, etc. The free spectral range is the bandwidth in a given order for which overlap of bandwidth from an adjacent order does not occur. It is equal to λ_1/m, where λ_1 is the short-wave end of the band, indicating that the free spectral range is longer for lower order. Actually, this is one reason that only the generated ±1st diffracted orders are needed in many applications.

2.1.3   High Order Diffraction Suppression for 1D Gratings

Traditionally, there are two possible ways to avoid the so-called overlapping of diffraction orders at a particular wavelength λ. One is to properly choose the planar grating period $D \in (\lambda, 2\lambda)$, which presents the major drawback of the wavelength range being limited to $(D, D/2)$. The other way is to use sinusoidal gratings with continuous relief [9]. However, the fabrication of continuous-relief diffractive optical elements is not compatible with the well-established silicon planar process, and still remains a challenge for mass production [10, 11].

Over the past decade, several methods have also been proposed to realize high order diffraction suppression for 1D gratings. In particular, Cao et al. proposed a binary grating composed of sinusoidal-shaped apertures to realize sinusoidal amplitude transmission at one direction [12]. After that, some new types of so-called single-order diffraction gratings, such as quantum-dot-array diffraction gratings [13], quasi-sinusoidal diffraction transmission gratings [14], modulated groove position diffraction gratings [15], zigzag diffraction transmission gratings [16], and trapezoidal transmission function [17], have been further developed. The key idea of all these work is to mimic a sinusoidal amplitude transmission function along one dimension by special structures. Single-order transmission diffraction gratings based on line-structure with rough edges [18] and dispersion engineered all-dielectric metasurfaces [19] have also been proposed. Besides, the order-sorting method using PbSe array detector [20] and thin films [21] have also been used to reduce the influence of higher-order diffractions.

2.2   Two-Dimensional Periodic Gratings

Thousands of papers have been published on various aspects of diffraction gratings, and most of them are devoted to 1D diffraction gratings consisting of a large number of equally spaced slits on an opaque screen. In recent years, two-dimensional (2D) diffraction gratings with complete order (periodic) structures have also received much attention due to their ability to simultaneously separate incident light into its constituent wavelength components in two perpendicular directions [22–27]. For example, in nearly all kinds of microscopes, 2D diffraction gratings are often used to calibrate the xy-plane [26]. In grating interferometry, 2D diffraction gratings are used to improve reconstruction of the wavefront phase [27].

Similar to the 1D case, 2D periodic structures also suffer from higher-order diffractions due to the diffractive nature of 2D structures. In order to get further insight into this, we begin our analysis from diffractions of a large number of identical and same oriented holes, as shown in Fig. 2.1. We denote the coordinate systems of the hole plane and the diffraction plane by (ξ, η) and (x, y), respectively. The coordinates of the hole center are (ξ_1, η_1), (ξ_2, η_2), ..., (ξ_N, η_N). We calculate the far-field diffraction pattern of N holes with area A by Fraunhofer diffraction formula [9, 33]

Fig. 2.1 Schematic of the
2D structures comprised of a
large number of identical and
same oriented holes

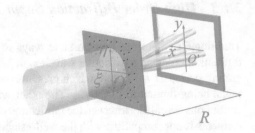

$$U(p,q) = C \sum_n e^{-ik(p\xi_n + q\eta_n)} \iint_A e^{-ik(p\xi' + q\eta')} \mathrm{d}\xi' \mathrm{d}\eta'. \tag{2.3}$$

Here $C = \sqrt{P}/(\lambda R)$, P is the power density incident on the hole array, R is the distance between the hole array plane and the observation plane, λ is the incident light wavelength, $k = 2\pi/\lambda$ is the wave vector in free space, and $p = x/R$, $q = y/R$.

From (2.3), the diffraction intensity pattern of a large number of holes in Fig. 2.1 can be expressed as

$$I(p,q) = I^0(p,q) \cdot \left| \sum_{n=1}^{N} e^{-ik(p\xi_n + q\eta_n)} \right|^2$$

$$= I^0(p,q) \cdot \sum_{m=1}^{M} \sum_{n=1}^{N} e^{-ik[p(\xi_n - \xi_m) + q(\eta_n - \eta_m)]}, \tag{2.4}$$

where $I^0(p,q)$ is the intensity distribution arising from a single hole and the remaining part denotes the interference effects of different holes. When we consider the effect of a large number of holes, we should obtain quite different results depending on whether the holes are distributed regularly or irregularly. When the holes are distributed irregularly, terms with different values of m and n will fluctuate rapidly between $+1$ and -1, and in consequence the sum of such terms will have zero mean value. Each remaining term ($m = n$) has the value unity. Hence the total intensity is N times the intensity of the light diffracted by a single hole: $I(p,q) \sim NI^0(p,q)$. The results are quite different when the holes are distributed regularly since the terms with $m \neq n$ will give appreciable contributions. For example, for some two dimensional array of holes, the phases of all the terms for which $m \neq n$ are exact multiple of 2π, their sum $I(p,q)$ will be equal to $N(N-1)$, and so for large N will be of the order of N^2.

Therefore, the locations of holes can be designed to create constructive interference leading to a subwavelength focus of prescribed size and shape [28–31]. The quasi-periodic distributed holes with special shape can acquire rich degrees of freedom (spatial position and geometric shape of holes) to realize complex functionalities, which are not achievable through periodic features of conventional grating with limited control in geometry. Thus, the suppression of high-order diffractions can be realized by the destructive interference of light from quasi-periodic array with specific distribution and specific shape of the holes. At the same time, the 1st order diffraction efficiency can be increased by the constructive interference of lights from the different holes.

2.3   Two-Dimensional Quasi-Periodic Gratings

In this section, the effects of the quasi-periodic grating structure parameters on the diffraction property will be evaluated. From (2.3) and (2.4), the diffraction intensity pattern of the quasi-triangle array with periods P_ξ and P_η in Fig. 2.2 can be expressed as

$$I(p, q) = \frac{\sin^2\left(N_\xi/2 \cdot kp2P_\xi/2\right) \cdot \sin^2\left(N_\eta kq\, P_\eta/2\right) \cdot \cos^2\left(kp2P_\xi/4 + kq\, P_\eta/4\right)}{\left(N_\xi/2\right)^2 \cdot \sin^2\left(kp2P_\xi/2\right) \cdot N_\eta^2 \cdot \sin^2\left(kq\, P_\eta/2\right)}$$

$$\cdot \left| C \iint_A e^{ik(p\xi + q\eta)} \mathrm{d}\xi\, \mathrm{d}\eta \right|^2 \cdot \left| \int \rho(s) \cdot e^{-ikps} \mathrm{d}s \right|^2$$

$$= I_1(p, q) \cdot I_2(p, q) \cdot I_3(p), \tag{2.5}$$

where s is the hole location deviation from the triangle lattice point along ξ axis, the holes are randomly shifted by s according to probability distribution function $\rho(s)$. Here, the hole location along η axis is fixed since we usually focus on the diffraction property along one direction. And the triangle array is selected rather than square array due to the more spacing between any two adjacent holes than the square one with the same period and hole size. There are three parts in (2.4). The first part $I_1(p, q) = \frac{\sin^2\left(N_\xi/2 \cdot kp2P_\xi/2\right)\cdot\sin^2\left(N_\eta kq\, P_\eta/2\right)\cdot\cos^2\left(kp2P_\xi/4 + kq\, P_\eta/4\right)}{\left(N_\xi/2\right)^2\cdot\sin^2\left(kp2P_\xi/2\right)\cdot N_\eta^2\cdot\sin^2\left(kq\, P_\eta/2\right)}$, only depends on P_ξ, P_η, N_ξ and N_η. It is the interference effect resulting from the triangle array.

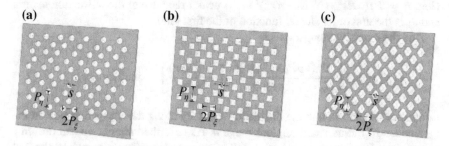

Fig. 2.2 Schematic of the quasi-triangle array gratings

The second part $I_2(p,q) = \left| C \iint_A e^{ik(p\xi+qn)} d\xi \, dn \right|^2$ denotes the effect of a single hole, and depends on the hole shape and size. It should be noted that the third part, $I_3(p) = \left| \int_S \rho(s) \cdot e^{-ikps} ds \right|^2$, is introduced by the location deviation of holes. Equation (2.4) shows that the geometric shape and spatial position of holes can be optimized to manipulate the diffraction intensity pattern $I(p,q)$.

Next we investigate effects of the hole shape and size on the diffraction intensity pattern. Here we only consider some simple shapes with less sides since the complicated structures are difficult to be theoretically analyzed and precisely fabricated. We will discuss quasi-period gratings of circles, rectangles, and hexagons in Fig. 2.2.

2.3.1 Two-Dimensional Quasi-Triangle Array of Circular Holes

Firstly two-dimensional gratings of circular holes, which are the simplest will be discussed shape and easiest to be fabricated [32]. At the same time, the simplest uniform distribution of hole location will be selected. That is to say, the holes are shifted by s from the lattice points along the ξ axis according to the probability distribution $\rho(s) = 1/(2a)$, $|s| \leq a$, where a is the shift range of circle holes along the ξ axis. From (2.5), for the quasi-triangle array of $N_\xi N_\eta$ circular holes with the radius r, the diffraction intensity pattern in the Fraunhofer diffraction is given by

$$I(p,q) = \frac{\sin^2(N_\xi/2 \cdot kp2P_\xi/2) \cdot \sin^2(N_\eta kq P_\eta/2) \cdot \cos^2(kp2P_\xi/4 + kq P_\eta/4)}{(N_\xi/2)^2 \cdot \sin^2(kp2P_\xi/2) \cdot N_\eta^2 \cdot \sin^2(kq P_\eta/2)}$$

$$\cdot I_0 \cdot \left[\frac{2J_1(kr\sqrt{p^2+q^2})}{kr\sqrt{p^2+q^2}} \right]^2 \cdot \text{sinc}^2(kpa/\pi). \tag{2.6}$$

Here $I_0 = P/(\lambda R)^2 \cdot (N_\xi N_\eta \cdot \pi r^2)^2$ is the peak irradiance of the diffraction pattern and J_1 is the first order Bessel function of the first kind.

And the intensity along x axis is

$$I(p) = I_0 \cdot \left[\frac{\sin(kp P_\xi N_\xi/4)}{N_\xi \cdot \sin(kp P_\xi/4)} \right]^2 \left[\frac{2J_1(kpr)}{kpr} \right]^2 \cdot \text{sinc}^2(kpa/\pi). \tag{2.7}$$

Equation (2.7) shows that the diffraction intensity along the x axis depends on the parameters of radius r and random range a. As described above, we can design r to make the first zero crossing of J_1 fall at some order diffraction such as the 2nd or 3rd order diffraction along x axis, and thus make it disappear. The third part $I_3(p) = \text{sinc}^2(kpa/\pi)$, is introduced by the location deviation of holes. It should be noted that the normalized sinc has zero crossings occurring periodically at non-zero integers. Thus we can optimize a to make these zero crossings of $\text{sinc}^2(kpa/\pi)$ fall at even order diffractions.

Now we investigate the dependence of all order diffractions on the radius r and random range a. According to (2.7), the m-th order diffraction intensity along the ξ axis is

$$I(m) = I_0 \cdot \left[\frac{2J_1(2\pi mr/P_\xi)}{2\pi mr/P_\xi} \right]^2 \cdot \text{sinc}^2(2ma/P_\xi). \tag{2.8}$$

In the real spectral measurement, only the adjacent diffractions (such as the 2nd and 3rd order diffractions) will overlap the 1st order diffraction. The higher order diffractions are usually very small and have little effects on the 1st order diffraction. Thus we pay the utmost attention to the structure parameters which lead to the vanishing of the 2nd and 3rd order diffractions. According to (2.8), the 2nd and 3rd order diffractions simultaneously disappear as $(r/P_\xi, a/P_\xi)$ take some special values: (0.203, 1/4), (0.305, 1/6), (0.305, 1/3), (0.372, 1/4). Considering the fabrication tolerance, we select $r/P_\xi = 0.203$ and $a/P_\xi = 0.25$ since the smallest spacing $\sqrt{(P_\xi - 2a)^2 + (P_\eta/2)^2} - 2r = 0.3011P_\xi$ (for $P_\xi = P_\eta$) of arbitrary adjacent holes is the largest one in the four cases of $(r/P_\xi, a/P_\xi)$. Here, it should be noted that the sinc function with $a/P_\xi = 0.25$ eliminates not only the 2nd order diffraction but also all the even order diffractions since it has periodic zero crossings.

Figure 2.3 presents the diffraction intensity pattern of $r/P_\xi = 0.203$ and $a/P_\xi = 0.25$ according to (2.6) and (2.7). As expected, Fig. 2.3a shows that the 0th and 1st order diffractions are kept along x axis, and high order diffractions disappear. Insets in Figs. 2.3b show clearly intensity distributions of the 0th and 1st order diffractions. The diffraction intensity along x axis in Fig. 2.3b presents clearly the complete suppression of the 2nd, 3rd 4th and 6th order diffractions. The 5th order diffraction of 5.370×10^{-5} is as low as 0.02% of the 1st order diffraction of 0.2637. As a result, it will be submerged in the background noise, i.e., it will decay to a negligible value in real experimental measurements.

In order to verify the validity of the theoretical analysis, the diffraction intensity pattern of the quasi-triangle array of 301×301 circular holes is simulated, according to (2.6) from Fraunhofer approximation. The locations of holes are determined by generating uniformly distributed pseudorandom numbers. The two-dimensional grating has the period. $P_\xi = P_\eta = 10 \, \mu\text{m}$ and the area 3.01 mm $\times$ 3.01 mm. Figure 2.4 shows that there exist the 0th and 1st order diffractions along x axis, and the 2nd, 3rd, 4th and 6th order diffractions disappear. Insets in Figs. 2.4b show clearly intensity distributions of the 0th and 1st order diffractions. The 5th order diffraction of 5.273×10^{-5} is as low as 0.02% of the 1st order diffraction of 0.2638. These agree very well with the theoretical results. Different from the theoretical results, the noise is introduced between any adjacent diffraction. Fortunately, the noise is much smaller than the 5th order diffraction and can be submerged in the background noise.

In order to further confirm the feasibility of the high order diffraction suppression of our theoretical predictions, a binary transmission grating comprised of quasi-

Fig. 2.3 a The far-field diffraction intensity pattern of the quasi-triangle array of circular holes. **b** The diffraction intensity along the x axis

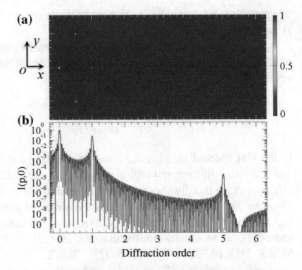

Fig. 2.4 a The far-field diffraction intensity pattern of the quasi-triangle array of 90,601 circular holes. **b** The diffraction intensity along the x axis

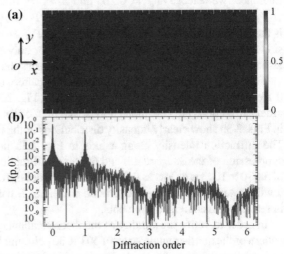

triangle array of 4000×4000 circular holes over area of $4\,cm \times 4\,cm$ on a glass substrate was fabricated for the visible light region by laser write lithography. Firstly, the chromium layer with thickness of $110\,nm$ was deposited onto the soda glass with $2.286\,mm$ thickness with an electron beam evaporation system, and $500\,nm$ thick AZ1500 photoresist was spin coated onto the chrome layer. Secondly, GDSII data of the designed binary gratings were imported into DESIGN WRITE LAZER 2000 (Heidelberg Instruments Mikrotechnik GmbH). Laser exposure and resist development were performed to pattern the binary gratings onto the resist. The exposure wavelength is $413\,nm$ and the exposure power is $80\,mW$. After that, wet etching

Fig. 2.5 Microphotograph of the fabricated quasi-triangle array of circular holes

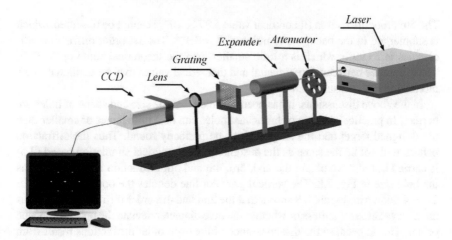

Fig. 2.6 .

technique was used to transfer the resist pattern onto the chrome layer. Finally, the residual resist was removed by using plasma ashing followed by acetone rinse.

The microphotograph of the fabricated structure is illustrated in Fig. 2.5. Periods $2P_\xi$ and P_η of the quasi-triangle array along the x and y axes are respectively 20 μm and 10 μm. The hole diameter is $0.406P_\xi \approx 4$ μm. It is shown clearly that the spacing between any two adjacent holes is larger than $0.3011P_\xi \approx 3$ μm. The experimental setup for optical demonstration is shown in Fig. 2.6. A collimated laser beams from Sprout (Lighthouse Photonics) with the wavelength of 532 nm was used to illuminate the two-dimensional grating, and the far-field diffraction pattern from the grating is focused by a lens and then recorded on a charge coupled device (CCD) camera (ANDOR DU920P-BU2) with 1024×256 pixels.

The measurement results were performed at low temperature of -85 °C, the results are shown in Fig. 2.7. It is clearly shown that only the 0th and the 1st orders exist along the x axis, which agrees well with the theoretical and simulation results.

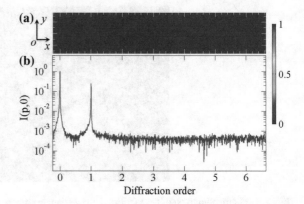

Fig. 2.7 a The far-field diffraction intensity pattern of the quasi-triangle array of circular holes. **b** The diffraction intensity along x axis

The 5th order diffraction (theoretical value 5.37×10^{-5}) cannot be observed, which is submerged in the background noise of 5×10^{-4}. The 1st order diffraction efficiencies is 23.99%, which is a little smaller than the theoretical value of 26.37%. The difference between experimental and theoretical values may be attributed to the fabrication and measurement errors.

In the above discussions, it has been assumed that the size and shape of holes are perfect. In practice, the size of fabricated holes can be a little larger or smaller than the designed target and the shape cannot be perfectly round. Thus the diffraction pattern will not be the same as the designed one. Numerical simulation based (2.5) is carried out and we obtain the 2nd, 3rd, 4th and 5th diffraction intensities versus the hole size in Fig. 2.8. The vertical grey dot line denotes the optimized diameter $2r = 4.0656\,\mu m$. Figure 2.8 shows that the 2nd and 4th order diffraction intensities are always zeros regardless of whether the hole diameter deviate the optimized value or not. This is because the disappearance of the even order diffractions result from zero crossings of the normalized sinc function, which is from the location randomness of holes. As $2r \in (3.6, 4.8)\,\mu m$, the 3rd order diffraction intensity will not be larger than 5×10^{-4}, even though it increases with the deviation of hole diameter from the optimized value. Similarly, the 5th order diffraction intensity is smaller than 3×10^{-4} as $2r \in (3.25, 4.9)\,\mu m$. Therefore, the quasi-periodic two-dimensional grating comprised of circular holes can, at least, tolerate $\pm10\%$ deviation of hole size. This large tolerance makes our structure can be easily fabricated by the current planar silicon technology.

2.3.2 Two-Dimensional Quasi-Triangle Array of Rectangular Holes

Two-dimensional gratings of rectangular holes will now be discussed. The holes are shifted by s along the ξ axis (Fig. 2.2b) according to the probability distribution

Fig. 2.8 The 2nd, 3rd, 4th and 5th order diffraction intensities versus the hole diameter

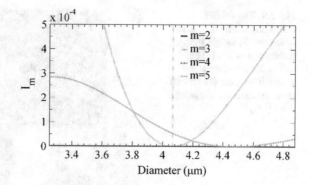

$\rho(s) = (\pi/P_\xi) \cdot \cos(2\pi s/P_\xi)$, $|s| \leq P_\xi/4$, where $2P_\xi$ is the period of the triangle array along the ξ axis [33].

For the quasi-triangle array of $N_\xi N_\eta$ rectangular holes of sides $2a = P_\xi/2$ and $2b = P_\eta/2$ as shown in Fig. 2.2b, the diffraction intensity pattern is [33]

$$I(p, q) = \frac{\sin^2\left(N_\xi/2 \cdot kp2P_\xi/2\right) \cdot \sin^2\left(N_\eta kq P_\eta/2\right) \cdot \cos^2\left(kp2P_\xi/4 + kq P_\eta/4\right)}{\left(N_\xi/2\right)^2 \cdot \sin^2\left(kp2P_\xi/2\right) \cdot N_\eta^2 \cdot \sin^2\left(kq P_\eta/2\right)}$$

$$\cdot I_0 \cdot \left(\frac{\sin kpa}{kpa}\right)^2 \left(\frac{\sin kqb}{kqb}\right)^2 \cdot \frac{\cos^2(kp P_\xi/4)}{(1 - kp P_\xi/2/\pi)^2 (1 + kp P_\xi/2/\pi)^2}. \quad (2.9)$$

Here $I_0 = P/(\lambda R)^2 \cdot (N_\xi N_\eta \cdot 4ab)^2$ is the peak irradiance of the diffraction pattern.

Figure 2.9 presents the diffraction intensity pattern according to (2.6). As expected, the 0th and 1st order diffractions are kept along x axis, and the high-order diffractions disappear. The logarithm of diffraction intensity along x axis in Fig. 2.9b presents clearly the complete suppression of the high order diffractions. Insets in Fig. 2.9 show the intensity distributions of the 0th and 1st order diffractions, respectively. From Fig. 2.9, one can see that the diffraction pattern of the quasi-triangle array of rectangular holes along x axis is the same as that of the ideal sinusoidal transmission grating.

Numerical simulation based on (2.9) is carried out to evaluate the diffraction property of the quasi-triangle array of 100,000 rectangular holes. The logarithm of diffraction intensity along x axis is shown in Fig. 2.10, and high-order diffraction is much less than the noise of 10^{-5} between 0th and 1st diffraction (red dash line in Fig. 2.10), which agrees well with the theoretical prediction of (2.9) and Fig. 2.9.

A binary transmission grating comprised of quasi-triangle array of rectangular holes was fabricated and tested by the above described process and setup. Figure 2.11 presents the recorded diffraction pattern and it is obvious that high order diffractions of the quasi-triangle array of rectangular holes are effectively suppressed. The diffraction intensity along x axis in Fig. 2.11b is almost the same as the ideal sinusoidal transmission gratings. The ratio of 1st order diffraction intensity to the 0th order diffraction intensity is 73.56% and much larger than the theoretical prediction and

Fig. 2.9 The far-field diffraction intensity pattern of the quasi-triangle array of rectangular holes. **b** The diffraction intensity along *x* axis. Insets: the 0th and 1st order diffractions. (Reprinted from [33])

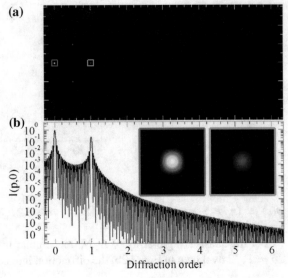

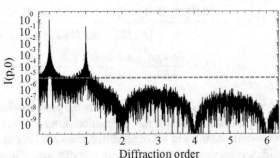

Fig. 2.10 The diffraction intensity along *x* axis of the quasi-triangle array with 100,000 rectangular holes. (Reprinted from [33])

numerical value 25%. This is because the CCD is saturated by the 0th order diffraction intensity. In addition, the red vertical lines in Fig. 2.11a are crosstalk along *y* direction due to our one-dimension CCD.

Now we focus on the complete suppression of high order diffractions along the *x* axis. As N_ξ is large enough, the intensity according to (2.9) along the *x* axis is given by [33]

$$
I(p) = \frac{I_0 \cdot \mathrm{sinc}^2(N_\xi kp8a/\pi)}{(1 - kp2a/\pi)^2(1 + kp2a/\pi)^2 \cos^2(kp4a)}
$$
$$
= \begin{cases} I_0, & p = 0 \\ \frac{1}{4}I_0, & p = \frac{\pm\pi}{2ka} \end{cases} \tag{2.10}
$$

Equation (2.10) demonstrates that the quasi-triangle array of infinite rectangular holes can generate the same diffraction pattern as sinusoidal transmission gratings along the *x* axis. Only three diffraction peaks (the 0th order and +1st/−1st orders)

Fig. 2.11 a The far-field diffraction intensity pattern of the quasi-triangle array of rectangular holes. b The diffraction intensity along the ξ axis. (Reprinted from [33])

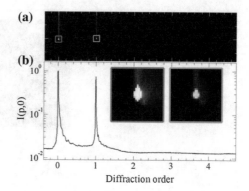

appear on the x-y plane. The above theoretical results are scalable from X-ray to far infrared wavelengths.

To obtain physical insight into the diffraction property of the quasi-triangle array of rectangular holes, the average transmission function along ξ axis is calculated by integrating the probability distribution over η axis [33]

$$T(\xi) = \int_{|\xi|-P_\xi/4}^{P_\xi/4} \rho(s) \mathrm{d}s = \frac{1}{2}(1 + \cos(\frac{2\pi}{P_\xi}\xi)). \tag{2.11}$$

Equation (2.11) shows that the quasi-triangle array of infinite rectangular holes has the same transmission function along the ξ axis as sinusoidal transmission gratings. It is the average diffractive effect similar to sinusoidal grating that eliminates high-order diffractions.

We can also understand the suppression of high-order diffractions by the interference weakening or strengthening. It is known that diffraction peaks is from the constructive interference of lights from the different holes. The interference of lights from different rectangular holes is controlled by the hole position. The desired diffraction pattern only containing the 0th order and +1st/−1st order diffractions can be tailored by the location distribution of holes according to some statistical law.

2.3.3 Two-Dimensional Quasi-Triangle Array of Hexagonal Holes

A grating with the quasi-triangle array of hexagonal holes to completely compressed the 2nd, 3rd, 4th, 5th and 6th order diffractions along x axis will now be addressed. The holes are shifted by s from the lattice points along the ξ axis according to the probability distribution $\rho(s) = 1/(2a)$, $|s| \leq a$, where a is the shift range of circle

Fig. 2.12 The far-field diffraction intensity pattern of the quasi-triangle array of hexagonal holes

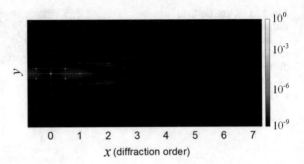

holes along the ξ axis. From (2.5), the diffraction intensity pattern $I(p, 0)$ can be described as

$$I(p, 0) = \frac{\sin^2 \left(N_\xi k p P_\xi /2\right)}{N_\xi^2 \cdot \sin^2 \left(k p P_\xi /2\right)}$$
$$\cdot \left(\frac{\sin(kp(a_2 + a_1)/2) \cdot \sin(kp(a_2 - a_1)/2)}{kp(a_2 + a_1)/2 \cdot kp(a_2 - a_1)/2}\right)^2 \cdot \text{sinc}^2(kpa/\pi). \tag{2.12}$$

Then the m-order diffraction intensity along x axis is

$$I(m) = \text{sinc}^2 \left(\frac{m(a_2 + a_1)}{P_\xi}\right) \cdot \text{sinc}^2 \left(\frac{m(a_2 - a_1)}{P_\xi}\right) \text{sinc}^2 (\frac{2ma}{P_\xi}). \tag{2.13}$$

Equation (2.13) shows that the m-order diffraction intensity $I(p, 0)$ is the product of three normalized sinc functions and depends on a, a_1 and a_2. Thus we can set $\frac{m(a_2+a_1)}{P_x} = n_1$, $\frac{m(a_2-a_1)}{P_x} = n_2$, and $\frac{2ma}{P_x} = n_3$ to suppress three kinds of the 2nd, 3rd, and 5th order diffractions.

In order to validate the theoretical analysis, numerical simulation for the case of $a = 1/10$, $a_1 = 1/12$ and $a_2 = 5/12$, has the lowest noise due to the smallest a. is carried out to evaluate the diffraction property of the quasi-triangle array with 100×100 hexagonal holes. As expected, the distribution of the diffraction intensity shown in Fig. 2.12 only has the 0th and $\pm$1st order diffractions along x axis, and the 2nd, 3rd, 4th, 5th and 6th order are completely suppressed. Different from the theoretical results, the noise is introduced due to the shift of the hole position. The 7th order diffraction of 7.473×10^{-6} is almost submerged in the noise, and as low as 0.003% of the 1st order diffraction of 0.2426.

A binary transmission grating comprised of hexagonal holes was fabricated and tested by the above described process and setup. The diffraction property is shown in Fig. 2.13. The 2nd, 3rd, 4th, 5th and 6th order diffractions disappear along x axis. This quantitatively agrees with the theoretical and simulation results. The 1st order diffraction efficiency is 24.65%, which is a little different from the theoretical value of 24.26%. The difference may result from the fabrication and measurement errors.

Fig. 2.13 The far-field
diffraction intensity pattern
of the quasi-triangle array
with hexagonal holes. **b** The
diffraction intensity along
the x axis

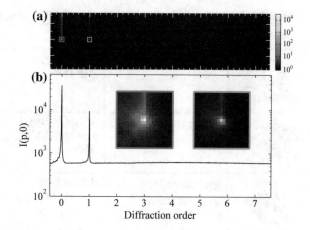

As expected, Fig. 2.13 also shows that the 7th order diffraction is submerged in
background noise.

2.3.4   Comparison of Three Types of Gratings

Table 2.1, summarizes the second part $I_2(p, 0)$ and $I_3(p)$ of (2.3), $I_2(p, 0)$ is the
envelope line of the diffraction intensity pattern [9] and $I_3(p)$ is introduced by the
location deviation of holes. $I_2(p, 0)$ only depends on one structure parameter for
circular or rectangular holes, which can make the 2nd or even order diffractions
zeros. Fortunately, for the hexagonal hole, $I_2(p, 0)$ is the product of two normalized
sinc functions depending on the two parameters a_2 and a_1, and thus has two kinds
of zero crossings, which can make both even and the 3rd, 6th, …, order diffractions
simultaneously disappear.

In other words, three types have their advantages and disadvantages. Two-
dimensional quasi-triangle array of circular holes the enough deviation tolerance
of hole size and the relatively large spacing of adjacent holes, which make our grating much easy to be fabricated than the rectangular or hexagonal holes. While it only
can completely suppress the 3rd and even order diffractions. The gratings composed
of rectangular holes with the special location probability distribution can completely
all the 2nd and higher order diffractions. However, the small gap from the probability
distribution and the right angle of hole make the fabrication difficult. The hexagonal
hole is the balance of the circular and rectangular holes. The complete suppression
of 2nd, 3rd, 4th, 5th and 6th order diffractions by the quasi-triangle array of the
hexagonal holes was demonstrated theoretically and experimentally. The 7th order
diffraction is as low as 2.2×10^{-5} of the 1st order, which can easily be submerged
in the background noise in real applications. At the same time, the smallest gap of

Table 2.1 The impact of the shape of the holes on the diffraction pattern

Shape	$I_2(p,0)$	$\rho(s)$	$I_3(p)$
	$\left(2J_1(krp)/krp\right)^2$	$1/(2a)$, $|s| \le a$, $a = P_\xi/4$	$\mathrm{sinc}^2(kpa/\pi)$, $a = P_\xi/4$
	$\mathrm{sinc}^2(kpa_2/\pi)$	$\rho(s) = (\pi/P_\xi) \cdot \cos(2\pi s/P_\xi), |s| \le P_\xi/4$	$\dfrac{\cos^2(kp\,P_\xi/4)}{(1-kp\,P_\xi/2/\pi)^2(1+kp\,P_\xi/2/\pi)^2}$
	$\mathrm{sinc}^2(\frac{kp(a_2+a_1)}{2\pi}) \cdot$ $\mathrm{sinc}^2(\frac{kp(a_2-a_1)}{2\pi})$	$1/(2a)$, $|s| \le a$, $a = P_\xi/10$	$\mathrm{sinc}^2(kpa/\pi)$, $a = P_\xi/10$

the hexagonal holes is larger than that of the rectangular gap, and thus the grating of hexagonal holes is easier to fabricate than the rectangular one.

2.4　Future Outlook for the Fabrication Method

Since the first diffraction grating was invented by the American astronomer David Rittenhouse in 1785 [6], great efforts have been devoted to the fabrication of high quality diffraction gratings [34, 35]. Until now, the difficulty in fabricating diffraction grating with nanometer feature size is still the major obstacle of advancing the dispersion performance. Various fabrication techniques, including mechanical ruling, interference lithography, e-beam lithography, laser beam lithography and nano-imprinting lithography, have been intensively studied to further reduce the feature size of diffraction grating.

2.4.1　The Mechanical Ruling Method

The first high quality diffraction grating was mechanically ruled [6]. After being polished carefully, the substrates (optical glass or fused silica) are coated with a thick aluminum or gold film. Such an Al or Au film acts as a functional material in the ruling process. Thousands of extremely fine and shallow grooves are faithfully produced by the ruling diamond with special shape in the cross section. For a ruled grating with 1000 grooves per mm and active area of $300 \times 300\,\mathrm{mm}^2$, the total groove number is 3.0×10^5, and the total length of these grooves is as large as 90 km. Thus, the ruling process is time consuming and the ultra-precision ruling engine may take above one month to work. Now only a few ruling engine are still operating for planar echelle gratings greater than 300 mm in width [36, 37].

2.4.2 Interference Lithography

Since the invention of the laser in the 1960s, interference lithography has proven to be an extremely useful technique. The principle of traditional interference lithography is simple, when two mutually coherent plane waves with same wavelength and equal intensity interfere each other, sinusoidal interference fringes will be produced and recorded by photoresist on the substrate.

The spatial resolution of interference lithography is fundamentally governed by the working wavelength due to the diffraction limit. Hence, to produce gratings and grids with feature sizes below 100 nm, shorter working wavelengths (deep ultraviolet and extreme ultraviolet spectral ranges) are required [38]. Several interference lithography methods have been proposed and developed, including Lloyd's mirror interferometer [39], amplitude division interferometers [40], grating-based interference lithography [41], etc.

Interference lithography has the advantage of generating high resolution periodic patterns over large area with extremely simple optical systems, large process latitude and large depth-of-focus. Also, no photomask is needed, one can easily manipulate the grating period by adjusting the angle between the two intersecting beams. A typical example of successful use of interference lithography is provided by MIT space nanotechnology laboratory, where thousands of X-ray transmission gratings have been produced by interference lithography with wavelength of 351.1 nm [42]. By carefully selecting the number of interfering beam or multiple exposures, interference lithography can produce 2D periodic arrays with arbitrary shaped nanomotifs. However, interference lithography is limited to patterning period features only, and the spatial-period of the grating is fundamentally governed by the working wavelength due to the diffraction limit.

2.4.3 Electron Beam Lithography

Electron beam lithography is a well-established lithographic technique for creating arbitrarily shaped patterns with resolution in the nanometer range. At present, in semiconductor industry, electron beam lithography is often used to expose mask patterns for optical lithography tools below 45 nm technology node. In some cases, it is also used for advanced prototyping of integrated circuits due to its flexibility and high resolution.

For larger exposure latitude, the commercial e-beam writer acceleration voltage is taken to be 100 kV. The size of the focused electron beam, which is an important factor directly influencing the resolution, can be reduced to round 1 nm at the expense of very low writing speed. Feature size below 5 nm is theoretically possible by exposing a thin (<30 nm) resist, but rarely demonstrated [43]. The actual patterning resolution is considerably larger and is limited mainly by the well-known proximity effect [44]. As charged particles, electrons undergo forward and backward scattering events when they penetrate through the resist into the substrate, resulting so-called proximity effect, i.e., unwanted dose to take place in the regions adjacent to those exposed by the focused electron beam.

Electron beam lithography is a promising approach for patterning 2D quasi-periodic gratings with high line density and high fidelity. In particular, it is convenient to perform pattern transfer from electron beam resist to a variety of materials. Obtaining high-quality grating patterns is always not straightforward, even using a state-of-the-art electron beam system. As a rule of thumb, for the 1D periodic gratings, the size of the focused electron beam can be one-sixth of the grating period for the purpose of higher writing speed. While for the 2D quasi-periodic gratings, this size should shrink to one-twelfth of the grating period to ensure sharp corners.

The drawback of the serial electron beam lithography is less practical in mass production due to the serial and slow scanning nature. To overcome this limitation, high-throughput multiple beam electron beam lithography, which is based on massively-parallel focused electron beams that can individually beam switched on and off, is now being developed.

2.4.4 Laser Beam Lithography

Laser beam lithography is performed by tightly focusing a laser beam instead of electron beam into a photoresist layer. Similar to serial electron beam lithography, the focused laser beam scans the patterned area to generate photoresist patterns pixel by pixel. On one hand, regarding the lithography costs, the less expensive laser beam lithography is now widely used for patterning mask above 45 nm technology node of semiconductor industry. On the other hand, it is also a powerful and widely used tool for the mask-free fabrication on various substrate materials.

Laser beam lithography is more flexible than the counterpart of electron beam lithography. For example, electron beam lithography must be performed on electrically conductive substrates with vacuum condition, while laser beam lithography is compatible with electrically insulating substrates and is performed under atmosphere condition. Furthermore, one-step laser beam lithography is capable of producing continuous-relief micro-structure with surface relief precision below 100 nm, which can be employed to increase the diffraction efficiency of the micro-optical elements.

In general, the spot size of the focused laser beam is fundamentally limited by the so-called Abbe's law, and is given by $\approx 1.22 \, \lambda/NA$, where NA is the numerical aperture of the light exposure system. This constitutes a major fabrication challenge in achieving sub-diffraction or nanometer resolution at visible wavelengths. To overcome this limitation, in recent years, numerous method have reported on how the resolution of laser direct writing scales into nanometer dimension [45–48]. Remarkably, feature size as small as 9 nm has been achieved by three-dimensional optical beam lithography [48].

2.4.5 Nano-imprint Lithography

As stated above, the processes of mechanical ruling, electron beam lithography and laser beam lithography are serial and hence less practical in high-volume production, and interference lithography is limited to period structures.

Since the first paper was published by Chou et al. [49], nano-imprint lithography, a nanomolding technique to transfer the topography of a template into a substrate, has spurred extensive research by many academic groups. It is now considered as a candidate for next generation lithography by the International Technology Roadmap for Semiconductors (ITRS) roadmap [50, 51], due to its potential for simple, large-area, high-throughput, high-resolution patterning. It also offers a promising way to replicate master gratings.

In nano-imprint lithography, the first and the most important step is the fabrication of master gratings on silicon or quartz substrates. Typically, electron beam lithography or laser beam lithography is used to pattern master gratings due to its ability to generate arbitrary structure with fine features, followed by inductively coupled plasma reactive ion etching to transfer the resist patterns onto the substrate. Once the master grating is fabricated, it can be replicated repeatedly by nano-imprint lithography. The cost-effective imprinting can be performed by the thermal curing method, ultraviolet visible assisted method and injection moulding method, yielding a resist mask [52]. It should be noted that the template on a transparent substrate such as quartz or glass must be used for the ultraviolet visible assisted method, and the use of the transparent substrate make optical alignment of the substrate feasible. In addition, based on this ultraviolet visible transparent template, step and flash imprint lithography that performed at low pressure and room temperature has been developed at wafer-scale [53]. After imprinting, anisotropic oxygen plasma ashing is used to remove the residual thin resist layer. The resist pattern is further transferred into a hard material by lift-off process, electroplate process or plasma process.

2.5 Conclusion

In conclusion, the optics community has witnessed great progress over the past 200 years in the development of one-dimensional diffraction gratings. In recent years, two-dimensional diffraction gratings with two duty cycles in two perpendicular directions are playing an increasingly important role in the optics community. Most previous work on two-dimensional grating diffraction is for 2D periodic structures, which result in so-called high-order diffraction contamination and limited free-spectral range. To overcome this limitation, 2D quasi-periodic gratings comprised of quasi-triangle array of holes have been proposed, and the effects of hole shape (circular, rectangular and hexagonal) and location distribution on the high-order diffraction suppression have been investigated analytically, numerically, and experimentally. These three types of quasi-periodic diffraction gratings have been demonstrated to be robust in suppressing high-order diffractions, and have their advantages and disadvantages.

While the demand for diffraction gratings is widespread, producing a large supply of high quality diffraction gratings at low cost and high speed still remains a challenge. Even the best-established optical lithography can meet the stringent technical conditions of gradual reduction in the minimum dimension of integrated circuits, following the well-known Moore's law, it cannot be applied to produce diffraction gratings. This is because the optical lithography tool is very expensive. Mechanical ruling, interference lithography, electron beam lithography and laser beam lithography have been successfully used to fabricate diffraction gratings. However, interference lithography tool with high throughput and low cost is limited to period structures, and the other methods suffer from low yield. Advances in nano-imprint lithography have made possible the high-volume production of 2D quasi-periodic gratings with nanostructures. In nano-imprint lithography, electron beam lithography or laser beam lithography is used to exposure various predesigned patterns of master gratings in a maskless process, taking advantage of their high resolution and ability to create patterns of arbitrary geometry. The replication process of using imprinting lithography is used for high volume manufacturing of the daughter gratings, taking advantage of its high resolution, low cost, high speed, high process latitude and process robustness. The development of gratings nanofabrication process will allow the proposed quasi-periodic diffraction gratings to find a wide variety of applications in areas as diverse as spectral analysis, imaging and microscopy and interferometry.

Acknowledgements The authors are particularly grateful to the noteworthy assistance of their colleagues. We also would like to thank L. Cao for helpful discussion over many years. This work was funded by National Key Research and Development Program of China (2017YFA0206002) and National Natural Science Foundation of China (61275170, 61107032).

References

1. J. Strong, The Johns Hopkins University and diffraction gratings. J. Opt. Soc. Am. A **50**(12), 1148–1152 (1960)
2. R.L.C. Filho, M.G.P. Homem, R. Landers, A.N. de Brito, Advances on the Brazilian toroidal grating monochromator (TGM) beamline. J. Electron Spectrosc. Relat. Phenom. **144–147**, 1125–1127 (2005)
3. A. Freise, A. Bunkowski, R. Schnabel, Phase and alignment noise in grating interferometers. New J. Phys. **9**, 433 (2007)
4. H. Zhang, J. Zhu, Z. Zhu, Y. Jin, Q. Li, G. Jin, Surface-plasmon-enhanced GaN-LED based on a multilayered M-shaped nano-grating. Opt. Express **21**(11), 13492–13501 (2013)
5. C. Palmer, E. Loewen, in *Diffraction Grating Handbook* (Newport Corp., 2005)
6. http://web.mit.edu/spectroscopy/history/nobel.html

7. N. Bonod, J. Neauport, Diffraction gratings: from principles to applications in high-intensity lasers. Adv. Opt. Photonics **8**(1), 156–199 (2016)
8. R.K. Heilmann, M. Ahn, E.M. Gullikson, M.L. Schattenburg, Blazed high-efficiency x-ray diffraction via transmission through arrays of nanometer-scale mirrors. Opt. Express **16**, 8658–8669 (2008)
9. M. Born, E. Wolf, in *Principles of Optics* (Pergamon, London, 1980)
10. P. Jin, Y. Gao, T. Liu, X. Li, J. Tan, Resist shaping for replication of micro-optical elements with continuous relief in fused silica. Opt. Lett. **35**(8), 1169–1171 (2010)
11. G. Vincent, R. Haidar, S. Collin, N. Guérineau, J. Primot, E. Cambril, J.-L. Pelouard, Realization of sinusoidal transmittance with subwavelength metallic structures. J. Opt. Soc. Am. B **25**(5), 834–840 (2008)
12. L. Cao, E. Förster, A. Fuhrmann, C. Wang, L. Kuang, S. Liu, Y. Ding, Single order x-ray diffraction with binary sinusoidal transmission grating. Appl. Phys. Lett. **90**(5), 053501 (2007)
13. C. Wang, L. Kuang, Z. Wang, S. Liu, Y. Ding, L. Cao, E. Foerster, D. Wang, C. Xie, T. Ye, Characterization of the diffraction properties of quantum-dot-array diffraction grating. Rev. Sci. Instrum. **78**, 053503 (2007)
14. L. Kuang, L. Cao, X. Zhu, S. Wu, Z. Wang, C. Wang, S. Liu, S. Jiang, J. Yang, Y. Ding, C. Xie, J. Zheng, Quasi-sinusoidal single-order diffraction transmission grating used in x-ray spectroscopy. Opt. Lett. **36**(20), 3954–3956 (2011)
15. N. Gao, C. Xie, High-order diffraction suppression using modulated groove position gratings. Opt. Lett. **36**(21), 4251–4253 (2011)
16. H. Zang, C. Wang, Y. Gao, W. Zhou, L. Kuang, L. Wei, W. Fan, W. Zhang, Z. Zhao, L. Cao, Y. Gu, B. Zhang, G. Jiang, X. Zhu, C. Xie, Y. Zhao, M. Cui, Elimination of higher order diffraction using zigzag transmission grating in soft x-ray region. Appl. Phys. Lett. **100**(11), 111904 (2012)
17. Q. Fan, Y. Liu, C. Wang, Z. Yang, L. Wei, X. Zhu, C. Xie, Q. Zhang, F. Qian, Z. Yan, Y. Gu, W. Zhou, G. Jiang, L. Cao, Single-order diffraction grating designed by trapezoidal transmission function. Opt. Lett. **40**(11), 2657–2660 (2015)
18. F.J. Torcal-Milla, L.M. Sanchez-Brea, E. Bernabeu, Diffraction of gratings with rough edges. Opt. Express **16**(24), 19757–19769 (2008)
19. S. Gupta, Single-order transmission diffraction gratings based on dispersion engineered all-dielectric metasurfaces. J. Opt. Soc. Am. A **33**(8), 1641–1647 (2016)
20. W. Lee, H. Lee, J. Hahn, Correction of spectral deformation by second-order diffraction overlap in a mid-infrared range grating spectrometer using a PbSe array detector. Infrared Phys. Technol. **67**, 327–332 (2014)
21. F. Quinn, D. Teehan, M. MacDonald, S. Downes, P. Bailey, Higher-order suppression in diffraction-grating monochromators using thin films. J. Synchrotron Radiat. **5**, 783–785 (1998)
22. R. BrZuer, O. Bryngdahl, Electromagnetic diffraction analysis of two-dimensional gratings. Opt. Commun. **100**, 1–5 (1993)
23. E. Grann, M. Moharam, D.A. Pommet, Artificial uniaxial and biaxial dielectrics with use of two-dimensional subwavelength binary gratings. J. Opt. Soc. Am. A **11**(10), 2695–2703 (1994)
24. M. Kagias, Z. Wang, P. Villanueva-Perez, K. Jefimovs, M. Stampanoni, 2D-omnidirectional hard-x-ray scattering sensitivity in a single shot. Phys. Rev. Lett. **116**, 093902 (2016)
25. S. Rutishauser, M. Bednarzik, I. Zanette, T. Weitkamp, M. Borner, J. Mohr, C. David, Fabrication of two-dimensional hard X-ray diffraction gratings. Microelectron. Eng. **101**, 12–16 (2013)
26. G. Dai, F. Pohlenz, T. Dziomba, M. Xu, A. Diener, L. Koenders, H. Danzebrink, Accurate and traceable calibration of two-dimensional gratings. Meas. Sci. Technol. **18**, 415–421 (2007)
27. Y. Kayser, S. Rutishauser, T. Katayama, T. Kameshima, H. Ohashi, U. Flechsig, M. Yabashi, C. David, Shot-to-shot diagnostic of the longitudinal photon source position at the SPring-8 Angstrom Compact Free Electron Laser by means of x-ray grating interferometry. Opt. Lett. **41**(4), 733–736 (2016)
28. L. Kipp, M. Skibowski, R.L. Johnson, R. Berndt, R. Adelung, S. Harm, R. Seemann, Sharper images by focusing soft X-rays with photon sieves. Nature **414**(6860), 184–188 (2001)

29. K. Huang, H. Liu, F. J. Garcia-Vidal, M. Hong, B. Luk'yanchuk, J. Teng, C. Qiu, Ultrahigh-capacity non-periodic photon sieves operating in visible light. Nat. Commun. **6**, 7059 (2015)
30. C. Xie, X. Zhu, H. Li, L. Shi, Y. Hua, M. Liu, Toward two-dimensional nanometer resolution hard X-ray differential-interference-contrast imaging using modified photon sieves. Opt. Lett. **37**(4), 749–751 (2012)
31. F. Huang, T. Kao, V. Fedotov, Y. Chen, N. Zheludev, Nanohole array as a lens. Nano Lett. **8**(8), 2469–2472 (2008)
32. J. Niu, L. Shi, Z. Liu, T. Pu, H. Li, G. Wang, C. Xie, High order diffraction suppression by quasi-periodic two-dimensional gratings. Opt. Mater. Express **7**, 366–375 (2017)
33. L. Shi, H. Li, Z. Liu, T. Pu, N. Gao, C. Xie, The quasi-triangle array of rectangular holes with the completely suppression of high order diffractions, in *Proceedings of the 5th International Conference on Photonics, Optics and Laser Technology—Volume 1: PHOTOPTICS*, Porto, Portugal, 2017, pp. 54–58
34. E. Di Fabrizio, S. Cabrini, D. Cojoc, F. Romanato, L. Businaro, M. Altissimo, B. Kaulich, T. Wilhein, J. Susini, M. De Vittorio, E. Vitale, G. Gigli, R. Cingolani, Shaping X-rays by diffractive coded nano-optics. Microelectron. Eng. **67–68**, 87–95 (2003)
35. H.I. Smith, 100 years of x-ray: impact on micro- and nanofabrication. J. Vac. Sci. Technol. B, **13**, 2323–2328 (1995)
36. G.R. Harrison, S.W. Thompson, H. Kazukonis, J.R. Connell, 750-mm ruling engine producing large gratings and echelles. J. Opt. Soc. Am. **62**, 751–756 (1972)
37. X. Li, H. Yu, X. Qi, S. Feng, J. Cui, S. Zhang, J. Tu, Y. Tang, 300 mm ruling engine producing gratings and echelles under interferometric control in China. Appl. Opt. **54**, 1819–1826 (2015)
38. B. Paivanranta, A. Langner, E. Kirk, C. David, Y. Ekinci, Sub-10 nm patterning using EUV interference lithography. Nanotechnology **22**, 375302 (2011)
39. A. Ritucci, A. Reale, P. Zuppella, L. Reale, P. Tucceri, G. Tomassetti, P. Bettotti, L. Pavesi, Interference lithography by a soft x-ray laser beam: nanopatterning on photoresists. J. Appl. Phys. **102**(3), 034313–034314 (2007)
40. P. Wachulak, M. Grisham, S. Heinbuch, D. Martz, W. Rockward, D. Hill, J. Rocca, C. Menoni, E. Anderson, M. Marconi, Interferometric lithography with an amplitude division interferometer and a desktop extreme ultraviolet laser. J. Opt. Soc. Am. B **25**, 104–107 (2008)
41. H. Shiotani, S. Suzuki, D. Gun Lee, P. Naulleau, Y. Fukushima, R. Ohnishi, T. Watanabe, H. Kinoshita, Dual grating interferometric lithography for 22-nm node. Jpn. J. Appl. Phys. **47**, 4881–4885 (2008)
42. M.L. Schattenburg, C.R. Canizares, D. Dewey, K.A. Flanagan, M.A. Hamnett, A.M. Levine, K.S.K. Lum, R. Manikkalingam, T.H. Markert, H.I. Smith, Transmission grating spectroscopy and the Advanced X-Ray Astrophysics Facility (AXAF). Opt. Eng. **30**(10), 1590–1600 (1991)
43. J.K.W. Yang, B. Cord, H. Duan, K.K. Berggren, J. Klingfus, S.W. Nam, K.B. Kim, M.J. Rooks, Understanding of hydrogen silsesquioxane electron resist for sub-5-nm-half-pitch lithography. J. Vac. Sci. Technol. B **27**(6), 2622–2627 (2009)
44. W. Chao, B.D. Harteneck, J.A. Liddle, E.H. Anderson. D.T. Attwood, Soft x-ray microscopy at a spatial resolution better than 15 nm. Nature **435**(7046), 1210–1213 (2005)
45. Y. Usami, T. Watanabe, Y. Kanazawa, K. Taga, H. Kawai, K. Ichikawa, 405 nm laser thermal lithography of 40 nm pattern using super resolution organic resist material. Appl. Phys. Express **2**, 126502 (2009)
46. L. Li, R.R. Gattass, E. Gershgoren, H. Hwang, J.T. Fourkas, Achieving λ/20 resolution by one-color initiation and deactivation of polymerization. Science **324**, 910–913 (2009)
47. Y. Cao, M. Gu, λ/26 silver nanodots fabricated by direct laser writing through highly sensitive two-photon photoreduction. Appl. Phys. Lett. **103**, 213104 (2013)
48. Z. Gan, Y. Cao, R.A. Evans, M, Gu, Three-dimensional deep sub-diffraction optical beam lithography with 9 nm feature size. Nat. Commun. **4**, 2061 (2013)
49. S.Y. Chou, P.R. Krauss, P.J. Renstrom, Imprint of sub-25 nm vias and trenches in polymers. Appl. Phys. Lett. **67**, 3114 (1995)
50. http://www.itrs.net/. Accessed 2017

51. J.V. Schoot, H. Schift, Next-generation lithography-an outlook on EUV projection and nanoimprint. Adv. Opt. Techn. **6**(3–4), 159–162 (2017)
52. M.T. Gale, C. Gimkiewicz, S. Obi, M. Schnieper, J. Sochtig, H. Thiele, S. Westenhofer, Replication technology for optical microsystems. Opt. Lasers Eng. **43**, 373–386 (2005)
53. M. Colburn, S. Johnson, M. Stewart, S. Damle, T. Bailey, B. Choi, M. Wedlake, T. Michaelson, S.V. Sreenivasan, J. Ekerdt, C.G. Willson, Step and flash imprint lithography: a new approach to high-resolution patterning. Proc. SPIE **3676**, 379–385 (1999)

Chapter 3
Two Dimensional Gratings of Connected Holes for High Order Diffraction Suppression

Lina Shi, Ziwei Liu, Tanchao Pu, Hailiang Li, Jiebin Niu and Changqing Xie

Abstract The use of two dimensional gratings comprised of connected holes for the high order diffraction suppression will be discussed. An analytical study of the diffraction property of the three kinds of gratings is described, and the dependence of the high order diffraction property on the transmission function is investigated. Notably, theoretical analysis reveals that the 2nd, 3rd and 4th order diffractions adjacent to the 1st order diffraction can be completely suppressed for all three kinds of gratings. The 5th order diffraction is as low as 0.16% of the 1st order diffraction, and thus can be submerged in the background noise for most practical applications. Experimental results are shown to be consistent with the theoretical predictions. Especially, the smallest characteristic size of the grating of the connected zigzag-profiled holes is the half of period P_x, which equals to the traditional 1:1 grating. For the two dimensional phase grating of the connected zigzag-profiled holes, the 1st order diffraction efficiency was 27.72%, which is much higher than 6.25% of the ideal sinusoidal transmission grating. At the same time, the phase grating can suppress completely the 0th order diffraction. These results are of great interest in the wide spectrum unscrambling from the infrared to the x-ray region.

3.1 Introduction

Grating is a fundamental component for spectral measurement equipment. For many years, spectroscopy has been important in physics, astronomical, biological, chemical and other analytical investigations. Usually, gratings disperse incident polychromatic light into its constituent monochromatic components and discrete orders. Generally, spectrum unscrambling only needs the +1st or −1st order diffraction of the grating. Unfortunately, the high order diffractions of the binary amplitude transmission grating always overlap the 1st one, and thus obstruct spectroscopic analysis [1–4].

L. Shi (✉) · Z. Liu · T. Pu · H. Li · J. Niu · C. Xie
Key Laboratory of Microelectronic Devices and Integrated Technology,
Institute of Microelectronics, Chinese Academy of Sciences, Beijing 100029, China
e-mail: shilina@ime.ac.cn

© Springer Nature Switzerland AG 2019
P. Ribeiro et al. (eds.), *Optics, Photonics and Laser Technology 2017*,
Springer Series in Optical Sciences 222,
https://doi.org/10.1007/978-3-030-12692-6_3

Consequently, the complicated unfolding process to filter the stray light is needed, thus may bring large errors in many calibrations, and even lead to wrong conclusions. It is known that sinusoidal amplitude transmission grating with continuous relief only presents the 0th and 1st order diffractions [3]. However, it is more difficult to fabricate the sinusoidal transmission gratings than the conventional binary ones for all wavelengths [5–7]. Another method to suppress high order diffractions is to decrease the grating period to the wavelength size [8–13]. An efficiency of 100% efficiency of the −1st order diffraction for the visible wavelengths has already been reported [12]. Unfortunately, such nano-structures are very expensive to fabricate by e-beam lithography [14], or difficult to be fabricated by the current nanofabrication technology such as nanoimprint lithography [15]. Therefore, we are in great need of large period binary transmission gratings with suppression of high order diffractions.

Several binary amplitude two-dimensional gratings with the large period have been developed, which can suppress the high order diffractions [9, 16–20]. Torcal Milla et al. found that the intensity of each diffraction order was modified by the statistical properties of the grating rough edges, and thus it is possible to obtain an amplitude binary grating with only diffraction orders −1, 0 and +1 [9]. At the same time, the two-dimensional gratings with amplitude sinusoidal transmission function along one direction were proposed [16–20]. This average transmission effect along one direction were obtained by integrating the transmission function along the other direction. There are two main methods: one is to modulate the hole position and the other is to introduce structures with complicated shapes. E.g. Combining the rectangular hole and the special probability distribution of hole location, the exact sinusoidal transmission function can be obtained along one direction [16–18]. However, the smallest gap between any two adjacent holes is zero due to the modulation of the hole positions. Thus, these structures can not be accurately fabricated by the state-of-the-art nanofabrication technology. Moreover, the background noise arising from the modulation of the hole position may interfere with the 1st order diffraction [16–18]. Similarly, the complicated shape such as sinusoidal-shaped aperture are also more difficult to fabricate than the traditional 1:1 grating [19, 20].

In order to increase the smallest gap between any adjacent holes, two dimensional gratings of quasi-array of circular holes were proposed [21, 22]. Considering that only the adjacent diffractions (such as the 2nd and 3rd order diffractions) will overlap the 1st order diffraction, these gratings based on quasi-array of holes decreased the shift range of hole location or hole size. This leads to larger gap between any two adjacent holes and smaller background noise. Unfortunately, the background noise can not be completely eliminated and the decrease of hole size leads to the low diffraction efficiency. Therefore, it has been our goal to design and fabricate a grating without the complicated shape or the modulation of the hole position, which has the high 1st order diffraction efficiency (not lower than the sinusoidal grating) and large characteristic size (comparable to traditional 1:1 grating).

In this chapter, we discuss the two-dimensional gratings with the special transmission function along one direction, which can completely suppress the 2nd, 3rd, and 4th order diffractions. The two-dimensional grating avoids the background noise, which is introduced by the modulation of the hole position [16–18]. Moreover, it has the opening area 50% of the total area, which leads to the large absolute diffraction efficiency. In addition, the characteristic size $P_x/2$ of the two dimensional gratings of connected zigzag-profiled holes is the same as the traditional 1:1 grating and thus the gratings are much easier to fabricate than the reported ones [16–22]. And the gratings can be scalable from soft x-ray to far infrared wavelengths. First, we theoretically analyze the dependence of the diffraction property on the structure parameters of the two-dimensional grating. The special transmission function results in our desired diffraction pattern. Then, we theoretically and experimentally demonstrate the diffraction pattern of the two-dimensional grating with the effective suppression of high order diffractions and the large absolute diffraction efficiency. These results will find an application in the wide spectrum unscrambling from the infrared to the x-ray region.

3.2 Diffractions of the Two Dimensional Grating

One starts the analysis from diffractions of two dimensional gratings in Fig. 3.1a. As shown in Fig. 3.1c, we denote the coordinate systems of the grating plane and the diffraction plane by xOy and $\xi O'\eta$, respectively. The grating periods along the x and y axes are respectively P_x and P_y as shown in Fig. 3.1b. The shadow area A_0 is a single period of the grating, and A is the opening area of the period.

For the square array of $N_x \times N_y$ periods in Fig. 3.1, the light field distribution in the Fraunhofer diffraction pattern is given by [4]

$$U(p, q) = C \cdot \iint_{A_0} F(x, y)e^{-ik(px+qy)}\mathrm{d}x\mathrm{d}y \cdot \sum_{n_x=0}^{N_x-1}\sum_{n_y=0}^{N_y-1} e^{-ik(pn_x P_x+qn_y P_y)} . \quad (3.1)$$

Here $C = \sqrt{P}/(\lambda R)$, P is the power density incident on the grating, λ is the incident light wavelength, R is the distance between the grating array plane and the diffraction plane. $F(x, y)$ is the complex transmission function, the area of a single period is $A_0 = P_x P_y$, k is the wave vector in free space, $p = \xi/R$ and $q = \eta/R$. The integration extends over a single period area of grating, and the integral expresses the effect of a single period. The sum represents the superposition of the coherent diffraction patterns.

Generally, the transmittance coefficient outside A is not zero but a complex value $\alpha e^{-i\beta}$ including the phase shift effect. Thus the transmission function is as follows:

$$F(x, y) = \begin{cases} \alpha e^{-i\beta}, & \text{outside } A, \\ 1, & \text{inside } A. \end{cases} \quad (3.2)$$

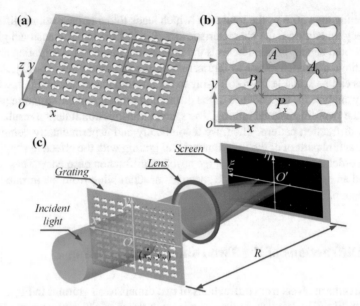

Fig. 3.1 **a** Schematic of the two dimensional grating. **b** The grating periods along x and y are respectively P_x and P_y. The area of a single period is A_0. The opening area of the period is A. **c** The coordinate systems of the grating plane and the diffraction plane

Then the diffraction intensity pattern is:

$$
\begin{aligned}
I(p,q) &= |U(p,q)|^2 \\
&= |C|^2 \cdot \left|(1 - \alpha e^{-i\beta})\iint_A e^{-ik(px+qy)}\mathrm{d}x\mathrm{d}y + \alpha e^{-i\beta}\iint_{A_0} e^{-ik(px+qy)}\mathrm{d}x\mathrm{d}y\right|^2 \\
&\quad \cdot \left|\frac{1 - e^{-iN_x kP_x p}}{1 - e^{-ikP_x p}} \cdot \frac{1 - e^{-iN_y kP_y q}}{1 - e^{-ikP_y q}}\right|^2 \\
&= C^2 \cdot \left[(1 + \alpha^2 - 2\alpha\cos\beta)|U_A|^2 + 2\mathrm{real}[\alpha(e^{i\beta} - \alpha)U_A U_{A0}^*] + \alpha^2|U_{A0}|^2\right] \\
&\quad \cdot N_x^2 N_y^2 \frac{\sin^2(N_x kpPx/2)}{N_x^2 \sin^2(kpP_x/2)} \frac{\sin^2(N_y kpPy/2)}{N_y^2 \sin^2(kpP_y/2)}.
\end{aligned}
\tag{3.3}
$$

Here $U_{A0} = U_{A0}^* = \iint_{A_0} e^{-ik(px+qy)}\mathrm{d}x\mathrm{d}y = A_0\mathrm{sinc}(kpP_x/2/\pi)\mathrm{sinc}(kqP_y/2/\pi)$, is the effect of a single period, and $U_A = \iint_A e^{-ik(px+qy)}\mathrm{d}x\mathrm{d}y$ denotes the effect of the opening area. The normalized interference function $\frac{\sin^2(N_x kpPx/2)}{N_x^2 \sin^2(kpP_x/2)} \frac{\sin^2(N_y kpPy/2)}{N_y^2 \sin^2(kpP_y/2)}$ denotes the array effects, and determines that the classical diffraction peaks locate at $p = m\lambda/P_x$ and $q = n\lambda/P_y$, here m and n are integers. As described in Sect. 3.1, we pay more attention on the diffractions along one direction, thus here we consider the case of $m = 0, 1, 2, \ldots$, and $n = 0$. According (3.3), the diffractions along the ξ axis is

$$
I(m, 0) = \begin{cases} S_0 \cdot \left[\frac{A^2}{A_0^2} + \left(1 - \frac{A}{A_0}\right)^2 \alpha^2 + 2\frac{A}{A_0}\left(1 - \frac{A}{A_0}\right)\alpha\cos\beta \right], & m = 0 \\ S_0 \cdot \left[(1 + \alpha^2 - 2\alpha\cos\beta)\frac{|U_A|^2}{A_0^2} \right], & m \neq 0 \end{cases} \tag{3.4}
$$

Here $S_0 = C^2 \cdot (N_x N_y A_0)^2$ is a constant relating to the power density and the grating area. Equation (3.4) shows that the 0th order diffraction only depends on the transmission function $F(x, y)$, the opening area A, and not depends on the integral U_A. While the m-th order diffraction ($m \neq 0$) depends on the integral U_A, and thus the shape of the opening region. Moreover, for all m values ($m \neq 0$), $I(m, 0)$ has the same coefficient $S_0(1 + \alpha^2 - 2\alpha\cos\beta)/A_0^2$, that is to say, the grating transmission function $F(x, y)$ cannot change the contrast of each order diffraction intensity ($m \neq 0$). Therefore, we can design the transmission function $F(x, y)$ to minimize the 0th order diffraction and maximize the 1st order one. At the same time, we optimize the shape of the opening region to increase the 1st order diffraction and suppress the 2nd and higher order diffractions. The following we focus on two cases: one is the amplitude grating ($\alpha = 0$) and the other is the phase grating ($\alpha = 1$ and $\beta = \pi$).

3.3 Two Dimensional Amplitude Gratings of Connected Hexagonal Holes

3.3.1 Theoretical Analysis

Discussion starts from the amplitude grating with $\alpha = 0$ of connected hexagonal holes (cyan) as shown in Fig. 3.2 [23, 24]. The dash white lines denote the square lattice and the crossings are the lattice points at which the hexagonal holes locate. The hexagonal hole has the side $2a_1$ along the x axis, the diagonal $2a$ along the x axis, and the height $2b = P_y$ along y axis. The opening area A (surrounded by dash red curve) of a single period A_0 (surrounded by dash black curve) is $2b(a + a_1) = P_y(a + a_1)$.

The transmission function $F(x, y)$ is

$$
F(x, y) = \begin{cases} 0, & \text{outside the hexagonal hole} \\ 1, & \text{inside the hexagonal hole} \end{cases} \tag{3.5}
$$

For the square array of $N_x \times N_y$ connected hexagonal holes, the diffraction intensity pattern from (3.3) is [23, 24]

Fig. 3.2 Schematic of the two dimensional grating of connected hexagonal holes

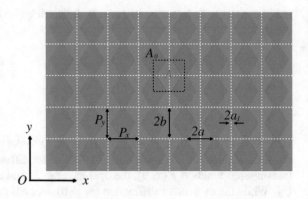

$$I(p,q)$$
$$= C^2 \cdot |U_A|^2 \cdot N_x^2 N_y^2 \frac{\sin^2(N_x k p P x/2)}{N_x^2 \sin^2(k p P_x/2)} \frac{\sin^2(N_y k p P y/2)}{N_y^2 \sin^2(k p P_y/2)}$$
$$= S_0 \left(\frac{a+a_1}{P_x}\right)^2 \cdot \frac{\sin^2(N_x k p P x/2)}{N_x^2 \sin^2(k p P_x/2)} \frac{\sin^2(N_y k p P y/2)}{N_y^2 \sin^2(k p P_y/2)}$$
$$\cdot \left| \frac{\cos(kpa_1 - kqb) - \cos kpa}{kp(a+a_1) \cdot (kp(a-a_1) + kqb)} + \frac{\cos kpa - \cos(kpa_1 + kqb)}{kp(a+a_1) \cdot (-kp(a-a_1) + kqb)} \right|^2 . \quad (3.6)$$

Now one focus on the diffraction intensity along the ξ axis since spectral measurement is usually at one direction. The diffraction intensity according to (3.6) along the ξ axis is given by

$$I(p,0) = S_0 \left(\frac{a+a_1}{P_x}\right)^2 \cdot \frac{\sin^2(N_x k p P x/2)}{N_x^2 \sin^2(k p P_x/2)} \cdot \text{sinc}^2 \frac{kp(a-a_1)}{2\pi} \text{sinc}^2 \frac{kp(a+a_1)}{2\pi}.$$
$$(3.7)$$

On the other hand, we can obtain the diffraction intensity along the ξ axis from the average transmission function along the x axis, which can be obtained by integrating the transmission function $F(x, y)$ along the y axis:

$$F(x) = \begin{cases} N_y \cdot 2b, & |x| \le a_1, \\ N_y \cdot 2b(a - |x|)/(a - a_1), & a_1 < |x| \le a, \\ 0, & a \le |x| \le P_x/2. \end{cases} \quad (3.8)$$

And according to Fraunhofer diffraction formula, we can also obtain the the same intensity pattern along the ξ axis as (3.7) directly from (3.5):

$$I(p) = \left| C \cdot \int_{-P_x/2}^{P_x/2} F(x) e^{-ikpx} dx \cdot \sum_{n_x=0}^{N_x-1} e^{-ikpn_x P_x} \right|^2$$

$$= S_0 \left(\frac{a+a_1}{P_x} \right)^2 \cdot \frac{\sin^2(N_x kp P_x/2)}{N_x^2 \sin^2(kp P_x/2)} \cdot \text{sinc}^2 \frac{kp(a-a_1)}{2\pi} \, \text{sinc}^2 \frac{kp(a+a_1)}{2\pi}. \quad (3.9)$$

And thus according to (3.7), the m-th order diffraction locating at $p = m\lambda/P_x$ along the ξ axis is

$$I(m) = S_0 \left(\frac{a+a_1}{P_x} \right)^2 \cdot \text{sinc}^2 \frac{m(a-a_1)}{P_x} \, \text{sinc}^2 \frac{m(a+a_1)}{P_x}. \quad (3.10)$$

Equation (3.10) shows that the m-th order diffraction is the product of two normalized sinc functions and depends on the two parameters a/P_x and a_1/P_x. It should be noted that the normalized sinc function has zero crossings occurring periodically at non-zero integers. Thus we can select the suitable values of a/P_x and a_1/P_x to make two kinds of zero crossings fall at two kinds of diffractions along the ξ axis. In real spectral measurement, only the near diffractions (such as the 2nd and 3rd order diffractions) will overlap the 1st order diffraction. The far diffractions are usually very small and have little effects on the 1st order diffraction. Thus we will consider the values of a and a_1 which lead to the zeros of the 2nd and 3rd order diffractions. According to (3.10), we can obtain $a_1/P_x = 1/12$ and $a/P_x = 5/12$ as the 2nd and 3rd order diffractions disappear simultaneously. And now, all the even and $3n$-th order diffractions disappear due to the periodic zero crossings of the sinc function. The 0th order diffraction is $25\% S_0$ and the 1st one is $6.93\% S_0$. The nearest order diffraction from the 1st one is the 5th order diffraction, which equals to $0.011088\% S_0$. This value is much smaller than $0.4\% S_0$ of the 5th order diffraction of conventional 1:1 grating.

Figure 3.3 presents the logarithm of the normalized diffraction intensity of the two dimensional grating of hexagonal holes with $a_1/P_x = 1/12$ and $a/P_x = 5/12$ according to (3.6). The intensity is normalized by the 0th order diffraction intensity

Fig. 3.3 The logarithm of the far-field diffraction intensity pattern of the two dimensional grating of connected hexagonal holes

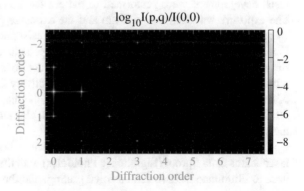

Fig. 3.4 The diffraction
intensity along the ξ axis of
the two dimensional grating
of connected hexagonal
holes (black) and that of 1:1
transmission gratings (red)

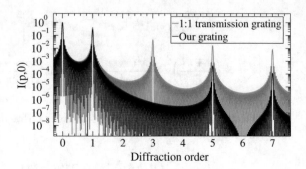

$I(0, 0) = 25\% S_0$. As expected, the 0th and 1st order diffractions are kept along the
ξ axis, and the 2nd, 3rd and 4th diffractions disappear. The 5th order diffraction is
smaller than the noise between the 0th and 1st order diffractions.

We compare the diffraction intensity along the ξ axis of our two dimensional
grating with that of the conventional 1:1 grating in Fig. 3.4. It is shown that the 5th
order diffraction of our grating is much suppressed, which is weak enough for real
application. Figure 3.4 also presents clearly the complete suppression of the 2nd, 3rd
and 4th order diffractions.

3.3.2 Experimental Results

In order to further confirm the feasibility of the high order diffraction suppression
of our theoretical predictions, a binary transmission grating with area 4 cm × 4 cm
on a glass substrate was fabricated for the visible light region by laser write lithography. Firstly, the chromium layer with thickness of 110 nm was deposited onto the
soda glass with 2.286 mm thickness with an electron beam evaporation system, and
500 nm thick AZ1500 photoresist was spin coated onto the chrome layer. Secondly,
GDSII data of the designed binary gratings were imported into DESIGN WRITE
LAZER 2000 (Heidelberg Instruments Mikrotechnik GmbH). Laser exposure and
resist development were performed to pattern the binary gratings onto the resist.
The exposure wavelength is 413 nm and the exposure power is 80 mw. After that,
wet etching technique was used to transfer the resist pattern onto the chrome layer.
Finally, the residual resist was removed using plasma ashing followed by acetone
rinse. The microphotograph of the fabricated structure is illustrated in Fig. 3.5. Periods P_x and P_y along the x and y axes are respectively 24 and 28 μm. The side $2a_1$
of the hexagonal hole along the x axis is 4 μm, the diagonal $2a$ along the x axis is
20 μm, and the height $2b$ along y axis is 28 μm. Our smallest size of fabrication is
the side $2a_1 = P_x/6 = 4$ μm.

The experimental setup for optical measurement is shown in Fig. 3.6. A collimated
laser beams from Sprout (Lighthouse Photonics) with the wavelength of 532 nm was
used to illuminate the two-dimensional grating, and the far-field diffraction pattern

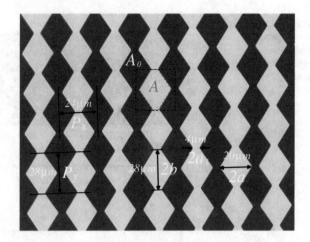

Fig. 3.5 Microphotograph of the fabricated two-dimensional grating of hexagonal holes

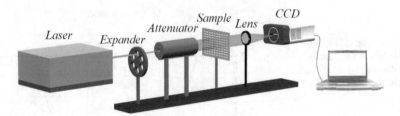

Fig. 3.6 Experimental setup for the optical measurement

from the grating is focused by a lens and then recorded on a charge coupled device (CCD) camera (ANDOR DU920P-BU2) with 1024×256 pixels. To decrease the background noise, CCD was cooled to minus 85 °C.

The measurement results are shown in Fig. 3.7. Each diffraction spot on the camera includes about 12 pixels. It is presented that only 0th and the 1st order diffractions exist along the ξ axis, which agrees well with the theoretical results. The 5th order diffraction cannot be observed since it is submerged in the background noise. The counts of the 0th and 1st order diffractions are 6.462×10^4 and 1.825×10^4.

3.3.3 Discuss of Fabrication Error

In the above discussions, it was assumed that the size and shape of connected hexagonal holes are perfect. In practice, the size of fabricated holes can be a little larger or smaller than the designed target and the angle of hexagonal hole can not be perfect. Thus the diffraction pattern will not be the same as the designed one. In the following, we analyze and evaluate the influence of the size and shape deviation of

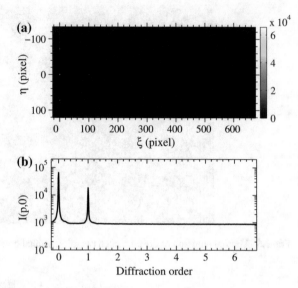

Fig. 3.7 **a** The far-field diffraction intensity pattern of the two dimensional grating of hexagonal holes. **b** The diffraction intensity along the ξ axis

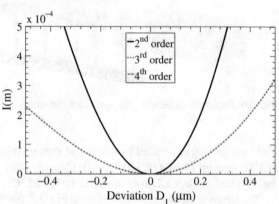

Fig. 3.8 The diffraction intensities of the 2nd, 3rd, and 4th order diffractions versus the deviation D_1

connected hexagonal holes on the practical performance. We define a size deviation $D_1 = a - a_f = a_1 - a_{1f}$, where a_f and a_{1f} are the actual parameters of the fabricated hexagonal holes. Numerical simulation based (3.10) is carried out and we obtain the 2nd, 3rd, 4th order diffraction intensities versus D_1 as shown in Fig. 3.8. From Fig. 3.8 one can see that the 3rd order diffraction intensity is always zero regardless of whether the size of hexagonal holes is changed. This is because the suppression of 3rd order diffraction depends on $a - a_1$, which has been described in (3.10). As $D_1 \in (-0.15, 0.15)\,\mu m$, the 2nd order diffraction intensity will not larger than 1×10^{-4}. And the 4th order one is not larger than 2.5×10^{-5}. The deviation D_1 is $\pm 3.75\%$ of the smallest fabrication size $2a_1$. The fabrication error at $10\,\mu m$ level of our laser write lithography is 2%. This tolerance makes our structure can be fabricated by the current technology.

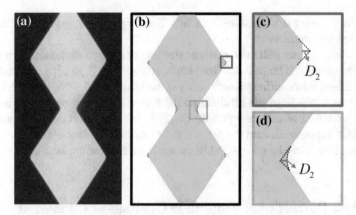

Fig. 3.9 **a** The fabricated hexagonal hole. **b** The simulated hexagonal hole. **c** The round corner 1 of hexagonal hole. **d** The round corner 2 of hexagonal hole

Fig. 3.10 The diffraction intensities of the 2nd, 3rd, and 4th order diffractions versus the deviation D_2

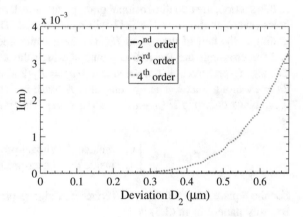

It should be noted that the sharp corner cannot be achieved in real fabrication process. Therefore, the effects of fabrication error of the corners on the diffraction pattern should also be discussed. For this the 2nd, 3rd and 4th order diffraction intensities were calculated pattern of the gratings of connected hexagonal holes with the round corner as shown in Fig. 3.9. A deviation D_2, which is the distance between the round corner and the vertex of the sharp corner was defined, as shown in Fig. 3.9c, d.

Figure 3.10 presents that the 2nd and 4th order diffraction intensities are always zero whether connected hexagonal holes have round corners. This is because the disappearance of the 2nd and 4th order diffractions depends on $\mathrm{sinc}^2 \frac{m(a+a_1)}{P_x}$, and both the concave and convex corners shown in Fig. 3.9b make $a + a_1$ keep constant with D_2 increasing. The 3rd order diffraction intensity increase with D_2. As $D_2 = 0.36\,\mu\mathrm{m}$, the 3rd order diffraction intensity is 1×10^{-4}. The tolerance of deviation

D_2 of sharp corner is at least $\pm9\%$ of $2a_1$. This large tolerance greatly relaxes the fabrication requirement.

In a word, the smallest characteristic size $2a_1$ of the two-dimensional grating of connected hexagonal holes is $P_x/6$, which is smaller than $P_x/2$ of the traditional 1:1 grating. The absolute diffraction efficiency of the 1st order is $6.93\% S_0$, which is a little higher than $6.25\% S_0$ of the ideal sinusoidal transmission grating. Both theoretical and experimental results demonstrate the 2nd, 3rd and 4th order diffractions are completely suppressed. And the 5th order diffraction is as low as 0.16% of the 1st order, and thus it can be submerged in the noise for the real application.

3.4 Two Dimensional Amplitude Gratings of Connected Zigzag-Profiled Holes

In this section, the two dimensional grating with $\alpha = 0$ of connected zigzag-profiled holes (cyan) as shown in Fig. 3.11 will be addressed. The characteristic size of the grating is the half of the period P_x. The dash white lines denote the square lattice and the crossings are the lattice points at which the zigzag-profiled holes locate. The zigzag-profiled hole has the side $a + a_1 = P_x/2$ along the x axis and the height $2b = P_y$ along y axis. The opening area A (dash red curve) of a single period A_0 (dash black curve) is $2b(a + a_1) = P_y(a + a_1)$. The transmission function $F(x, y)$ is

$$F(x, y) = \begin{cases} 0, & \text{outside the zigzag-profiled hole} \\ 1, & \text{inside the zigzag-profiled hole} \end{cases} \tag{3.11}$$

For the square array of $N_x \times N_y$ connected zigzag-profiled holes, the diffraction intensity pattern from (3.3) is

Fig. 3.11 Schematic of the two dimensional grating of connected zigzag-profiled holes

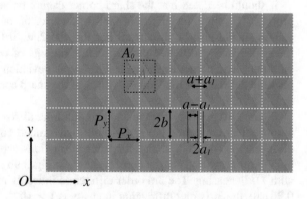

$$I(p, q)$$

$$= C^2 \cdot |U_A|^2 \cdot N_x^2 N_y^2 \frac{\sin^2(N_x kp Px/2)}{N_x^2 \sin^2(kp P_x/2)} \frac{\sin^2(N_y kp Py/2)}{N_y^2 \sin^2(kp P_y/2)}$$

$$= S_0 \left(\frac{a + a_1}{2P_x} \right)^2 \cdot \frac{\sin^2(N_x kp Px/2)}{N_x^2 \sin^2(kp P_x/2)} \frac{\sin^2(N_y kp Py/2)}{N_y^2 \sin^2(kp P_y/2)}$$

$$\cdot \operatorname{sinc}^2 \frac{kp(a + a_1)}{2\pi} \cdot \left[\operatorname{sinc}^2 \frac{kqb + kp(a - a_1)}{2\pi} + \operatorname{sinc}^2 \frac{kqb - kp(a - a_1)}{2\pi} \right.$$

$$\left. + 2 \cos kqb \cdot \operatorname{sinc} \frac{kqb + kp(a - a_1)}{2\pi} \cdot \operatorname{sinc} \frac{kqb - kp(a - a_1)}{2\pi} \right]. \qquad (3.12)$$

And according to (3.12), the diffraction intensity along the ξ axis is given by

$$I(p, 0) = S_0 \left(\frac{a + a_1}{P_x} \right)^2 \cdot \operatorname{sinc}^2 \frac{kp(a - a_1)}{2\pi} \operatorname{sinc}^2 \frac{kp(a + a_1)}{2\pi} \cdot \frac{\sin^2(N_x kp Px/2)}{N_x^2 \sin^2(kp P_x/2)}.$$
$$(3.13)$$

It can be seen that (3.13) is the same as (3.7). It means that the diffraction property along the ξ axis of the two dimensional grating of connected zigzag-profiled holes is the same as that of connected hexagonal holes. And thus the same structure parameter $a_1/P_x = 1/12$ and $a/P_x = 5/12$ can be obtained as the 2nd and 3rd order diffractions disappear simultaneously. This is due to the same average transmission function $F(x)$ along the x axis obtained by integrating the transmission function $F(x, y)$ along the y axis. And also the influence of the size and shape deviation of connected zigzag-profiled holes on the practical performance is similar as that of the hexagonal ones. Fortunately, the characteristic size is three times of that of the hexagonal ones and thus the tolerance becomes three times of that of the hexagonal ones.

Figure 3.12 presents the logarithm of the normalized diffraction intensity of the two dimensional grating of connected zigzag-profiled holes with $a_1/P_x = 1/12$ and $a/P_x = 5/12$ according to (3.12). The intensity is normalized by the 0th order diffraction intensity $I(0, 0) = 25\% S_0$. It is shown that the diffraction property of zigzag-profiled holes along the ξ axis is the same as that of hexagonal holes although the diffraction property along the η axis is different.

A binary transmission grating of connected zigzag-profiled holes was also fabricated and tested by the above process and the experiment setup in Fig. 3.6. The microphotograph of the fabricated structure is illustrated in Fig. 3.13. The structure parameters such as $P_x = 24\,\mu m$, $P_y = 28\,\mu m$, $2a_1 = 4\,\mu m$, $a - a_1 = 8\,\mu m$ and $2b = 28\,\mu m$ are the same as that of hexagonal holes. The only difference is the smallest size of fabrication is $a + a_1 = P_x/2 = 12\,\mu m$.

The measurement results are shown in Fig. 3.14. Each diffraction spot on the camera includes about 12 pixels. It is presented that only 0th and the 1st order diffractions exist along the ξ axis, which agrees well with the theoretical results. The 5th order diffraction cannot be observed since it is submerged in the background noise. The counts of the 0th and 1st order diffractions are 5.413×10^4 and 1.426×10^4.

Fig. 3.12 The logarithm of the far-field diffraction intensity pattern of the two dimensional grating of connected zigzag-profiled holes

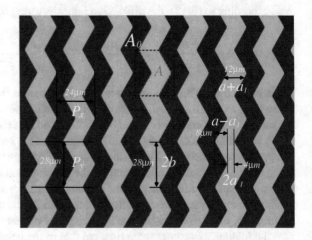

Fig. 3.13 Microphotograph of the fabricated two-dimensional grating of connected zigzag-profiled holes

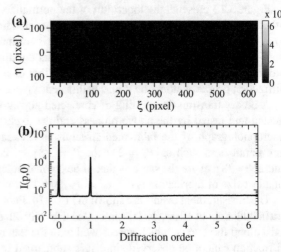

Fig. 3.14 **a** The far-field diffraction intensity pattern of the two dimensional grating of connected zigzag-profiled holes. **b** The diffraction intensity along the ξ axis

In a word, the smallest characteristic size $a + a_1$ of the two-dimensional grating of connected zigzag-profiled holes is $P_x/2$, which equals to that of the traditional 1:1 grating and is much larger than $P_x/6$ of the two-dimensional grating of connected hexagonal holes. At the same time, the diffraction property along the ξ axis is the same as that of the two-dimensional grating of connected hexagonal holes.

3.5 Two Dimensional Phase Gratings of Connected Zigzag-Profiled Holes

This section focuses on the increase of the 1st order diffraction efficiency. According to the discuss in Sect. 3.2, we know that the grating transmission function $F(x, y)$ can change the absolute efficiency of each order diffraction. We select $\alpha = 1$ and $\beta = \pi$ since the coefficient $(1 + \alpha^2 - 2\alpha \cos \beta)$ of the m-th ($m \neq 0$) diffraction can obtain the maximum 4 from (3.4). And now the diffraction peaks along the ξ axis are

$$I(m, 0) = \begin{cases} S_0 \cdot (2A/A_0 - 1)^2, & m = 0 \\ S_0 \cdot 4\frac{|U_A|^2}{A_0^2}, & m \neq 0 \end{cases} \tag{3.14}$$

Equation (3.14) shows that the 0th order diffraction disappears as $A = A_0/2$. Here we consider the grating of connected zigzag-profiled holes (cyan) with $a_1/P_x = 1/12$, $a/P_x = 5/12$ and $2b = P_y$ as shown in Fig. 3.11, whose opening area is $A = 2b(a + a_1) = A_0/2$.

From (3.3), the diffraction intensity pattern of the phase grating of $N_x \times N_y$ connected zigzag-profiled holes is

$$I(p, q) = C^2 \cdot N_x^2 N_y^2 \frac{\sin^2(N_x kp P x/2)}{N_x^2 \sin^2(kp P_x/2)} \frac{\sin^2(N_y kp P y/2)}{N_y^2 \sin^2(kp P_y/2)}$$
$$\left[4|U_A|^2 - 4\text{real}(U_A)U_{A0} + |U_{A0}|^2 \right]. \tag{3.15}$$

Here

$$U_{A0} = U_{A0}^* = \iint_{A_0} e^{-ik(px+qy)} dx dy = P_x P_y \text{sinc}\left(\frac{kp P_x}{2\pi}\right) \text{sinc}\left(\frac{kq P_y}{2\pi}\right), \tag{3.16}$$

$$U_A = \iint_A e^{-ik(px+qy)} dx dy$$
$$= \frac{e^{-ikpa_1} - e^{ikpa}}{-ikp} \cdot \left[\frac{1}{ikp(a-a_1)/b + ikq} + \frac{1}{ikp(a-a_1)/b - ikq} \right.$$
$$+ \frac{e^{-ikp(a-a_1)-ikqb}}{-ikp(a-a_1)/b - ikq} + \left. \frac{e^{-ikp(a-a_1)+ikqb}}{-ikp(a-a_1)/b + ikq} \right], \tag{3.17}$$

Fig. 3.15 The logarithm of
the far-field diffraction
intensity pattern of the two
dimensional grating of
connected zigzag-profiled
holes

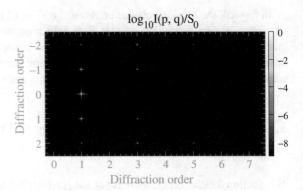

and

$$|U_A|^2 = \left(\frac{a+a_1}{2P_x}\right)^2 \cdot \text{sinc}^2 \frac{kp(a+a_1)}{2\pi}$$
$$\cdot \left[\text{sinc}^2 \frac{kqb+kp(a-a_1)}{2\pi} + \text{sinc}^2 \frac{kqb-kp(a-a_1)}{2\pi}\right.$$
$$\left. +2\cos kqb \cdot \text{sinc}\frac{kqb+kp(a-a_1)}{2\pi} \cdot \text{sinc}\frac{kqb-kp(a-a_1)}{2\pi}\right]. \quad (3.18)$$

Then the diffraction intensity according to (3.15) along the ξ axis is given by

$$I(p,0) = S_0 \left(\frac{a+a_1}{P_x}\right)^2 \cdot \frac{\sin^2(N_x kp P_x/2)}{N_x^2 \sin^2(kp P_x/2)}$$
$$\cdot \left[2\,\text{sinc}\frac{kp(a-a_1)}{2\pi}\,\text{sinc}\frac{kp(a+a_1)}{2\pi} - \frac{P_x}{(a+a_1)}\,\text{sinc}\frac{kp P_x}{2\pi}\right]^2. \quad (3.19)$$

Thus according to (3.19), the m-th ($m \neq 0$) order diffraction locating at $p = m\lambda/P_x$ along the ξ axis is

$$I(m) = 4S_0 \left(\frac{a+a_1}{P_x}\right)^2 \cdot \text{sinc}^2 \frac{m(a-a_1)}{P_x}\,\text{sinc}^2 \frac{m(a+a_1)}{P_x}. \quad (3.20)$$

Equation (3.20) shows that the m-th order diffraction intensity is four times of that of the amplitude gratings in Sects. 3.3 and 3.4. The 1st order diffraction efficiency attains $27.72\% S_0$ for the case of $a_1/P_x = 1/12$ and $a/P_x = 5/12$. The nearest order diffraction from the 1st one is the 5th order diffraction, which equals to $0.044352\% S_0$.

Figure 3.15 presents the logarithm of the diffraction intensity of the phase grating of connected zigzag-profiled holes with $a_1/P_x = 1/12$ and $a/P_x = 5/12$ according to (3.15). The intensity is normalized by the parameter $S_0 = C^2 \cdot (N_x N_y P_x P_y)^2$. Different from the amplitude grating, the 0th order diffraction disappears. The 1st

Fig. 3.16 The diffraction intensity along the ξ axis of the two dimensional grating of connected zigzag-profiled holes

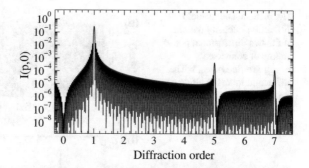

Fig. 3.17 Surface Contour photograph of the fabricated two-dimensional grating of connected zigzag-profiled holes

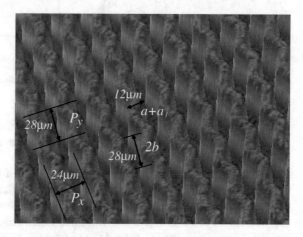

order diffraction is kept along the ξ axis, and the 2nd, 3rd and 4th diffractions disappear. The 5th order diffraction is smaller than the noise around the 1st order diffraction.

Figure 3.16 presents the diffraction intensity along the ξ axis. It is shown that the 5th order diffraction is much suppressed, the 0th, 2nd, 3rd and 4th order diffractions are completely suppressed.

A phase grating of connected zigzag-profiled holes was also fabricated by the above process followed by the soda glass etching and the chrome removing. The surface contour photograph of the fabricated structure is illustrated in Fig. 3.17. The structure parameters such as $P_x = 24\,\mu\text{m}$, $P_y = 28\,\mu\text{m}$, $2a_1 = 4\,\mu\text{m}$, $a - a_1 = 8\,\mu\text{m}$ and $2b = 28\,\mu\text{m}$ are the same as that of the amplitude grating of connected zigzag-profiled holes. The smallest size of fabrication of phase grating is also $a + a_1 = P_x/2 = 12\,\mu\text{m}$.

The phase grating was tested by the experiment setup in Fig. 3.6. The measurement results are shown in Fig. 3.18. Each diffraction spot on the camera includes about 12 pixels. It is presented that only the 1st order diffractions exists along the ξ axis, which agrees well with the theoretical results. The counts of the 1st and 5th order diffractions are 5.462×10^4 and 8.2×10^2. The 0th order diffraction is 1.016×10^3,

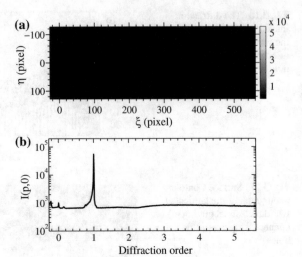

Fig. 3.18 **a** The far-field diffraction intensity pattern of the two dimensional phase grating of connected zigzag-profiled holes. **b** The diffraction intensity along the ξ axis

which is different from the theoretical result. This difference is from the phase shift $\beta \neq \pi$ since the etching height of the soda glass may be a little larger or smaller than the designed one.

In a word, the smallest characteristic size $a + a_1$ of the two-dimensional phase grating of connected zigzag-profiled holes is $P_x/2$, which equals to that of the traditional 1:1 grating. The contrast of each order diffraction (for $m \neq 0$) intensity along the ξ axis is the same as that of the two-dimensional amplitude gratings in Sects. 3.3 and 3.4. Fortunately, the 1st order diffraction absolute efficiency attains $27.72\% S_0$, and the 0th order one is zero.

3.6 Conclusion

The diffraction property of the two dimensional gratings of connected holes has been analysed. The complete suppression of 2nd, 3rd, and 4th order diffractions by the two dimensional gratings was demonstrated theoretically and experimentally. The 5th order diffraction is as low as 0.16% of the 1st order one, and thus it can be submerged in the background noise in real applications. The smallest characteristic size of the grating of the connected zigzag-profiled holes is the half of period P_x, which equals to that of the traditional 1:1 grating. For the two dimensional phase grating of the connected zigzag-profiled holes, we obtain the 1st order diffraction efficiency $27.72\% S_0$, which is much higher than $6.25\% S_0$ of the ideal sinusoidal transmission grating. At the same time, the phase grating can suppress completely the 0th order diffraction. Therefore, the two-dimensional grating with connected holes offers an opportunity for high-accuracy spectral measurement and will possess broad potential applications in optical science and engineering fields.

Acknowledgements We would like to thank L. Cao for helpful discussions over many years. This work was supported by National Key Research and Development Program of China (2017YFA0206 002), Major National Scientific Instruments Developed Special Project (2013YQ1508290602), National Natural Science Foundation of China (61107032, 61275170), and the Opening Project of Key Laboratory of Microelectronic Devices and Integrated Technology, Institute of Microelectronics of Chinese Academy of Sciences.

References

1. C. Palmer, *Diffraction Grating Handbook* (Richardson Grating Laboratory, 2005)
2. R. Petit, *Electromagnetic Theory of Gratings* (Springer, Berlin, 1980)
3. E.G. Loewen, E. Popov, *Diffraction Gratings and Applications* (Marcel Dekker, New York, 1997)
4. M. Born, E. Wolf, *Principle of Optics*, 6th edn. (Pergamon, Oxford, 1980), pp. 370–427
5. H.X. Miao, A.A. Gomella, N. Chedid, L. Chen, H. Wen, Fabrication of 200 nm period hard X-ray phase gratings. Nano Lett. **14**(6), 3453–3458 (2014)
6. P. Jin, Y. Gao, T. Liu, X. Li, J. Tan, Resist shaping for replication of micro-optical elements with continuous relief in fused silica. Opt. Lett. **35**(8), 1169–1171 (2010)
7. G. Vincent, R. Haidar, S. Collin, N. Gurineau, J. Primot, E. Cambril, J.-L. Pelouard, Realization of sinusoidal transmittance with subwavelength metallic structures. J. Opt. Soc. Am. B **25**(5), 834–840 (2008)
8. S. Gupta, Single-order transmission diffraction gratings based on dispersion engineered all-dielectric metasurfaces. J. Opt. Soc. Am. A **33**(8), 1641–1647 (2016)
9. F.J. Torcal-Milla, L.M. Sanchez-Brea, E. Bernabeu, Diffraction of gratings with rough edges. Opt. Express **16**(24), 19757–19769 (2008)
10. Z. Li, E. Palacios, S. Butun, K. Aydin, Visible-frequency metasurfaces for broadband anomalous reflection and high-efficiency spectrum splitting. Nano Lett. **15**(3), 1615–1621 (2015)
11. L. Zhu, J. Kapraun, J. Ferrara, C.J. Chang-Hasnain, Flexible photonic metastructures for tunable coloration. Optica **2**(3), 255–258 (2015)
12. T. Clausnitzer, T. Kämpfe, E.-B. Kley, A. Tnnermann, A.V. Tishchenko, O. Parriaux, Highly-dispersive dielectric transmission gratings with 100% diffraction efficiency. Opt. Express **16**(8), 5577–5584 (2008)
13. C. Zhou, T. Seki, T. Kitamura, Y. Kuramoto, T. Sukegawa, N. Ishii, T. Kanai, J. Itatani, Y. Kobayashi, S. Watanabe, Wavefront analysis of high-efficiency, large-scale, thin transmission gratings. Opt. Express **22**(5), 5995–6008 (2014)
14. T.R. Groves, D. Pickard, B. Rafferty, N. Crosland, D. Adam, G. Schubert, Maskless electron beam lithography: prospects, progress, challenges. Microelectron. Eng. **61**, 285–293 (2002)
15. C.-H. Lin, Y.-M. Lin, C.-C. Liang, Y.-Y. Lee, H.-S. Fung, B.-Y. Shew, S.-H. Chen, Extreme UV diffraction grating fabricated by nanoimprint lithography. Microelectron. Eng. **98**, 194–197 (2012)
16. N. Gao, C. Xie, High-order diffraction suppression using modulated groove position gratings. Opt. Lett. **36**(21), 4251–4253 (2011)
17. L. Kuang, L. Cao, X. Zhu, S. Wu, Z. Wang, C. Wang, S. Liu, S. Jiang, J. Yang, Y. Ding, C. Xie, J. Zheng, Quasi-sinusoidal single-order diffraction transmission grating used in X-ray spectroscopy. Opt. Lett. **36**(20), 3954–3956 (2011)
18. L. Shi, L. Li, Z. Liu, T. Pu, N. Gao, C. Xie, The quasi-triangle array of rectangular holes with the completely suppression of high order diffractions, in *Proceedings of the 5th International Conference on Photonics, Optics and Laser Technology - Volume 1: PHOTOPTICS, Porto, Portugal* (2017), pp. 54–58
19. L. Cao, E. Förster, A. Fuhrmann, C. Wang, L. Kuang, S. Liu, Y. Ding, Single order X-ray diffraction with binary sinusoidal transmission grating. Appl. Phys. Lett. **90**(5), 053501 (2007)

20. H. Zang, C. Wang, Y. Gao, W. Zhou, L. Kuang, L. Wei, W. Fan, W. Zhang, Z. Zhao, L. Cao, Y. Gu, B. Zhang, G. Jiang, X. Zhu, C. Xie, Y. Zhao, M. Cui, Elimination of higher-order diffraction using zigzag transmission grating in soft x-ray region. Appl. Phys. Lett. **100**, 111904 (2012)
21. J. Niu, L. Shi, Z. Liu, T. Pu, H. Li, G. Wang, C. Xie, High order diffraction suppression by quasi-periodic two-dimensional gratings. Opt. Mater. Express **7**, 366–375 (2017)
22. H. Li, L. Shi, L. Wei, C. Xie, L. Cao, Higher-order diffraction suppression of free-standing quasiperiodic nanohole arrays in the X-ray region. Appl. Phys. Lett. **110**, 041104 (2017)
23. Z. Liu, L. Shi, T. Pu, H. Li, J. Niu, G. Wang, C. Xie, Two-dimensional gratings of hexagonal holes for high order diffraction suppression. Opt. Express **25**, 1339–1349 (2017)
24. Z. Liu, L. Shi, T. Pu, C. Xie, H. Li, J. Niu, High order diffraction suppression of the membrane with hexagonal hole array, in *Proceedings of the 5th International Conference on Photonics, Optics and Laser Technology - Volume 1: PHOTOPTICS, Porto, Portugal* (2017), pp. 59–64

Chapter 4
Polarization-Independent Tunable Spectral Filter by the Use of Liquid Crystal

Mitsunori Saito

Abstract Liquid crystals are useful for creating spectral filters, since they exhibit a large birefringence and an efficient electro-optic effect. Liquid crystal devices, however, usually require a polarizer that halves light intensity. This polarization dependence is unfavorable particularly in the long-wavelength infrared region in which neither efficient polarizers nor strong light sources is available due to opaqueness of ordinary optical materials. Polarizer-free devices are therefore desired for extending the application fields of liquid crystals. A solution is to split a light signal into two beams with orthogonal polarization directions and superpose them after transmission through a liquid crystal cell. This optical system, however, has to be constructed with multiple optical components, which require a complicated alignment process and induce high insertion losses for infrared light. It is therefore beneficial to integrate multiple components in a single device. A compact optical device that integrates a polarization beam splitter and two retarders can be constructed by using silicon pentaprisms. The silicon prisms act as a substrate and an electrode for a nematic liquid crystal that achieves both polarization division and retardation. A function of a tunable spectral filter (Lyot filter) is attainable in the 2–8 μm wavelength range. Another solution for creating a polarizer-free device is to eliminate an optical anisotropy of the liquid crystal. Unique optical properties of cholesteric liquid crystals realize a polarization-independent index change. Whereas visible light changes its polarization direction as it passes through a chiral structure of the cholesteric liquid crystal, infrared light keeps its polarization state since their wavelength is too long to recognize the micro chiral structure. Consequently, the cholesteric liquid crystal exhibits an isotropic index of refraction that is tunable by voltage application. This isotropic property is useful to create a polarizer-free liquid crystal device. Interference filters (Fabry-Perot filters) for the infrared measurements can be constructed by enclosing a cholesteric liquid crystal in a narrow gap of two silicon plates. A polarization-independent index change between 1.52 and 1.61 is attainable by application of ~10 V.

M. Saito (✉)
Department of Electronics and Informatics, Ryukoku University, Seta, Otsu 520-2194, Japan
e-mail: msaito@rins.ryukoku.ac.jp

© Springer Nature Switzerland AG 2019
P. Ribeiro et al. (eds.), *Optics, Photonics and Laser Technology 2017*,
Springer Series in Optical Sciences 222,
https://doi.org/10.1007/978-3-030-12692-6_4

4.1 Introduction

Applications of liquid crystals (LCs) are not limited to displays but extend to a variety of technical fields including optical communications and measurements [1–4]. Whereas LCs are used widely in the visible spectral range, they are seldom used in the infrared range beyond 2 μm wavelength [5, 6]. The long-wavelength infrared range is important for conducting spectrometry and thermometry [7, 8]. Recent development of low-cost infrared cameras has promoted the research in the imaging spectrometry [9–12]. Night vision techniques for security and self-driving systems will further promote the infrared measurement technology in future.

A problem that hinders development of infrared LC devices is the polarization dependence. A polarizer usually halves light intensity and accordingly reduces the device efficiency. This problem is serious in the infrared measurements, since infrared light sources, e.g., heaters and human bodies, emit weaker radiation than visible lamps do. In addition, infrared wire-grid polarizers are expensive and less efficient in comparison with dichroic plates or prisms that are used in the visible spectral range [13]. The author has been conducting researches on infrared measurements by the use of LCs and semiconductors [14–16]. When creating infrared LC devices, silicon (Si) plates act as a transparent window as well as a conductive substrate (electrode). This performance is advantageous over ordinary glass plates that need an oxide-film coating for creating a transparent electrode. In the following sections, recent development of the polarization-independent LC devices is introduced with special emphasis on tunable spectral filters [17–22].

4.2 Principle of Polarization Selection

Since the early days of the LC device development, polarization-independent devices have been created by the use of a polarization beam splitter [23–27]. In those devices, a beam splitter divides non-polarized light into two beams whose polarization directions are orthogonal to one another. These beams are controlled separately by different LC cells and then superposed to reproduce a single beam. In the visible spectral region, polarization beam splitters are created by coating multiple films on a glass plate or prism [28]. Polarization separation is also achievable with calcite prisms, which exhibit birefringence if suitable crystallographic surfaces are selected. These devices, however, cannot be used in the infrared region, since ordinary oxide glasses and crystals are opaque in the wavelength range beyond 2 μm. If the birefringence of LC is combined with the infrared transparency of Si, a polarization beam splitter for the infrared light can be prepared in a simple process. The electric conductivity of Si is also useful to control the LC orientation [15]. In addition, the high refractive index of Si ($n = 3.4$) is advantageous for creating a polarization beam splitter, since a total internal reflection takes place at the Si-LC boundary due to the large difference in their indices.

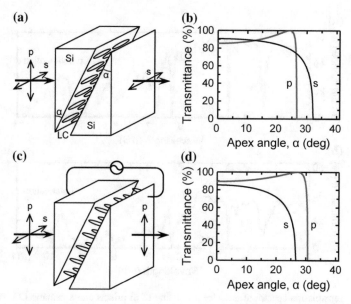

Fig. 4.1 **a** Structure of a polarization switch in which a nematic LC is aligned parallel to the Si prisms [17]. The polarization direction is denoted as p or s with reference to the slope of the prism. If the apex angle, α, is selected suitably, only an s-polarized beam emerges from the prism. **b** Theoretical transmittance at the Si-LC boundary for the LC orientation in (**a**). The assumed indices of refraction are $n_o = 1.53$ for p polarization and $n_e = 1.82$ for s polarization. **c** Reorientation of the LC molecules by application of an electric field. The output beam turns to p-polarized light due to the refractive index change. **d** Theoretical transmittance at the Si-LC boundary during the voltage application process in (**c**)

Figure 4.1 shows a principle of the polarization separation at the Si-LC boundary [17]. As Fig. 4.1a shows, a nematic LC is sandwiched between two Si prisms. The LC is aligned parallel to the Si surface so that its director is perpendicular to the slope direction. The assumed refractive indices are $n_o = 1.53$ for the ordinary light and $n_e = 1.82$ for the extraordinary light, corresponding to the LC that is used in the following experiment. A light beam that is normally incident on the outer surface of the Si prism impinges on the Si-LC boundary at an incident angle, α, that is determined by the apex angle of the prism. Figure 4.1b shows the theoretical transmittance at the Si-LC boundary, which is based on the Fresnel's reflection formula [14, 17]. If the apex angle is $\alpha = 0°$, i.e., if a light beam is normally incident on a flat Si plate, the transmittance for p- or s-polarization is 84 or 90%, respectively, corresponding to the index of $n_o = 1.53$ or $n_e = 1.82$. As the apex angle increases, the transmittance of p-polarized light increases gradually, and becomes 100% at $\alpha = 24°$ since the incident angle at the Si-LC boundary coincides with the Brewster's angle. As the apex angle increases further, the transmittance decreases rapidly. If the apex angle exceeds 26°, no light enters the LC layer, since the total internal reflection takes place at the boundary. By contrast, the transmittance of s-polarized light decreases monotonically as the apex

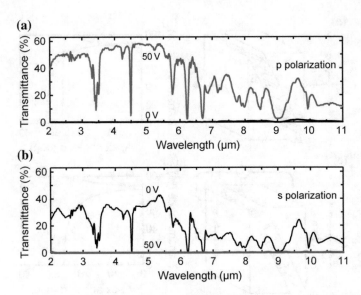

Fig. 4.2 Transmission spectra of a device consisting of Si prisms and a nematic LC. The probe light of the spectrometer was linearly polarized in the direction of **a** p or **b** s polarization that is shown in Fig. 4.1. The black lines show the spectra that were measured with no voltage application. The gray lines show the spectra during a 50 V (peak voltage, 1 kHz) application process

angle increases. The total internal reflection takes place at 32°, which is larger than the critical angle for p-polarized light since the refractive index for s polarization (n_e) is closer to that of Si. As Fig. 4.1c shows, voltage application induces reorientation of the LC layer, and accordingly, the refractive index for s polarization decreases to n_o. By contrast the p-polarized light takes an index between n_o and n_e. Consequently, the total internal reflection occurs at a smaller angle for s polarization than p polarization, as shown in Fig. 4.1d. Figure 4.1b, d predict that a polarization switch is attainable if the apex angle is adjusted at around 28°; i.e., at this angle, only p-polarized light passes through the LC layer during the voltage application process, whereas only s-polarized light passes when no voltage is applied.

A polarization switch was fabricated on the basis of the theoretical prediction above [17]. The input surface of the Si prism was a 20 mm square, and the thickness was adjusted to 11 mm so that the apex angle became $\alpha = 28°$. A polyimide film (Chisso, PIA-X107-G01) was coated on the inclined surface of the prism for aligning the LC molecules in the direction shown in Fig. 4.1a. Two prisms were adhered to one another with an epoxy adhesive containing glass rods of 5 μm diameter. A nematic LC (Merck, BL006) was put into this gap (5 μm) between the prisms. Figure 4.2 shows transmission spectra of the sample that were measured by using a Fourier transformation infrared (FTIR) spectrometer. A wire-grid polarizer was used to create a linearly-polarized probe beam. The sample transmittance was evaluated by taking the ratio of light intensities that were measured before and after placing the sample in the spectrometer. The spectral transmittance of the polarizer, therefore, did

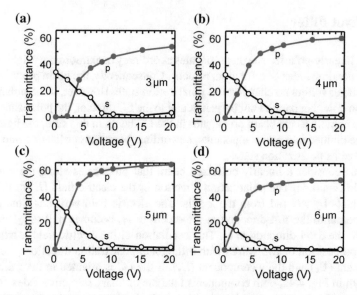

Fig. 4.3 Transmittance change during the voltage application process. The transmittances at **a** 3, **b** 4, **c** 5, or **d** 6 μm wavelength were taken from the spectra that were measured for each polarization by using the FTIR spectrometer. The horizontal axis shows the peak voltage of the driving signal (1 kHz)

not affect the evaluated sample transmittance. As the black lines show, the sample transmitted only s-polarized light in the original state (0 V). When the LC was reoriented by application of a sinusoidal signal of 50 V (peak voltage), s-polarized light was suppressed and p-polarized light emerged from the sample (the gray lines). The transmittance decreased at specific wavelengths (3.4 and 4.5 μm) as well as in the long wavelength range (>7 μm), since either the LC or the polyimide film was absorptive at these wavelengths. The maximum transmittance for p polarization was about 60% because of the high index of Si (3.4), which induced a high reflection loss at the outer surfaces of the prisms (the Si-air boundaries). The maximum transmittance for s polarization was lower than that for p polarization, since, as Fig. 4.1b shows, s-polarized light suffered a reflection loss due to the lack of the Brewster's angle.

Figure 4.3 shows the voltage dependence of the spectral transmittance. The transmittance change or the LC reorientation seemed to take place below 10 V. The transmittance, however, continued to increase after the voltage exceeded 10 V. The reorientation of the LC in the vicinity of the alignment coating probably needed a high electric field.

4.3   Lyot Filter

When a linearly-polarized beam propagates as ordinary or extraordinary light, it takes a corresponding index of refraction, n_o or n_e. Consequently, the beam propagating in the LC layer suffers no change in its polarization state. However, if the polarization direction is neither parallel nor perpendicular to the LC director, the polarization state changes notably as the beam propagates in the LC layer. In this case, the light beam has to be divided into two components, i.e., ordinary and extraordinary components, to analyze its polarization state.

Let us consider a linearly polarized beam that propagates along the z axis, as shown in Fig. 4.4a. The polarization direction or the electric field (E) of this light beam slants by $\pi/4$ rad from the x axis. The electric field varies as time passes; i.e., it points to the $\pi/4$ direction at a moment of t_1, becomes null at t_2, and then points to the $5\pi/4$ direction at t_3. This oscillation of the electric field is expressed by a superposition of two waves with orthogonal polarization directions, i.e., the x component (E_x) and the y component (E_y). If the LC is oriented in the z direction, as shown in Fig. 4.4a, both components take the ordinary refractive index (n_o) and propagate with the same wavelength, λ/n_o, in the LC layer ($z = 0$ to d), as shown in Fig. 4.4b. As Fig. 4.4c shows, therefore, the two waves come out of the LC cell in the same phase, creating a beam with the same polarization state as that of the input beam.

The LC orientation in Fig. 4.4a corresponds to the voltage application process between the substrates (The substrates are placed at $z = 0$ and d). If the alignment coating on the substrates is oriented in y direction, the voltage reduction causes the LC to tilt toward y direction, as shown in Fig. 4.4d. Since the LC director rotates in the y-z plane (from z direction toward y direction), the refractive index for the x component (E_x) is unchanged, whereas the other component (E_y) takes an index that is larger than n_o ($n_o < n' < n_e$). This causes the wavelength reduction to λ/n', which in turn induces the phase difference between the x and y components, as shown in Fig. 4.4e. Consequently, the output beam exhibits a circular or elliptical polarization, as shown in Fig. 4.4f.

As Fig. 4.4g shows, the LC orientation becomes closer to y direction by further decreasing the application voltage. Then the refractive index for the y component becomes even larger ($n' < n'' < n_e$), and accordingly, the wavelength (λ/n'') becomes shorter than before, as shown in Fig. 4.4h. Since the wavelength of the x component is unchanged, the phase difference becomes larger at the exit, yielding a linear polarization in the $3\pi/4$ direction, as shown in Fig. 4.4i.

The polarization state of the output beam is affected by the optical wavelength as well. Figure 4.4j, k show the same LC orientation as that in Fig. 4.4g, h, but correspond to a different wavelength, λ'. Accordingly, the wavelengths in the LC layer become λ'/n_o and λ'/n'' for the x and y components, respectively. As Fig. 4.4l shows, a combination of these components yields a linear polarization in the $\pi/4$ direction for a certain wavelength, λ'.

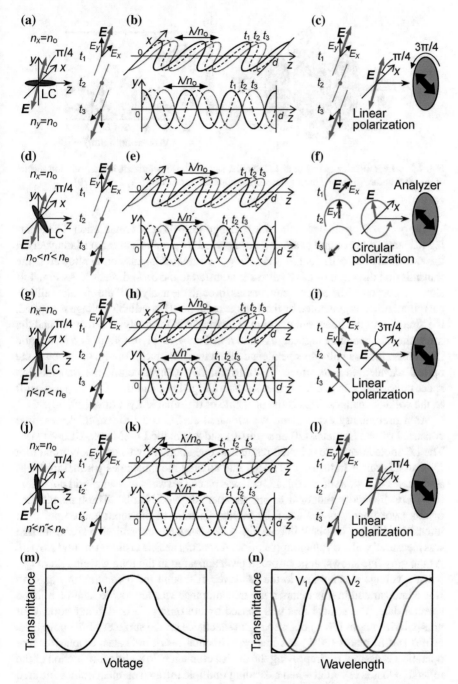

Fig. 4.4 a–l Schematic illustration for the change in the polarization state (retardation) inside a LC layer of thickness d. **m, n** Graphical illustration for the principle of the Lyot filter. The transmittance of the analyzer (polarizer at the output end) varies depending on the applied voltage and the wavelength

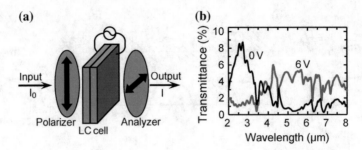

Fig. 4.5 **a** Lyot filter consisting of an LC retarder and wire-grid polarizers. **b** Transmission spectra of the filter before (0 V) and during the voltage application process (6 V, 1 kHz) [19]

The polarization change or the phase change above is called retardation. The polarization state of the output beam can be analyzed by placing a polarizer (analyzer) at the exit of the LC cell. As Fig. 4.4c shows, no light transmission is allowed if the transmission direction of the analyzer is adjusted to the $3\pi/4$ direction. As Fig. 4.4f shows, a portion of the output beam passes through the analyzer if the circular (elliptical) polarization is generated by the voltage reduction. If a suitable voltage is applied, the direction of the linear polarization coincides with the analyzer direction, and the transmittance becomes highest, as shown in Fig. 4.4i. Figure 4.4m is a schematic illustration for this voltage dependence of the transmittance. As Fig. 4.4l shows, the optical wavelength also affects the transmittance. Consequently, the transmission spectrum exhibits an oscillation, as shown in Fig. 4.4n, and the spectral peaks shift as the voltage changes. This is the principle of the tunable Lyot filter [29, 30].

As a preliminary experiment, we prepared an LC cell (20 μm thickness) that consisted of two Si plates (20 mm square) and a nematic LC (Merck, BL006) [19]. The LC molecules were oriented parallel to one another by using a polyimide film (Nissan Chemical Industries, SE-410) that aligned the molecules parallel to the Si surface. As Fig. 4.5a shows, this LC cell was placed between two wire-grid polarizers that were directed orthogonal to one another (crossed Nicols). The transmittance of this Lyot filter was evaluated by taking the ratio of the input and output light intensities (I/I_0). The black line in Fig. 4.5b shows the transmission spectrum that was measured with no voltage application. A spectral peak is visible at around 2.6 μm. As the gray line shows, another broad peak appeared in the long wavelength range, when a voltage was applied to the LC layer. Although the prediction in Fig. 4.4n was demonstrated in this manner, the transmittance was too low to attain a reliable spectral data. The optical loss was caused by not only polarization adjustment for non-polarized light (50%) but also high reflectance on the surfaces of the polarizers (10%) and Si plates (30%). As this experimental result indicates, a polarizer-free operation is desired for improving the device efficiency. In addition, reduction of the reflection loss is essential when cascading multiple infrared elements, since infrared transmitting materials usually possess a high refractive index that induces a high reflectance.

These problems, i.e., the polarization dependence and the reflection loss, were solved by integration of the LC elements [18, 19]. Figure 4.6a shows the structure of the fabricated device. The device consists of two pentaprisms, three LC layers, and two mirrors. As Fig. 4.6b shows, a Si pentaprism receives an input beam at an incident angle of $\theta_0 = \alpha = 39°$, and refracts it toward the central LC layer so that the incident angle at the Si-LC boundary becomes $\theta = 28°$. As Fig. 4.6c shows, a nematic LC (Merck, BL006) is enclosed between the two prisms (5 μm gap), being oriented perpendicular to the prism surface by voltage application (40 V). When the input beam impinges on this boundary at the incident angle of $\theta = 28°$, the s-polarized component is reflected (total internal reflection), whereas the p-polarized component passes through the LC layer (Brewster's angle), as shown in Fig. 4.1d. In this manner, this LC layer acts as a polarization beam splitter.

The s-polarized beam that is reflected by the central LC layer makes a normal incidence on the lower left surface of the prism, since this surface is inclined at the angle of $\beta = \theta = 28°$. Similarly, the p-polarized beam that passes through the central LC layer impinges normally on the lower right surface of the other prism. At this surface, another LC layer is formed, as shown in Fig. 4.6d. A gap of ~20 μm between the Si prism and a gold-coated glass mirror is filled with the same LC as used in the central layer. The surfaces of the prism and the mirror are coated with a polyimide film (Nissan Chemical Industries, SE-410) to align the LC molecules parallel to the surfaces. These films are rubbed in the $\pi/4$ direction ($\pi/4$ rad apart from the polarization direction) so that the LC layer acts as a retarder. An incident p-polarized beam, therefore, changes its polarization state depending on the wavelength as it makes a round trip in the LC layer. A voltage application to this LC layer also changes the polarization state due to reorientation of the LC molecules. The other LC layer at the lower left surface (retarder 1) also induces a similar polarization change for the s-polarized light.

The reflected beams return to the central LC layer that reflects the s-polarized component and transmits the p-polarized component. Consequently, the central LC layer acts as a crossed-Nicol polarizer for the retarders 1 and 2. That is, the three LC layers exhibit a function of Lyot filter. The output beam, which is a superposition of the two beams, propagates along the same optical axis as that of the input beam. Figure 4.6e shows a photograph of this integrated device. Electric wires are soldered on the prisms and mirrors to control the LC orientation. In this manner, the combination of Si and LC realizes a device that is free from polarizers and large reflection losses.

The transmission characteristics of the fabricated device were examined by using the FTIR spectrometer. First, an s-polarized probe beam was used to confirm the function of each LC layer. Figure 4.7a–j show the transmission spectra that were measured during a voltage application process to the retarder 1 (V_1). No voltage was applied to the retarder 2 ($V_2 = 0$). Three peaks at 2.6, 3.6, and 5.3 μm in the low voltage range (the gray, white, and black arrows) shifted to shorter wavelengths as the voltage rose. This result indicated that the retardation became smaller as the LC molecules were oriented perpendicular to the prism surface by the voltage application. By contrast, no peak shift was visible when the voltage was applied to

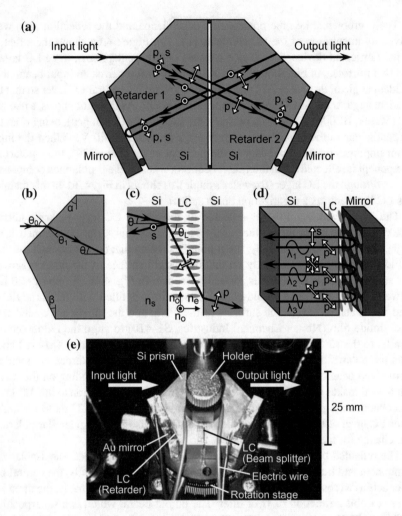

Fig. 4.6 **a** Operation principle of the integrated Lyot filter. The device consists of **b** Si pentaprisms, **c** a homeotropic LC layer (polarization beam splitter), and **d** plane-parallel LC layers (retarders). **e** Photograph of the device that is mounted on a rotation stage [18, 19]

the retarder 2, as shown in Fig. 4.7k–t. As Fig. 4.6a shows, no light propagated to the retarder 2 when the s-polarized light was incident on the central LC layer. Therefore the voltage application to the retarder 2 induced no spectral change. This fact verified that the central layer acted correctly as a polarization beam splitter.

Spectral change is difficult to recognize in Fig. 4.7, since the transmission spectra are disturbed by the absorption bands that are caused by the LC or the alignment coating. Figure 4.8a–f show the voltage dependence of the transmittance (T), which is taken from the spectral data for s polarization (Fig. 4.7). The transmittance at each wavelength changes notably as the voltage V_1 changes, whereas V_2 induces

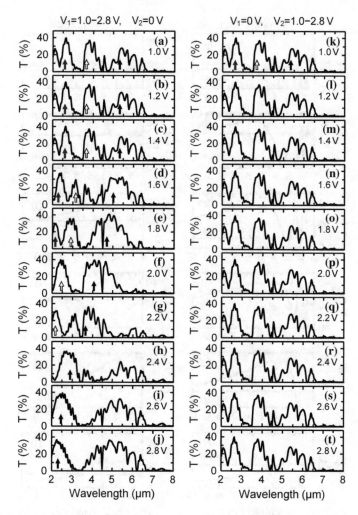

Fig. 4.7 Transmission spectra of the integrated Lyot filter for s polarization. Voltage was applied only to **a–j** the retarder 1 or **k–t** the retarder 2. The arrows denote the peak wavelengths [19]

no change. Figure 4.8g–l show the results for p polarization. The transmittance is affected by the voltage V_2, since, as Fig. 4.6a shows, the p-polarized light propagates to the retarder 2. The transmittance, however, does not decrease to 0% and is affected a little by the voltage V_1. Both the prism design and the fabrication technique (precision) need improvement for attaining a better performance.

Figure 4.9a–j show the transmission spectra that were measured by using a non-polarized probe beam. A numeral in the figures denotes the voltage V_1 that was applied to the retarder 1. The voltage for the retarder 2 was adjusted at $V_2 = 0.86V_1 + 0.09$, since, as Fig. 4.8g–l show, the retarder 2 exhibited a smaller retardation than the retarder 1 [19]. As the arrows in the graphs indicate, the spectral peaks shifted to

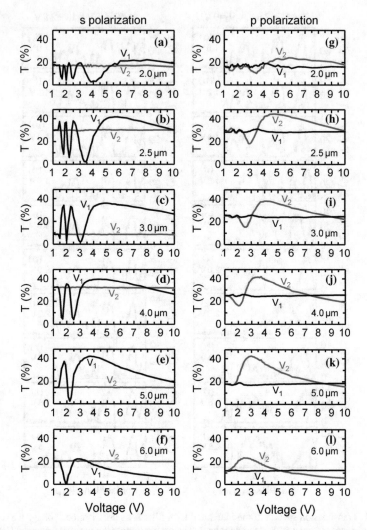

Fig. 4.8 Voltage dependence of the transmittance. The voltage was applied to the retarder 1 (the black lines) or the retarder 2 (the gray lines). Measurements were conducted for **a–f** the s- or **g–l** p-polarized probe beam. The probe light wavelengths are denoted in the figures [19]

shorter wavelengths with the increase of the applied voltage. Figure 4.9k–p show the voltage dependence of the transmittance. In comparison with the conventional Lyot filter in Fig. 4.5, this integrated device realized five-fold increase of the maximum transmittance (>40%) due to the removal of polarizers and the reduction of the reflection loss. In addition to these advantages, the integration of the LC elements reduced the device size, facilitated handling, and dispensed with a troublesome alignment process for invisible infrared beams.

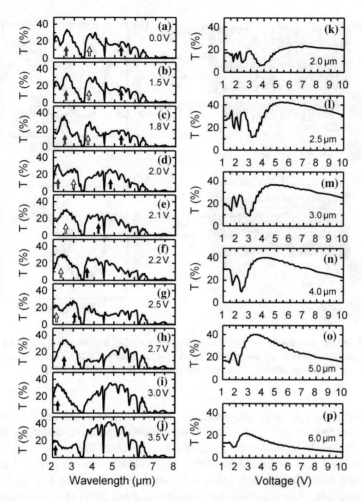

Fig. 4.9 Non-polarized light transmittance through the polarizer-free Lyot filter. **a–j** Spectral change by voltage application. The voltage that is denoted in the figures (V_1) was applied to the retarder 1 and $V_2 = (0.86V_1 + 0.09)$ was applied to the retarder 2. The arrows denote the transmission peaks. **k–p** Voltage dependence of the transmittance at each wavelength [19]

4.4 Fabry-Perot Filter

Fabry-Perot interferometers are usually used for creating tunable filters [3, 12, 25, 31]. Figure 4.10 shows a typical filter structure and the principle of spectral tuning. As Fig. 4.10a shows, a nematic LC is sandwiched between two cavity mirrors with its director being parallel to them. If the input light is linearly polarized in the direction parallel to the director (extraordinary light), its wavelength in the LC layer becomes λ/n_e. As it propagates back and forth in the layer (thickness: d), a constructive

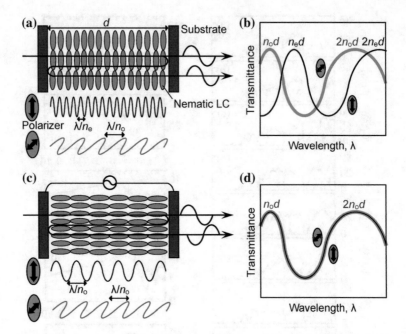

Fig. 4.10 **a** Structure of a Fabry-Perot filter that uses nematic LC. **b** Interference peaks in the transmission spectra of the ordinary (the gray line) or extraordinary light (the black line). **c** Reorientation of the LC by voltage application. **d** Interference spectra for the orientation in (**c**)

interference takes place if a phase-matching condition is satisfied, i.e., $2d = m(\lambda/n_e)$ or $\lambda = 2n_e d/m$ ($m = 1, 2, 3 \ldots$). The transmission spectrum, therefore, exhibits interference peaks at $2n_e d$ and $n_e d$, as shown by the black line in Fig. 4.10b.

When the LC is reoriented by voltage application, as shown in Fig. 4.10c, the light beam takes the ordinary index (n_o) and propagates in the LC layer with a wavelength λ/n_o. Consequently, the interference peak shifts to $\lambda = (2n_o d)/m$, as shown in Fig. 4.10d. This is the principle of the spectral tuning. As the gray waves and the gray spectra in Fig. 4.10 show, however, no peak shift occurs for the linear polarization in the orthogonal direction (ordinary light). The LC Fabry-Perot filters, therefore, usually need a polarizer.

Various methods have been proposed for eliminating the polarization dependence of LC devices [25, 27, 32–37]. If the anisotropy of the LC itself is eliminated, a polarizer-free operation is attainable. It is known in the scientific field of composite materials that mixtures exhibit uniform optical properties if the component size is smaller than optical wavelength [38, 39]. That is, if two materials with a dielectric constant of ε_1 or ε_2 are mixed at the ratio of $q:(1 - q)$, the dielectric constant of the mixture becomes $\varepsilon = q\varepsilon_1 + (1 - q)\varepsilon_2$. Since the refractive index is proportional to the square-root of the dielectric constant, the index of the mixture becomes $n = [qn_1^2 + (1 - q)n_2^2]^{1/2}$, in which $n_1^2 = \varepsilon_1$ and $n_2^2 = \varepsilon_2$ (the effective medium theory). The refractive index of a saline solution, i.e., a mixture of water and salt, is an

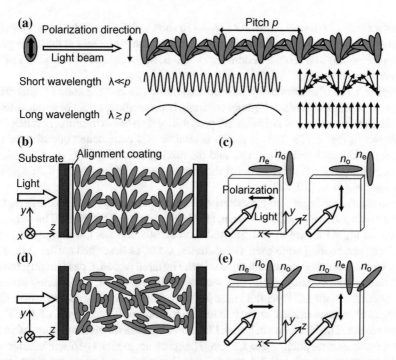

Fig. 4.11 **a** Orientation of a cholesteric LC and the polarization directions of light beams propagating in it. **b** Cholesteric LC that is aligned parallel to the substrates. **c** Two typical directors in the x-y plane that are encountered by a light beam. **d** Orientation of cholesteric LC in a cell with no alignment coating. **e** Three typical directors in the random domain texture in (**d**)

example that is represented by this expression. This principle seems to hold true for LC. In a twisted nematic LC, for example, long-wavelength light propagates with no polarization change, whereas short-wavelength light rotates its polarization direction according to the twist of the LC molecules [40]. This fact indicates that the long-wavelength light does not recognize a microstructure but grasps an average of the entire structure. Therefore, if the twist pitch of the LC is sufficiently small, a propagating beam will take a polarization-independent refractive index between the ordinary and extraordinary indices ($n_o < n < n_e$).

Chiral nematic LC, which is commonly called cholesteric LC, possesses a microstructure that is suitable for inducing isotropic optical properties in the long-wavelength range [41–43]. Figure 4.11a illustrates an orientation texture of the cholesteric LC. The LC director rotates with a chiral pitch, p, which is typically 0.1–10 μm. As a linearly-polarized beam propagates in this LC layer, its polarization direction rotates with the same period (p), if the wavelength is sufficiently shorter than the chiral pitch ($\lambda < p$). A light beam with a long wavelength ($\lambda > p$), however, cannot recognize this microstructure, and hence, propagates with no polarization rotation. Figure 4.11b is a schematic illustration of a cholesteric LC that is oriented parallel to the substrates by the use of alignment coatings. In this alignment, the LC director rotates in the x-y plane. As Fig. 4.11c shows, therefore, a beam propagating in this LC layer encounters the ordinary and extraordinary indices at the same probability ($q = 0.5$) and takes a certain index n_{AV} between n_o and n_e regardless of the polarization direction.

Figure 4.11d illustrates a random domain texture of the cholesteric LC that arises when the substrates have no alignment coating. In this texture, a light beam encounters the ordinary index (n_o) at twice higher probability than the extraordinary index (n_e), as shown in Fig. 4.11e. This situation is similar to a composite consisting of two materials with an index of n_o or n_e and the mixing ratio of $q:(1 - q) = 0.67:0.33$. Consequently, the refractive index of this texture (n_R) become smaller than n_{AV}; i.e., $n_o < n_R < n_{AV} < n_e$ [20, 22].

Polarization-independent characteristics were examined by using cholesteric LCs with a chiral pitch of $p = 1.5$ or 4.8 μm (JNC Corporation, JD-1036L). The refractive indices of these LCs are $n_o = 1.516$ and $n_e = 1.760$. Si plates (20 mm square) with no alignment coating were used as substrates, since, as described earlier (Fig. 4.2), the alignment coating caused absorption in the infrared region. Consequently, the LC took the random domain texture that was shown in Fig. 4.11d. The spacing between the substrates or the LC layer thickness (d) was adjusted by using a glass spacer [22].

Spectral measurements were conducted for non-polarized light of the FTIR spectrometer. The black line in Fig. 4.12a shows the transmission spectrum of an LC sample with a chiral pitch of $p = 1.5$ μm. The spectrum exhibits periodic interference peaks together with dips at 3.4 and 5.7 μm that originate from absorption by the LC. The gray line shows the interference peak wavelengths that were calculated by assuming a suitable index ($n = 1.61$) and thickness ($d = 20.8$ μm); i.e., $\lambda = (2nd)/m$ [22]. This evaluated index (the best fit value for the measured peak wavelengths) agreed with the prediction of the effective medium theory.

As Fig. 4.12b shows, the transmittance decreased over the entire spectral range when 60 V (peak voltage) was applied to the LC layer. This phenomenon was caused by optical scattering during the reorientation process [20–22]. As Fig. 4.12c shows, the transmittance recovered when the voltage rose to 100 V. The maximum transmittance (~50%) was determined by the reflection losses at the outer surfaces of the sample (the Si-air boundaries). Figure 4.12d shows the refractive indices that were evaluated by drawing the fitting curves (the gray lines). As the voltage rose, the LC orientation approached the homeotropic phase that was shown in Fig. 4.10c. Consequently, the index decreased gradually in the range below 90 V, and then dropped rapidly to the ordinary index (n_o) at 93 V. The polarization-independent index change was demonstrated in this manner. This sample, however, required a high operation voltage and exhibited a notable scattering loss on the midway of the index change.

These problems seemed to be solved by reducing the sample thickness. Therefore an LC cell of ~4 μm thickness was prepared by the use of smaller glass spacers. Figure 4.12e shows the transmission spectra of this thin sample. The spacing between the interference peaks is wider than that of the former sample because of the thickness reduction. The refractive index is evaluated to be 1.64 by the peak wavelength fitting (the gray line). This value is higher than that of the thick sample. The possible reason for this index increase is a plane-parallel orientation of the LC; i.e., although the

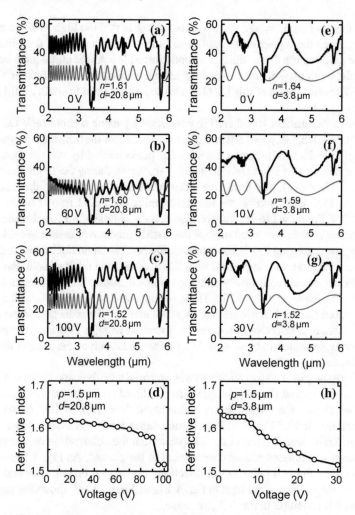

Fig. 4.12 Transmission spectra and evaluated index of refraction for the LC samples with a chiral pitch of 1.5 μm and a thickness of **a–d** ~20 or **e–h** ~4 μm. The black lines show the measured spectra and the gray lines show the fitting curves that were drawn by assuming a suitable index (*n*) and thickness (*d*). The voltage indicates the amplitude (peak value) of a sinusoidal electric signal (1 kHz) for inducing LC reorientation [22]

substrates have no alignment coating, they affect the LC orientation more strongly in the narrow gap, and hence, the LC molecules possibly tend to be oriented parallel to the substrate, as shown in Fig. 4.11b. When the voltage was applied to the sample, the transmittance decreased slightly at around 10 V, as shown in Fig. 4.12f. As Fig. 4.12g shows, the transmittance returned to the original level when the voltage was raised to 30 V. Figure 4.12h shows the voltage dependence of the refractive index. The index decreases gradually as the voltage increases exceeding 6 V. Although the thin sample reduces the operation voltage to 30 V, this voltage is still too high to attain a stable function, since the high-voltage application to the narrow gap (<4 μm) induces a substrate deformation due to the electrostatic force [44]. As the chiral pitch becomes shorter, the reorientation process tends to require a stronger electric field or a higher voltage. The chiral pitch of this LC (1.5 μm) seems too short to attain useful optical functions.

The next experiment was therefore conducted by using a cholesteric LC with a longer chiral pitch, i.e., $p = 4.8$ μm. Figure 4.13a–c show the transmission spectra of the sample of ~20 μm thickness. Interference peaks shifted by voltage application. However, this sample also exhibited a heavy scattering during the reorientation process (10–20 V range). The refractive index was evaluated by fitting the theoretical curves. As Fig. 4.13d shows, the refractive index decreased from 1.61 to 1.51 at around 10 V. Figure 4.13e–h show the results for a sample of ~5 μm thickness. The reorientation voltage decreased to ~5 V due to the reduction of the sample thickness. However, the scattering loss in the reorientation process could not be reduced by the thickness reduction. In comparison with the result for the thick sample, the refractive index decreased more slowly with the voltage increase. The short-pitch LC also exhibited this tendency (Fig. 4.12). These results clarified that the LC with a longer pitch required a lower electric voltage for reorientation. The reduction of the sample thickness was also effective to reduce the operation voltage. However, the problem of scattering remained in these samples. Further reduction of the sample thickness was desired.

Finally a sample of ~3 μm thickness was prepared with the same LC ($p = 4.8$ μm). As Fig. 4.14a–c show, the voltage application induced no transmittance decrease. As Fig. 4.14d shows, the refractive index decreased from 1.61 to 1.52 in the 2–6 V range, and reached 1.51 at 12 V. Polarization dependence was examined by placing a wire-grid polarizer in front of the sample so that the polarization direction of the probe beam was either horizontal or vertical to the ground. As Fig. 4.14e–g show, the spectra for the two polarization directions overlapped one another and agreed with those for non-polarized light in Fig. 4.14a–c. Figure 4.14h shows the tunability of the peak wavelength in the 2–3 μm range.

The interference spectra in Fig. 4.14 oscillate between 30 and 55%. An ideal spectral filter has to exhibit a maximum transmittance of 100% and a minimum transmittance of 0%. The maximum transmittance of the current Fabry-Perot filter is determined by the reflection loss at the outer surfaces (the Si-air boundaries). The antireflection coating on the outer surface of the Si substrates will enhance the maximum transmittance of the current filter. On the other hand, the minimum transmittance is determined by the reflectance of the inner surface (the Si-LC boundary). A high-reflection coating on the inner surface of the Si substrate will reduce the minimum transmittance [20]. Further, a function of a band-pass filter is also required to select

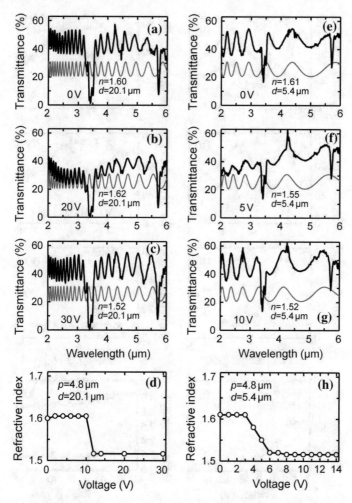

Fig. 4.13 Transmission spectra and refractive indices of the LC with a chiral pitch of 4.8 μm. The sample thickness was **a–d** ~20 or **e–h** ~5 μm. The black and gray lines show the measured and calculated values, respectively

a single transmission wavelength from a series of periodic interference peaks. The author's group is currently designing such multilayer coatings on the Si surfaces that realize anti-reflection, high-reflection, and band-pass properties. Combination of these Si substrates and the cholesteric LC will realize an efficient tunable filter for the infrared measurements.

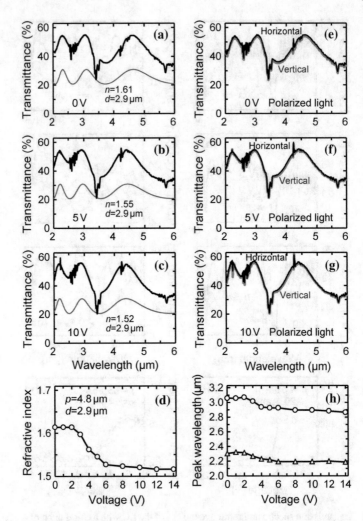

Fig. 4.14 **a–c** Transmission spectra of non-polarized light that passed through a LC layer with a chiral pitch of $p = 4.8\,\mu m$ and a thickness of $d = 2.9\,\mu m$ (the black lines). The gray lines are drawn for the index evaluation. **d** Refractive index change during the voltage application process. **e–g** Transmission spectra of the same sample for linearly-polarized light. The black and gray lines (overlapping) correspond to the polarization directions that are horizontal or vertical to the ground, respectively. **h** Peak wavelength tuning by the voltage application

4.5   Summary

Spectral measurements in the infrared region are important for chemical analyses, gas sensings, and thermal imagings. Compact measurement instruments can be created if LCs are used for spectral tuning. Polarizer-free LC devices are useful particularly for the infrared measurements, since efficient optical elements for the visible light are unusable in the long wavelength region. The isotropic index change in a cholesteric LC layer as well as the element integration capability of Si-based devices will extend the application field of LCs in the infrared technology.

Acknowledgements The author thanks the students in his laboratory for their help in conducting experiments of this research. A part of this research was supported by a grant from Japan Society for Promotion of Science.

References

1. I.C. Khoo (ed.), *Liquid Crystals*, 2nd edn. (Wiley, New York, 2007)
2. J.W. Doane, A. Golemme, J.L. West, J.B. Whitehead, B.-G. Wu, Polymer dispersed liquid crystals for display applications. Mol. Cryst. Liq. Cryst. **165**, 511–532 (1988)
3. M.W. Maeda, J.S. Patel, C. Lin, J. Horrobin, R. Spicer, Electronically tunable liquid-crystal-etalon filter for high-density WDM systems. IEEE Photon. Technol. Lett. **2**, 820–822 (1990)
4. W.A. Crossland, I.G. Manolis, M.M. Redmond, K.L. Tan, T.D. Wilkinson, M.J. Holmes, T.R. Parker, H.H. Chu, J. Croucher, V.A. Handerek, S.T. Warr, B. Robertson, I.G. Bonas, R. Franklin, C. Stace, H.J. White, R.A. Woolley, G. Henshall, Holographic, optical switching: the "ROSES" demonstrator. J. Lightwave Technol. **18**, 1845–1854 (2000)
5. A.F. Fray, C. Hilsum, D. Jones, Some properties of liquid crystals as infrared modulators. Infrared Phys. **18**, 35–41 (1978)
6. D.P. Resler, D.S. Hobbs, R.C. Sharp, L.J. Friedman, T.A. Dorschner, High-efficiency liquid-crystal optical phase-array beam steering. Opt. Lett. **21**, 689–691 (1996)
7. D.P. Dewitt, G.D. Nutter, *Theory and Practice of Radiation Thermometry* (Wiley, New York, 1988)
8. E. Klimov, M. Fuelleborn, H.W. Siesler, Electric-field-induced reorientation of liquid crystalline p-cyanophenyl-p-n-alkylbenzoates: a time-resolved study by Fourier transform infrared transmission and attenuated total reflection spectroscopy. Appl. Spectrosc. **57**, 499–505 (2003)
9. L.H. Kidder, I.W. Levin, E.N. Lewis, V.D. Kleiman, E.J. Heilweil, Mercury cadmium telluride focal-plane array detection for mid-infrared Fourier-transform spectroscopic imaging. Opt. Lett. **22**, 742–744 (1997)
10. M. Saito, K. Kikuchi, C. Tanaka, H. Sone, S. Morimoto, T. Yamashita, J. Nishii, Spectroscopic gas-flow imaging through an infrared optical fiber bundle. Rev. Sci. Instrum. **70**, 4308–4312 (1999)
11. M. Saito, Y. Okubo, Time-resolved infrared spectrometry with a focal plane array and a galvano-mirror. Opt. Lett. **32**, 1656–1658 (2007)
12. H. Zhang, A. Muhammmad, J. Luo, Q. Tong, Y. Lei, X. Zhang, H. Sang, C. Xie, Electrically tunable infrared filter based on the liquid crystal Fabry-Perot structure for spectral imaging detection. Appl. Opt. **53**, 5632–5639 (2014)
13. I. Yamada, K. Kintaka, J. Nishii, S. Akioka, Y. Yamagishi, M. Saito, Mid-infrared wire-grid polarizer with silicides. Opt. Lett. **33**, 258–260 (2008)
14. M. Saito, T. Yasuda, Complex refractive-index spectrum of liquid crystal in the infrared. Appl. Opt. **42**, 2366–2371 (2003)

15. M. Saito, R. Takeda, K. Yoshimura, R. Okamoto, I. Yamada, Self-controlled signal branch by the use of a nonlinear liquid crystal cell. Appl. Phys. Lett. **91**, 141110-1–3 (2007)
16. M. Saito, K. Yoshimura, K. Kanatani, Silicon-based liquid crystal cell for self-branching of optical packets. Opt. Lett. **36**, 208–210 (2011)
17. M. Saito, T. Yasuda, An infrared polarization switch consisting of silicon and liquid crystal. J. Opt. **12**, 015504-1–6 (2010)
18. M. Saito, K. Hayashi, Tunable retarder made of pentaprisms and liquid crystal. Proc. SPIE **8114**, 811412-1–12 (2011)
19. M. Saito, K. Hayashi, Integration of liquid crystal elements for creating an infrared Lyot filter. Opt. Express **21**, 11984–11993 (2013)
20. M. Saito, A. Maruyama, J. Fujiwara, Polarization-independent refractive-index change of a cholesteric liquid crystal. Opt. Mater. Express **5**, 1588–1597 (2015)
21. M. Saito, H. Uemi, A spatial light modulator that uses scattering in a cholesteric liquid crystal. Rev. Sci. Instrum. **87**, 033102-1–6 (2016)
22. M. Saito, J. Fujiwara, Reduction of an optical rotation and scattering in a cholesteric liquid crystal layer, in *Proceedings of 5th International Conference on Photonics, Optics and Laser Technology* (Porto, Feb 2017), pp. 22–30
23. R.E. Wagner, J. Cheng, Electrically controlled optical switch for multimode fiber applications. Appl. Opt. **19**, 2921–2925 (1980)
24. E.G. Hanson, Polarization-independent liquid-crystal optical attenuator for fiber-optics applications. Appl. Opt. **19**, 2921–2925 (1980)
25. J.S. Patel, M.W. Maeda, Tunable polarization diversity liquid-crystal wavelength filter. IEEE Photon. Technol. Lett. **3**, 739–740 (1991)
26. W.C. Yip, H.C. Huang, H.S. Kwok, Efficient polarization converter for projection displays. Appl. Opt. **36**, 6453–6457 (1997)
27. J. Moore, N. Collings, W.A. Crossland, A.B. Davey, M. Evans, A.M. Jeziorska, M. Komarčević, R.J. Parker, T.D. Wilkinson, H. Xu, The silicon backplane design for an LCOS polarization-insensitive phase hologram SLM. IEEE Photon. Technol. Lett. **20**, 60–62 (2008)
28. L. Li, J.A. Dobrowolski, High-performance thin-film polarizing beam splitter operating at angles greater than the critical angle. Appl. Opt. **39**, 2754–2771 (2000)
29. B. Lyot, Le filtre monochromatique polarisant et ses applications en physique solaire. Ann. Astrophys. **7**, 31–79 (1939)
30. J.W. Evans, Solc birefringent filter. J. Opt. Soc. Am. **48**, 142–145 (1958)
31. E. Hecht, *Optics* (Addison-Wesley, Reading, MA, 1998). Chapter 9
32. K. Hirabayashi, H. Tsuda, T. Kurokawa, Tunable liquid-crystal Fabry-Perot interferometer filter for wavelength-division multiplexing communication systems. J. Lightwave Technol. **11**, 2033–2043 (1993)
33. J.-H. Lee, H.-R. Kim, S.-D. Lee, Polarization-insensitive wavelength selection in an axially symmetric liquid-crystal Fabry-Perot filter. Appl. Phys. Lett. **75**, 859–861 (1999)
34. Y.-H. Lin, H. Ren, Y.-H. Wu, Y. Zhao, J. Fang, Z. Ge, S.-T. Wu, Polarization-independent liquid crystal phase modulator using a thin polymer-separated double-layered structure. Opt. Express **13**, 8746–8752 (2005)
35. A. Alberucci, M. Peccianti, G. Assanto, G. Coschignano, A.D. Luca, C. Umeton, Self-healing generation of spatial solitons in liquid crystals. Opt. Lett. **30**, 1381–1383 (2005)
36. M. Ye, B. Wang, S. Sato, Polarization-independent liquid crystal lens with four liquid crystal layers. IEEE Photon. Technol. Lett. **18**, 505–507 (2006)
37. C. Provenzano, P. Pagliusi, G. Cipparrone, Highly efficient liquid crystal based diffraction grating induced by polarization holograms at the aligning surfaces. Appl. Phys. Lett. **89**, 121105-1–3 (2006)
38. O. Wiener, Die Theorie des Mischkörpers für das Feld der Stationären Strömung. Abh. Math. Phys. Kl. Sächs. Akad. Wiss. **32**, 507–604 (1912)
39. M. Born, E. Wolf, *Principles of Optics* (Pergamon, Oxford, 1980). Chapter 2
40. J.S. Patel, S.-D. Lee, Electrically tunable and polarization insensitive Fabry-Perot étalon with a liquid-crystal film. Appl. Phys. Lett. **58**, 2491–2493 (1991)

41. H.-S. Kitzerow, C. Bahr (eds.), *Chirality in Liquid Crystals* (Springer, New York, 2001)
42. K.-H. Kim, H.-J. Jin, K.-H. Park, J.-H. Lee, J.C. Kim, T.-H. Yoon, Long-pitch cholesteric liquid crystal cell for switchable achromatic reflection. Opt. Express **18**, 16745–16750 (2010)
43. Y.-C. Hsiao, C.-Y. Tan, W. Lee, Fast-switching bistable cholesteric intensity modulator. Opt. Express **19**, 9744–9749 (2011)
44. M. Saito, H. Furukawa, Infrared tunable filter by the use of electrostatic force. Appl. Phys. Lett. **79**, 4283–4285 (2001)

Chapter 5
Polyethylene Glycol as a Bistable Scattering Matrix for Fluorescent Materials

Mitsunori Saito

Abstract Polyethylene glycol (PEG) exhibits a strong scattering in the solid phase whereas it is transparent in the liquid phase. Since various molecules and ions, e.g., organic dyes and fluorescent lanthanides, are soluble in bipolar PEG, it acts as a tunable matrix for those guest materials. When a fluorescent dye (rhodamine 6G) is dispersed in PEG, the strong scattering in the solid phase induces random lasing that exhibits a narrow peak in the fluorescence spectrum. This stimulated emission peak, however, reduces to a weak, broad spectrum of a spontaneous emission through the phase transition process to the transparent liquid. Regarding the lanthanide ions (europium, erbium, etc.), which generally require a long optical path for absorbing excitation light, a stronger photoluminescence occurs in the solid phase than the transparent liquid phase, since the scattering extends the optical path of the excitation light. In addition to this scattering characteristic, PEG exhibits a bistability (hysteresis) in the phase transition process; i.e., the melting point is higher than the freezing point. Consequently, both the solid and liquid phases are stable in a specific temperature range. This bistable range extends by mixing PEGs with different molecular weights; e.g., a mixture of PEGs with molecular weights of 300 and 2000 takes both the solid and liquid phases in the range between 2 and 38 °C. This bistable scattering matrix is useful to realize controllable optical devices including tunable lasers and rewritable signboards.

5.1 Introduction

Optical bistability is useful to realize various optical functions. One of the most successful application fields of the optical bistability is the rewritable memory, in which a bistable isomerization process of photochromic dyes was used in early days (compact disk, CD) [1, 2] and a bistable phase transition of chalcogenides is used

M. Saito (✉)
Department of Electronics and Informatics, Ryukoku University,
Seta, Otsu 520-2194, Japan
e-mail: msaito@rins.ryukoku.ac.jp

© Springer Nature Switzerland AG 2019
P. Ribeiro et al. (eds.), *Optics, Photonics and Laser Technology 2017*,
Springer Series in Optical Sciences 222,
https://doi.org/10.1007/978-3-030-12692-6_5

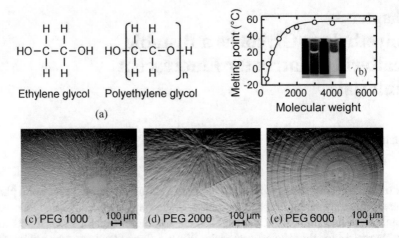

Fig. 5.1 a Molecular structures of ethylene glycol and polyethylene glycol (PEG). **b** Molecular weight dependence of the melting point. The inset shows a photograph of PEG in the liquid (left) or solid (right) phase [18]. **c–e** Micrographs of solid PEGs with a molecular weight of 1000, 2000, or 6000

currently (digital versatile disk, DVD) [3]. Liquid crystals are also used widely to create bistable devices, since they occasionally exhibit a hysteresis in the phase transition process [4–12]. Besides these phase transition characteristics of specific materials, various optical effects have been examined to attain bistable functions; e.g., micro-cavity resonance [13, 14] and electro-, thermo-, or acousto-optic effects [15–17]. These optical functions, however, require a complicated device structure, a high-voltage electric source, or a high-frequency instrument.

Polyethylene glycol (PEG) has rarely been used in the optical technology, although it is a well-known chemical agent. As Fig. 5.1a shows, ethylene glycol (EG) contains both polar and nonpolar groups, and accordingly, chemists usually use it as a bipolar solvent. PEG also acts as a solvent for various dye molecules or fluorescent ions. PEG is usually denoted together with its molecular weight, e.g., PEG 1000. The melting point of PEG is adjustable at around room temperature, since, as Fig. 5.1b shows, it varies notably depending on the molecular weight [18]. As the photograph (inset) shows, PEG is transparent in the liquid phase but becomes milky-white in the solid phase. (Imagine a white fat that turns to transparent oil on a frying pan.) Figure 5.1c–e show the micrographs of PEGs in the solid phase. Radial or circular patterns are visible. These microstructures, which seem to originate from self-alignment of PEG molecules, induce a strong scattering.

Recently, scattering materials have attracted interests, since they exhibit useful optical functions including coherent scattering and random lasing [19–23]. Experiments on the scattering phenomena are usually conducted by either dispersing nanoparticles in a matrix or infiltrating a solution into a porous material [24]. If PEG is used as a solvent or a solid matrix for dyes, neither nanoparticles nor porous materials is needed for attaining a strong scattering. In addition, PEG provides tunability of the scattering strength due to the phase transition at around room temperature. In the following sections, optical characteristics and functions of PEG are described on the basis of recent experiments in the author's laboratory.

5.2 Bistable Transmission Properties

The transmittance change during the phase transition process was measured by putting PEG in a sample cell of 10 mm thickness [25]. The sample was heated or cooled at ~10 deg/min by applying a suitable voltage (−4 to +4 V) to the Peltier element that was attached to the sample cell. The sample temperature was measured by inserting a sheathed thermocouple (1 mm diameter) into the PEG. Figure 5.2a shows the transmission spectra of PEG 1000 in the heating process. The sample became transparent at 41 °C, indicating that the sample melted at this temperature. The sample was still in the solid phase at 38 °C, and hence, it exhibited no light transmission over the entire visible spectral range. Interestingly, the sample exhibited a high transmittance at the same temperature (38 °C) in the cooling process, as shown in Fig. 5.2b. In other words, both the transparent and opaque states (the liquid and solid phases) were stable at around 38 °C. The reproducibility of these results was confirmed by repeating the heating and cooling processes several times.

Other PEGs also exhibited a similar hysteresis in the phase transition process. Figure 5.3a shows the transmittance at 500 nm wavelength during the heating (circles) and cooling (triangles) processes. The bistable ranges of PEG 300, PEG 600, and PEG 2000 were −13 to −6 °C, 15–21 °C, and 40–52 °C, respectively. This bistability was caused by notable stability of the supercooling state. It is generally known that non-equilibrium states are stabilized by mixing different materials. This rule holds true for PEG; i.e., the bistable range extends by mixing PEGs with different molecular weights. As Fig. 5.3b shows, for example, the bistable range extends to 32–49 °C, when PEG 300 and PEG 2000 are mixed at a weight ratio of 30/70 (the gray marks). The bistable range extends further to 2–38 °C, when the mixing ratio was 95/5 (the black marks). Another mixture of PEG 600 (99.5 wt%) and PEG 2000 (0.5 wt%)

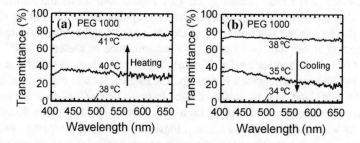

Fig. 5.2 Transmission spectra of PEG 1000 during the **a** heating or **b** cooling process. The sample thickness was 10 mm

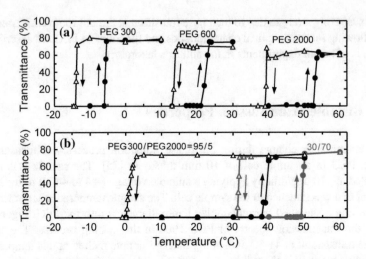

Fig. 5.3 Transmittance change of PEG during the phase transition process (500 nm wavelength). The circles and triangles correspond to the heating and cooling processes, respectively. The samples were **a** PEGs with a different molecular weight and **b** mixtures of PEG 300 and PEG 2000 (mixing ratio: 95/5 or 30/70)

exhibits the bistability in the 25–50 °C range. As these results indicate, the bistable range extends notably when a small amount of heavy PEG (large molecular weight) is added to light PEG (small molecular weight).

The bistable characteristics of PEG were examined carefully by using a small sample cell in order to improve the thermal response. As Fig. 5.4a shows, two acrylate fibers (1 mm diameter) were fixed in a groove (1 mm width) of an aluminum plate so that a gap of ~1 mm is created between them. This gap was filled with a sample that was prepared by mixing PEG 300 and PEG 2000 at a ratio of 30/70. A Peltier element (30 mm square) was attached to the aluminum plate for heating or cooling the sample. A thermocouple was inserted into a deep hole in the aluminum plate to measure temperature in the close vicinity of the sample. A HeNe laser beam (633 nm wavelength) was transmitted through these fibers to measure the optical transmittance of the PEG sample in the gap. Figure 5.4b shows the transmittance change that was triggered by a positive voltage application to the Peltier element (heating). The sample was originally in the solid phase, and hence, no output light was detected at the start (−10 min). The transmittance increased to ~60% at the moment of voltage application (at −5 min). The coupling loss of ~40% was caused mainly by the beam divergence in the gap. The electric voltage was removed 5 min later (at 0 min), and then the sample cooled to room temperature in a few minutes. The transmittance, however, kept the high level (~60%) for longer than 30 min. The bistability was confirmed in this manner. Figure 5.4c shows the repeated switching processes, in which a positive or negative voltage was applied to the Peltier element to induce the transmittance change. The probe light beam was controlled reproducibly by application of these trigger signals.

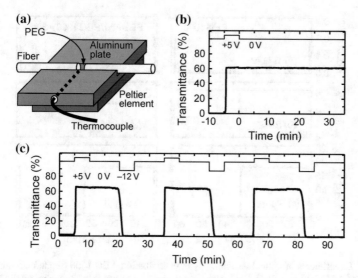

Fig. 5.4 **a** Structure of the sample cell. PEG was put into the gap (~1 mm) between the two optical fibers (1 mm diameter). **b** Transmittance increase of the PEG (the lower line) that was triggered by voltage application to the Peltier element (+5 V, 5 min). Probe light was a HeNe laser beam of 633 nm. **c** Switching of the probe light by alternate application of +5 (heating) and −12 V (cooling) to the Peltier element

As the gray marks in Fig. 5.3b shows, the 30/70 mixture exhibited the bistability in the 32–49 °C range when it was put in a large cell. Interestingly, the same mixture in the narrow gap (Fig. 5.4) stayed in the liquid phase for longer than 30 min after the voltage removal, although the sample cooled to below 30 °C in a few minutes according to the thermocouple measurement. In the switch-off process of Fig. 5.4c as well, the sample had to be cooled to below 10 °C for solidification by application of −12 V. Recent experiments in our laboratory have indicated that the phase transition temperature of PEG is affected heavily by the surface condition or the surrounding material. Similar phenomenon has been reported for a liquid metal; i.e., although the melting point of gallium (Ga) is said to be ~30 °C in literature, a thin Ga film (or particle) stays in the liquid phase at ~0 °C [26, 27]. Further experiments and theoretical studies are needed to clarify the mechanism of this super-cooling phenomenon.

The phase transition was controlled by the Peltier element in the above experiments. Although the sample (PEG) volume was small (~1 mm³), the entire sample cell consisting of the aluminum plate and the fibers had to be heated. To realize a local heating of the sample cell, a laser irradiation method was examined. Since PEG had no strong absorption band in the visible spectral range (Fig. 5.2), PEG was colored with a cyanine dye (trimethyl-indolium iodide) that induced a strong absorption at around 450 nm. PEG was molten at 70 °C to dissolve this dye (10^{-3} mol/l), and then poured into the sample cell of Fig. 5.4a to attain a solid sample at room temperature. A blue laser diode of 450 nm wavelength was used to heat the sample. The laser beam was directed downward and focused on the top of the PEG

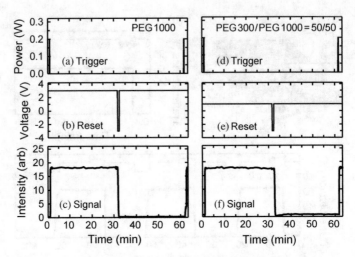

Fig. 5.5 Experiments of optical switching by laser irradiation. PEG 1000 (the left column) or the mixture of PEG 300 and PEG 1000 (the right column) was put into the sample cell shown in Fig. 5.4a. **a, d** A blue laser beam (450 nm) of 0.2 W was irradiated for 10 s as a trigger signal. **b, e** A negative voltage (−3 V) was applied to the Peltier element for 30 s as a reset signal. The voltage was kept at +3 or +1 V in the ordinary state to keep the sample temperature at ~36 (the left column) or ~29 °C (the right column). **c, f** Temporal change of the transmitted light intensity (550 nm wavelength)

sample. A lamp light beam was transmitted through the fibers as a probe signal, and the signal intensity was monitored by a spectrometer. Figure 5.5a–c show the experimental results for PEG 1000. Since this sample exhibited the bistability at 36–38 °C (Fig. 5.2), +3 V was continuously applied to the Peltier element, as shown in Fig. 5.5b, to keep the sample temperature in this range. As Fig. 5.5a shows, the laser beam of 0.2 W was irradiated for 10 s as a trigger. This laser irradiation caused the phase transition (melting), and accordingly, the signal intensity (550 nm wavelength) increased immediately, as shown in Fig. 5.5c. This transmission state continued for 30 min. It changed to the opaque state, when a reset signal of −5 V was applied to the Peltier element for 30 s, as shown in Fig. 5.5b. This opaque state continued after the voltage returned to the original level (+3 V). The signal emerged again, when the trigger laser beam was irradiated 30 min later. Similar results were attained with a 50/50 mixture of PEG 300 and PEG 1000, as shown in Fig. 5.5d–f. Since the bistable temperature became lower by addition of PEG 300, the application voltage to the Peltier element was reduced to +1 V to keep the sample temperature at ~29 °C. The reset signal was the same as that in the former experiment (−5 V, 30 s).

The experiments of Fig. 5.5 required a continuous voltage application (+3 or +1 V) for keeping the sample temperature in the bistable range. Additional experiments are in progress by the use of other PEG mixtures to realize an optical control that needs no electric signal. As Figs. 5.4 and 5.5 show, the transmission state was preserved for longer than 30 min in the current experiments. One might think that this memory period is too short in comparison with those of other memory devices, e.g., semiconductor chips or optical disks. Note that the dynamic random access memories (DRAMs) in computers have to be refreshed several times in 1 s lest they should lose their memory. The transmission state of PEG can be preserved for a long time if a refresh signal (voltage application or laser irradiation) is provided every 30 min. In addition to applications to the fluorescent devices, which are described in the following sections, this function of memory switching will be useful for creating rewritable signboards or reconfigurable 3D ornaments.

5.3 Absorption Enhancement

Photoluminescence of lanthanide ions, e.g., europium (Eu^{3+}) and erbium (Er^{3+}), are used extensively in illuminators, optical amplifiers, and fiber lasers. In comparison with organic dyes, e.g., rhodamine and coumarin, the optical absorbance of the lanthanide ions are smaller by several orders of magnitude. Therefore the excitation light for inducing fluorescence is absorbed only weakly by the lanthanide-based phosphors. As Fig. 5.6 shows, for example, the absorbance at the pump laser wavelength is 10^4-fold smaller in Eu^{3+} than rhodamine 6G. This property is unfavorable for inducing strong fluorescence or generating a population inversion that creates a laser beam. Lanthanide-doped phosphors generally require a long optical path for efficient absorption of pump light energy.

PEG is expected to be a suitable matrix for those fluorescent materials with a small absorption coefficient, since the optical path of pump light extends by scattering [18]. This prediction was examined by using the optical setup shown in Fig. 5.7a. The pump light source was a violet laser diode, whose emission peak is shown in Fig. 5.7b. PEG that contained $EuCl_3$ of 10^{-2} mol/l was put into a glass cell of 50 mm length. The solution was readily attainable with PEG 300 that took the liquid phase at room temperature. Since PEG 1000 took the solid phase at room temperature, it was molten at 70 °C to dissolve $EuCl_3$, and then cooled to room temperature for

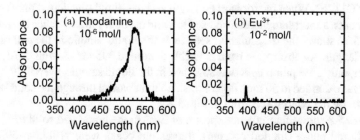

Fig. 5.6 Absorption spectra of the aqueous solutions of **a** rhodamine 6G (10^{-6} mol/l) and **b** $EuCl_3$ (10^{-2} mol/l). The vertical axis shows the optical density, $\log T$, where T is the sample transmittance for 10 mm thickness

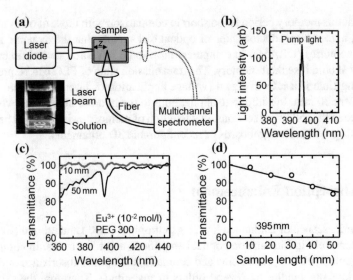

Fig. 5.7 a Optical system for the fluorescence measurements [18]. A pump laser beam (100 mW), whose emission spectrum is shown in (**b**), was focused into the sample cell of 50 mm length. Fluorescence spectra were measured from the cell side by using a multichannel spectrometer. The pickup position (distance from the input end, z) was varied by moving the probe fiber (400 μm diameter). As the dotted lines illustrate, fluorescence was measured at the exit end in the later experiment (Fig. 5.9). **c** Transmission spectra of Eu^{3+}-doped PEG 300 (liquid) in the sample cells of 10 or 50 mm length [18]. **d** Transmittance of the liquid sample at the pump laser wavelength (395 nm). The horizontal axis shows the sample cell length

attaining the solid sample. The photograph in Fig. 5.7a shows the optical path of the pump laser beam or the fluorescence of Eu^{3+} along the path. As this photograph shows, fluorescence of the liquid sample was detectable in all directions. In the solid sample, however, the strong scattering attenuated the emitted light heavily before it emerged from the output end of the long sample cell. Fluorescence was therefore collected from the cell side. The pickup position, z (the distance from the entrance), was changed by moving the probe fiber of the multichannel spectrometer (B & W Tek, BTC112E). When the forward emission was measured (the dotted lines) in later experiments, a short sample cell was used to avoid the heavy fluorescence attenuation. Figure 5.7c shows the transmission spectra of the liquid sample in the cell of 10 or 50 mm length. An absorption band is visible at around 395 nm. As Fig. 5.7d shows, however, only 15% pump light was absorbed in the liquid sample, even if the sample length was extended to 50 mm. By contrast, no light passed through the solid sample, as shown by the bottom spectrum in Fig. 5.2.

Figure 5.8 shows fluorescence spectra of Eu^{3+} in the liquid and solid PEGs [18]. In the liquid phase (the left column), fluorescent peaks were visible at 592, 613, and 698 nm. These peaks correspond to the electronic transitions of $^5D_0 \rightarrow {}^7F_1$, $^5D_0 \rightarrow {}^7F_2$, and $^5D_0 \rightarrow {}^7F_4$ in the Eu^{3+} ion [28]. A peak at 395 nm was caused by scattered pump light. Although the measurement position changed from $z = 0$ to 10 mm, no

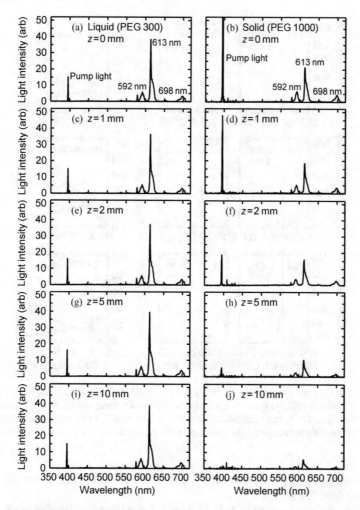

Fig. 5.8 Fluorescence spectra that were measured from the sample side. The samples were **a** PEG 300 in the liquid phase (the left column) or **b** PEG 1000 in the solid phase (the right column), which contained EuCl$_3$ at a concentration of 10^{-2} mol/l. The measurements were conducted at different positions ($z = 0$–10 mm) [18]

notable change was visible in the spectra. This result indicated that Eu^{3+} ions were excited equally at all positions along the optical axis. In other words, attenuation of the pump light is negligible due to the small absorbance of Eu^{3+}. By contrast, the sample in the solid phase (the right column) exhibited a notable attenuation of the pump light. Accordingly, the fluorescent peaks became smaller as the measurement position became distant from the entrance. This fact indicated that the pump light was confined in the vicinity of the entrance. The fluorescence was weaker in the solid

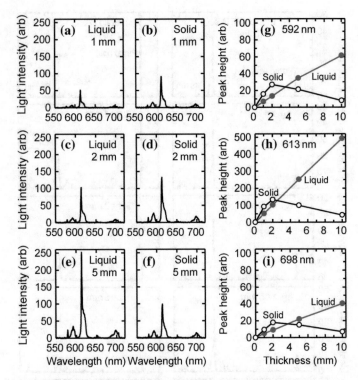

Fig. 5.9 a–f Forward fluorescence of the PEG 300 (liquid) or PEG 1000 (solid) that contained EuCl₃ (10^{-2} mol/l) [18]. As the dotted lines in Fig. 5.7 shows, these spectra were measured in the forward direction. The sample cell thickness was 1, 2, or 5 mm. **g–i** Fluorescent peak height as a function of the sample thickness. The numerals in the graphs denote the peak wavelengths

phase than the liquid phase, since fluorescent rays were also scattered heavily before emerging from the sample cell.

The experimental results in Fig. 5.8 revealed that the pump light penetrated only within 1–2 mm depth in the solid PEG. Fluorescence measurement in the forward direction was therefore conducted by using a thin sample cell, i.e., 1, 2, 5, or 10 mm. As Fig. 5.9a–f show, the liquid (the left column) and solid samples (the central column) exhibited a different dependence on the sample thickness [18]. When the thickness was 1 mm, the fluorescent peak of the solid sample was twice higher than that of the liquid sample. The samples of 2 mm thickness also exhibited a stronger fluorescence in the solid phase. Figure 5.9g–i show the thickness dependence of the fluorescent peak heights. In the liquid phase, the fluorescence became stronger in proportion to the sample thickness. By contrast, the solid sample exhibited the strongest fluorescence when the sample thickness was 2 mm. As these results indicate, the scattering in the solid PEG is effective to enhance the pump light absorption and accordingly the fluorescence intensity, if the sample is smaller than 2 mm. In the current samples, the scattered pump light diverged to a large volume around the optical axis. If a miniature cell is used, scattered pump light will be confined in a small volume to enhance the excitation efficiency, which will lead to a strong fluorescence emission.

5.4 Random Lasing

Laser oscillation is induced by the stimulated emission. To promote the stimulated emission, fluorescence emitters (dyes or ions) have to be excited strongly to induce a population inversion. At the same time emitted rays have to be confined in a cavity to stimulate other emitters. Ordinary laser oscillators use a pair of mirrors to create a resonant cavity. Droplets act as a mirror-less cavity, since fluorescent rays circulate along the periphery (whispering gallery mode) inducing a stimulated emission [29]. In comparison with solid spheres or disks, droplets are advantageous in tunability; i.e., the droplet deformation changes the wavelength, direction, polarization state of the laser emission. The deformability or fluidity of liquid, however, becomes a disadvantage when considering stability and handling capability. These problems are solved by enclosing a droplet in rubber. Silicone (polydimethyl siloxane, PDMS) rubber is a useful matrix for optical liquids, since it is transparent, flexible, nonreactive, and nontoxic. The rubber is creatable from silicone oil in a simple curing process at room temperature. A droplet can be embedded in silicone rubber easily by using an inkjet-method [30]. A droplet in this flexible matrix can be handled like a solid with its deformability being preserved.

In conventional experiments, an ethanol solution was usually used for creating droplet dye lasers that were dispersed in air [29]. In silicone rubber, however, an ethanol droplet emits no strong fluorescence, as shown by the lower spectrum in Fig. 5.10a. The spectral shape is similar to that of a spontaneous emission from an

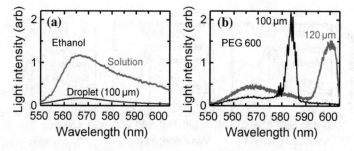

Fig. 5.10 a Fluorescence spectrum of an ethanol droplet (100 μm diameter) containing rhodamine 6G (10^{-3} mol/l). The droplet was enclosed in silicone rubber. A frequency-doubled Nd:YAG laser pulse (532 nm, 5 ns, 1 mJ/mm^2) was used for the dye excitation. The upper gray line shows a spectrum of the solution in an ordinary sample cell. **b** Fluorescence spectra of PEG droplets (100 or 120 μm diameter) containing rhodamine 6G (10^{-3} mol/l). The droplets were enclosed in silicone rubber. The excitation laser was the same as the above

ordinary dye solution (the upper line). This weak emission of the ethanol droplet is attributed to the fact that the refractive index of ethanol ($n = 1.36$) is too low to induce the total internal reflection at the boundary of the droplet and the silicone rubber ($n = 1.40$). PEG is a suitable solvent for creating a droplet laser, since its index of refraction ($n = 1.46$) is higher than that of the silicon rubber as well as it dissolves various dyes and ions [31]. As Fig. 5.10b shows, the stimulated emission takes place in dye-doped PEG droplets (~100 μm diameter) that are enclosed in silicone rubber [30]. Comb-like resonant peaks are visible on the fluorescent peak.

In a small droplet, however, pump light is not absorbed sufficiently, and hence, the excitation efficiency becomes low. The scattering in the PEG matrix is effective to enhance the pump light absorption. In addition, emitted rays (fluorescence) are also scattered strongly in the PEG matrix, which possibly promotes the stimulated emission. Laser emission in a scattering matrix is attracting interests, being called "random laser" [20, 21, 23, 24]. On the basis of these preceding researches on random lasers as well as the above experimental results on PEG, the author is thinking of creating a droplet random laser. As mentioned earlier, deformability is both an advantage (tunability) and disadvantage (handling) of droplets. This fact is also applied to the solid PEG, since it is a soft matter that is difficult to handle. Silicone rubber that contains a PEG particle seems useful for creating a droplet random laser.

As a first step toward the creation of a droplet random laser, scattering and fluorescence characteristics of dye-dispersed PEG were studied by using the small sample cell shown in Fig. 5.4a [32]. PEGs with a molecular weight of 1000, 2000, or 6000 were examined to compare the characteristics of the liquid and solid phases. Samples were prepared by dissolving rhodamine 6G at 10^{-3} mol/l in molten PEGs (70 °C). This sample was put into the gap (1 mm) between the fibers of the sample cell. The dye molecules dispersed uniformly in PEG through the solidification (cooling) process. Figure 5.11 shows the transmission loss spectra of the samples. The vertical axis shows the optical density, $-\log T$, which was calculated from the measured transmittance T. The optical density of the solid sample is related to absorption by the

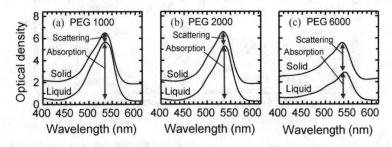

Fig. 5.11 Optical loss spectra of PEGs containing rhodamine 6G (10^{-3} mol/l). The PEG samples were contained in the small sample cell (1 mm thickness) shown in Fig. 5.4a. Measurements were conducted in the solid (30 °C) or liquid (70 °C) phase. The vertical axis shows the optical density, which is the sum of the absorption loss (rhodamine), the scattering loss (PEG), and the coupling loss (fibers)

dye molecules and scattering by the PEG matrix. The absorption peak of rhodamine 6G is located at around 530 nm. Green laser is therefore suitable for excitation. The absorbance decreases as the molecular weight increases, although the dye concentration is the same for all samples. The scattering by the PEG matrix, i.e., the difference of the upper and lower spectra, becomes larger as the molecular weight increases. It is assumed therefore that the effect of the scattering will be most notable in the sample of PEG 6000.

In the fluorescence measurements, the sample temperature was controlled by the Peltier element and the thermocouple, as shown in Fig. 5.4a. A green pulse (527 nm, 10 ns) of a frequency-doubled Nd:YLF laser was irradiated on the sample top. The laser beam diameter was 1 mm, and the pulse energy was adjusted between 0 and 190 μJ by using an attenuator. Fluorescence was collected by the optical fibers that sandwiched the sample, and measured by a multichannel spectrometer. Figure 5.12a, b show the fluorescence spectra of PEG 6000 containing rhodamine 6G (10^{-3} mol/l). The liquid sample exhibited weak, broad spectra of the spontaneous emission. The fluorescence intensity increased in proportion to the pump energy. In comparison with the liquid sample, the solid sample exhibited a strong fluorescent peak when pumped by a high-energy pulse. This strong fluorescence verifies the occurrence of the stimulated emission or the amplified spontaneous emission (ASE). Figure 5.12c shows the pump energy dependence of the peak height. In the liquid phase, the peak height is proportional to the pump energy. In the solid phase, however, the peak height increases nonlinearly with the pump energy. The threshold energy is about 50 μJ. Figure 5.12d shows the change in the peak width (full width at half maximum, FWHM). Rhodamine 6G usually exhibits a peak width of ~40 nm in the spontaneous emission process. As the pump energy increases, the peak width decreases slightly in the liquid phase. The stimulated emission seems to be at the initial stage in the dye solution. In the solid phase, the peak width decreases notably with the increase in the pump energy, which is the typical tendency in the stimulated emission process.

The reproducibility of the fluorescence emission was examined by repeating the heating and cooling processes. Figure 5.13a shows the fluorescence spectra of the PEG 6000 that were measured at 30 °C after the cooling process of ~3 deg/s, i.e., the same conditions as those of Fig. 5.12b. Similar spectra were attained in the measurements that were repeated three times. That is, no notable change was visible in both wavelength and height of the fluorescent peak, which verified a good reproducibility of the fluorescence emission. It was found, however, that the cooling rate notably affected the fluorescence. When the cooling rate was decreased to 0.5 deg/s, for example, the fluorescent peak shifted to a shorter wavelength, as shown in Fig. 5.13b. This fact indicates that the emission wavelength is tunable by the cooling rate. The reason for the peak shift has not been clarified yet, but the cooling rate seems to affect the microstructures that are shown in Fig. 5.1c–e.

Figure 5.14a shows the spectral change during the heating process of the PEG 6000. A strong fluorescence was visible until temperature reached 52 °C, since PEG 6000 stayed in the solid phase below that temperature. When the temperature exceeded 52 °C, the PEG began to melt, and accordingly, the fluorescent peak

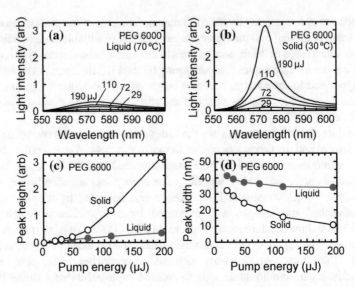

Fig. 5.12 **a, b** Fluorescence spectra of rhodamine 6G in PEG 6000. The PEG was in the liquid (70 °C) or solid (30 °C) phase. The numerals beside the spectra denote the pulse energies of the pump laser beam (527 nm wavelength). **c** Peak height and **d** width of the fluorescence as a function of the pump energy

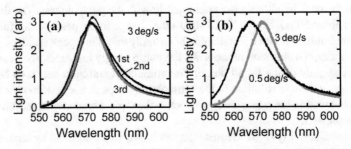

Fig. 5.13 **a** Reproducibility of the fluorescence emission. The sample (PEG 6000 containing rhodamine 6G of 10^{-3} mol/l) was heated (70 °C) and cooled (30 °C) repeatedly. The cooling rate was ~3 deg/s. The pump energy was 190 μJ. **b** Blue-shift of the fluorescent peak by the reduction of the cooling rate

became lower and disappeared at 65 °C. The peak shift to a long wavelength, i.e., so-called "red shift", was also visible in the melting process. In the cooling process that is shown in Fig. 5.14b, no strong fluorescence was visible until the temperature became 52 °C. Then the fluorescent peak extended as the temperature was lowered. At 40 °C, the fluorescent peak recovered to the original shape, i.e., the gray spectrum in Fig. 5.14a. As these data show, PEG 6000 exhibits different spectra at the same temperature (52 °C). That is, the fluorescence emission also exhibits a bistability during the phase transition process.

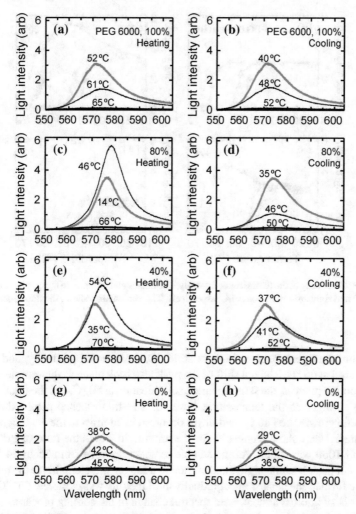

Fig. 5.14 Fluorescence spectra of rhodamine 6G (10^{-3} mol/l) in PEGs during the heating (the left column) and cooling (the right column) processes. The matrix was **a, b** PEG 6000, **c–f** mixtures of PEG 6000 and PEG 1000 (the ratio of PEG 6000: 80 or 40%), or **g, h** PEG 1000. The numerals beside the spectra denote temperatures at which measurement was conducted. The pump energy was 190 µJ

As mentioned in Sect. 5.2, the bistable function is facilitated by mixing PEGs with different molecular weights. A fluorescent sample was therefore prepared by mixing PEG 6000 and PEG 1000 at a ratio of 80/20. Figure 5.14c, d show the fluorescence spectra of this mixture. In the heating process, the sample stayed in the solid phase until 14 °C, and no spectral change was visible in the fluorescence. When the temperature exceeded 14 °C, the sample began to melt. Interestingly, the fluorescence became stronger as the temperature rose or the melting progressed.

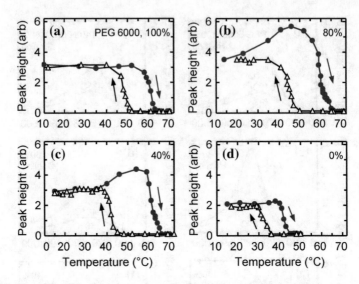

Fig. 5.15 Bistability of the fluorescence intensity. The circles and triangles in the graphs correspond to the left and right spectra in Fig. 5.14, respectively. The vertical axis shows the fluorescence peak height

The fluorescent peak became highest at 46 °C, then shrank gradually, and finally disappeared at 66 °C. The red shift of the peak also took place as the sample melted. In the cooling process, the sample started solidification at 50 °C, and the fluorescence became stronger as the temperature lowered. The fluorescence peak exhibited a monotonic growth, i.e., no up-and-down occurred in contrast to the heating process.

Figure 5.14e, f show the results for the mixture in which the ratio of PEG 6000 and PEG 1000 was 40/60. As the sample temperature rose from 35 to 54 °C, the fluorescent peak became higher and shifted to a longer wavelength. As the temperature rose further, the peak shrank gradually until melting was complete (70 °C). This sample also exhibited a monotonic spectral change in the cooling process.

Figure 5.14g, h show the spectral change of PEG 1000 (no PEG 6000 was contained). The fluorescence decreased monotonically in the heating process (from 38 to 45 °C), and increased monotonically in the cooling process (from 36 to 29 °C). This monotonic change in the melting process was similar to that of the PEG 6000 and contrasted with the peculiar changes of the mixtures. In addition, the peak shift in the phase transition process was negligible in PEG 1000.

Figure 5.15 show the change in the peak height during the heating (circles) and cooling (triangles) processes. As Fig. 5.15a shows, the fluorescence in the PEG 6000 decreases rapidly at around 60 °C in the heating process, and increases at around 50 °C in the cooling process. Consequently, a bistable fluorescence emission takes place in the 50–60 °C range. As Fig. 5.15b shows, the mixture containing 80% PEG 6000 exhibits a peculiar fluorescence change in the heating process. The fluorescence becomes strongest at 46 °C. The bistable fluorescence takes place in

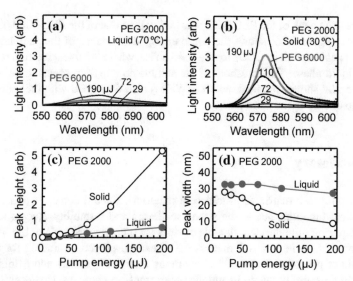

Fig. 5.16 a, b Fluorescence spectra of rhodamine 6G in PEG 2000. The PEG was in the liquid (70 °C) or solid (30 °C) phase. The numerals beside the spectra denote the pulse energies of the pump laser beam (527 nm wavelength). **c** Peak height and **d** width of the fluorescence as a function of the pump energy

the 20−60 °C range. As Fig. 5.15c shows, the mixture containing 40% PEG 6000 exhibits a bistability in the 40–60 °C range, and emits the strongest fluorescence at 54 °C. Finally, as Fig. 5.15d shows, PEG 1000 exhibits a monotonic change in both heating and cooling processes. The fluorescence is weaker and the bistable range is narrower in PEG 1000 than the other samples.

As Fig. 5.11 shows, the absorbance of rhodamine 6G is twice higher in PEG 1000 than PEG 6000. As Figs. 5.14 and 5.15 show, however, the fluorescent peak is higher in PEG 6000 than PEG 1000. The stimulated emission seems to take place more efficiently in PEG 6000, since, as Fig. 5.11 shows, the scattering is stronger in PEG 6000 than PEG 1000. The random lasing requires not only the strong excitation (absorption of pump light) but also the confinement (scattering) of the emitted rays. From this viewpoint, PEG 2000 seems to be a desirable matrix, since, as Fig. 5.11b shows, it provides a stronger absorbance than PEG 6000 and a larger scattering cross-section than PEG 1000. To examine this assumption, the fluorescence spectra of the dye-dispersed PEG 2000 were measured by using the same sample cell and the same pump laser. Figure 5.16a shows the fluorescence spectra in the liquid phase. A broad peak of the spontaneous emission emerged and grew in proportion to the pump energy. The gray line shows the fluorescence spectrum of PEG 6000 for a pump energy of 190 μJ. In comparison with this gray spectrum, the fluorescence intensity of the PEG 2000 was stronger, since the absorbance was twice stronger. Figure 5.16b shows the fluorescence spectra of the solid phase. A narrow peak grew nonlinearly with the increase of the pump energy. When the pump energy was 190 μJ, the fluorescent peak of the PEG 2000 was higher than that of the PEG 6000 (the gray spectrum). Figure 5.16c shows the pump energy dependence of the peak height. A nonlinear peak growth is visible for the solid phase, whereas a linear relation is visible for the liquid phase. The threshold of the stimulated emission seems to be ~50 μJ. As Fig. 5.16d shows, the peak width (FWHM) also decreased with the increase of the pump energy in the solid phase.

5.5 Summary

PEG is useful as a matrix for fluorescent materials, since it enhances and controls their optical functions due to the strong scattering and bistability. Recent technology is requiring miniaturized, flexible devices for creating wearable or bio-adaptive instruments. A variety of soft matters have been developed or studied for uses in electronic or photonic devices. PEG is among those soft matters and exhibits tunabililty and bistability due to its unique phase transition process. However, nothing has been clarified about the microstructure in the solid phase and the mechanism of scattering. Fundamental studies on the basis of physics and chemistry have to be achieved in addition to the application studies.

Acknowledgements The experiments that were described in this article were conducted by students in the author's laboratory at Ryukoku University. The author appreciates their contribution in the advancement of this research. A part of this research was supported by a grant from Japan Society for the Promotion of Science.

References

1. G.H. Brown (ed.), *Photochromism* (Wiley, New York, 1971)
2. C.J.G. Kirkby, R. Cush, I. Bennion, Optical nonlinearity and bistability in organic photochromic thin films. Opt. Commun. **56**, 288–292 (1985)
3. G.K. Ahluwalia, Applications of chalcogenides: S, Se, and Te, in *Data Storage Devices*, ed. by G.K. Ahluwalia (Springer, New York, 2017)
4. N.A. Clark, S.T. Lagerwall, Submicrosecond bistable electro-optic switching in liquid crystals. Appl. Phys. Lett. **36**, 899–901 (1980)
5. D.W. Berreman, W.R. Heffner, New bistable cholesteric liquid-crystal display. Appl. Phys. Lett. **37**, 109–111 (1980)
6. I.C. Khoo, R. Normandin, V.C.Y. So, Optical bistability using a nematic liquid crystal film in a Fabry-Perot cavity. J. Appl. Phys. **53**, 7599–7601 (1982)
7. J.D. Valera, B. Svensson, C.T. Seaton, G.I. Stegeman, Bistability and switching in thin-film waveguides with liquid-crystal cladding. Appl. Phys. Lett. **48**, 573–574 (1986)
8. A.D. Lloyd, B.S. Wherrett, All-optical bistability in nematic liquid crystals at 20 μW power levels. Appl. Phys. Lett. **53**, 460–461 (1988)
9. C.-Y. Huang, K.-Y. Fu, K.-Y. Lo, M.-S. Tsai, Bistable transflective cholesteric light shutters. Opt. Express **11**, 560–565 (2003)
10. J. Ma, L. Shi, D.-K. Yang, Bistable polymer stabilized cholesteric texture light shutter. Appl. Phys. Express **3**, 021702-1–3 (2010)

11. C.-Y. Wu, Y.-H. Zou, I. Timofeev, Y.-T. Lin, V.Y. Zyryanov, J.-S. Hsu, W. Lee, Tunable bi-functional photonic device based on one-dimensional photonic crystal infiltrated with a bistable liquid-crystal layer. Opt. Express **19**, 7349–7355 (2011)
12. C.-T. Wang, T.-H. Lin, Bistable reflective polarizer-free optical switch based on dye-doped cholesteric liquid crystal. Opt. Mater. Express **1**, 1457–1462 (2011)
13. J. Bravo-Abad, A. Rodriguez, P. Bermel, S.G. Johnson, J.D. Joannopoulos, M. Soljacic, Enhanced nonlinear optics in photonic-crystal microcavities. Opt. Express **15**, 16161–16176 (2007)
14. L.-D. Haret, T. Tanabe, E. Kuramochi, M. Notomi, Extremely low power optical bistability in silicon demonstrated using 1D photonic crystal nanocavity. Opt. Express **17**, 21108–21117 (2009)
15. G.R. Olbright, T.E. Zipperian, J. Klem, G.R. Hadley, Optical switching in $N \times N$ arrays of individually addressable electroabsorption modulators based on Wannier-Stark carrier localization in GaAs/GaAlAs superlattices. J. Opt. Soc. Am. B **8**, 346–354 (1991)
16. I.A. Temnykh, N.F. Baril, Z. Liu, J.V. Badding, V. Gopalan, Optical multistability in a silicon-core silica-cladding fiber. Opt. Express **18**, 5305–5313 (2010)
17. S. Ramchandran, M.R. Chatterjee, Nonlinear dynamics of a Bragg cell under intensity feedback in the near-Bragg, four-order regime. Appl. Opt. **41**, 6154–6167 (2002)
18. M. Saito, T. Koketsu, Fluorescence enhancement of europium ions in a scattering matrix, in *5th International Conference on Photonics, Optics and Laser Technology*, Porto, Feb 2017, Proceedings, pp. 15–21
19. Y. Kuga, A. Ishimaru, Retroreflectance from a dense distribution of spherical particles. J. Am. Opt. Soc. A **1**, 831–835 (1984)
20. N.M. Lawandy, R.M. Balachandran, A.S.L. Gomes, E. Sauvain, Laser action in strongly scattering media. Nature **368**, 436–438 (1994)
21. D.S. Wiersma, M.P. Albada, A. Lagendijk, N.M. Lawandy, R.M. Balachandran, Random laser? Nature **373**, 203–204 (1995)
22. A. Kurita, Y. Kanematsu, M. Watanabe, K. Hirata, T. Kushida, Wavelength- and angle-selective optical memory effect by interference of multiple-scattered light. Phys. Rev. Lett. **83**, 1582–1585 (1999)
23. R.C. Polson, A. Chipouline, Z.V. Vardeny, Random lasing in π-conjugated films and infiltrated opals. Adv. Mater. **13**, 760–764 (2001)
24. S. Wiersma, M. Colocci, R. Righini, F. Aliev, Temperature-controlled light diffusion in random media. Phys. Rev. B **64**, 144208-1–6 (2001)
25. M. Saito, Y. Nishimura, Bistable optical transmission properties of polyethylene-glycol. Proc. SPIE **8474**, 11-1–12 (2012)
26. M. Hida, A. Sakakibara, H. Kamiyabu, Surface tension and supercooling phenomenon of liquid Ga. J. Japan Inst. Metals **53**, 1263–1267 (1989)
27. T. Hashimoto, M. Saito, I. Yamada, J. Nishii, Flexible reflection grating that is made of liquid metal, in *62th Spring Meeting of Japan Society of Applied Physics*, Hiratsuka, Japan, Mar 2015, Extended Abstracts, pp. 04–430
28. S. Shionoya, W.M. Yen (eds.), *Phosphor Handbook* (CRC Press, Boca Raton, Florida, 1999)
29. H.M. Tzeng, K.F. Wall, M.B. Long, R.K. Chang, Laser emission from individual droplets at wavelengths corresponding to morphology-dependent resonances. Opt. Lett. **9**, 499–501 (1984)
30. M. Saito, K. Koyama, Spectral and polarization characteristics of a deformed droplet laser. J. Opt. **14**, 065002-1–6 (2012)
31. M. Saito, T. Hashimoto, Whispering gallery mode emission of a cylindrical droplet laser, in *5th International Conference on Photonics, Optics and Laser Technology*, Porto, Feb 2017, Proceedings, pp. 32–38
32. M. Saito, Y. Nishimura, Bistable random laser that uses a phase transition of polyethylene glycol. Appl. Phys. Lett. **108**, 131107-1–4 (2016)

Chapter 6
Advances in Fs-Laser Micromachining Towards the Development of Optofluidic Devices

João M. Maia, Vítor A. Amorim, D. Alexandre and P. V. S. Marques

Abstract In this chapter the developments made in femtosecond laser micromachining for applications in the fields of optofluidics and lab-on-a-chip devices are reviewed. This technology can be applied to a wide range of materials (glasses, crystals, polymers) and relies on a non-linear absorption process that leads to a permanent alteration of the material structure. This modification can induce, for instance, a smooth variation of the refractive index or generate etching selectivity, which can be used to form integrated optical circuits and microfluidic systems, respectively. Unlike conventional techniques, fs-laser micromachining offers a way to produce high-resolution three-dimensional components and integrate them in a monolithic approach. Recent advances made in two-photon polymerization have also enabled combination of polymeric structures with microfluidic channels, which can provide additional functionalities, such as fluid transport control. In particular, here it is emphasised the integration of microfluidic systems with optical layers and polymeric structures for the fabrication of miniaturized hybrid devices for chemical synthesis and biosensing.

J. M. Maia (✉) · V. A. Amorim · P. V. S. Marques
Department of Physics and Astronomy, Faculty of Sciences of University of Porto, Rua do Campo Alegre 687, Porto, Portugal
e-mail: joao.m.maia@inesctec.pt

V. A. Amorim
e-mail: vitor.a.amorim@inesctec.pt

P. V. S. Marques
e-mail: psmarque@fc.up.pt

D. Alexandre
Department of Physics, University of Trás-os-Montes e Alto Douro, Quinta de Prados, 5001-801 Vila Real, Portugal
e-mail: daniel@utad.pt

J. M. Maia · V. A. Amorim · D. Alexandre · P. V. S. Marques
CAP - Centre for Applied Photonics, INESC TEC, Rua Dr. Roberto Frias, Porto, Portugal

© Springer Nature Switzerland AG 2019   119
P. Ribeiro et al. (eds.), *Optics, Photonics and Laser Technology 2017*,
Springer Series in Optical Sciences 222,
https://doi.org/10.1007/978-3-030-12692-6_6

6.1 Introduction

Lab-on-a-chip systems correspond to miniaturized devices capable of performing several functions, which are typically realized in a full scale laboratory [1, 2]. These devices consist of a network of microfluidic channels (with a usual cross-section size of 10–100 μm) that are ended by reservoirs or inlet/outlet ports, which allow connection with tubes for fluid pumping. The small dimension of these microfluidic channels introduces unique properties in the kinematics of the fluid flow, which has strong implications in basic processes. In particular, these channels can be used for transport of fluids, for mixing of different reagents or for control of the concentration of particles present in the stream [3]. Microfluidic systems can also be combined with other components (electrical or optical) for active control of the flow or for sensing applications in medical and biochemical environments [4, 5]. Regarding the latter application, these systems can alleviate some issues related to long analysis time and high reagent consumption, while at the same time maintaining high accuracy and performance.

Photolithography and etching were the first technique used for the development of microfluidic devices on silicon and glass and are still one of the main technologies used in planar lightwave circuit technology [6, 7]. Despite its maturity, this technique presents some limitations: (a) it is a planar technique, meaning that the devices are made at the substrate's surface which limits the maximum density of components that can be integrated, and (b) it is a multistep technique that requires access to a specialized cleanroom facility. In particular, the high cost of this method has promoted the development of new microfabrication techniques (soft lithography, injection molding and hot embossing) which use polymers (PDMS—polydimethyl-siloxane, PMMA—polymethylmethacrylate) as a target material and rely on replication of a mold [8, 9]. Despite allowing rapid prototyping of microfluidic structures with submicrometer resolution, these methods possess some issues: (a) like all methods based on photolithography they are planar, and (b) integration of microfluidics with optical layers is pretty challenging given that polymers typically exhibit higher propagation losses and poorer light confinement than glasses, not to speak on the difficulties of the fabrication methods.

In order to overcome these issues, fs-laser micromachining has been adopted for the production of lab-on-a-chip devices. In this technique, a sample (glass, crystal or polymer) is scanned by a focused laser beam which gives rise to a non-linear interaction between the laser beam and the irradiated material. This interaction leads to a permanent modification of the material confined within the focal volume, thus submicrometer resolutions can be obtained [10]. Depending on the pulse characteristics (wavelength, repetition rate, laser fluence) and on the material used different modifications can occur, namely variation of the refractive index (either positive or negative), generation of etching selectivity, formation of voids, among other effects. These modifications have led to the formation of different systems: optical circuits [11], electrical layers [12], microfluidic systems [13], polymeric structures [14] that can be fabricated either separately or integrated in a single chip. Additionally, the inherent three-dimensional aspect of this technology accounts for a flexible system design and for high component integration.

The possibility of combining microfluidic systems with optical layers and polymeric structures in a monolithic glass block provides a unique opportunity for the design of optofluidic devices with far-reaching applications in basic science, chemical synthesis and drug development, and biochemical analysis [15, 16]. Thus, the goal of this chapter is to discuss recent advances made in fs-laser micromachining to fulfil the production of such devices. After this introduction, the fundamentals of fs-laser micromachining are discussed. The types of structural modifications that can be obtained are reviewed, alongside with the materials in which these modifications can be observed. Some examples of existent devices for applications in optical communications and in microfluidics are presented. Then, novel optofluidic devices for on-chip optical sensing and particle manipulation are reviewed. A summary of future developments expected for this field is also presented.

6.2   Femtosecond Laser Micromachining

In a typical laser direct writing setup, the fs-laser beam is focused inside a material transparent to the beam wavelength through a focusing lens, as shown in Fig. 6.1. The most common fs-laser systems used for micromachining are regeneratively amplified Ti:sapphire lasers (fundamental wavelength of 800 nm) and Ytterbium-based lasers (fundamental wavelength of 1030 nm); wavelengths other than the fundamental can be used by generation of second or third harmonic. Several free-space optics components can be used to, for example, control the beam polarisation (through a half-wave plate), or to adjust the beam spot size to the lens entrance diameter (through a beam expander). External components can also be used to control the pulse energy. The sample is positioned by three motorized stages and is scanned relative to the beam propagation direction across any spatial direction. In certain systems, beam scanning techniques that use galvanometric mirrors are combined with sample scanning for an improved exposure.

6.2.1   Fundamentals of Interaction Laser-Matter

The interaction between the laser beam and the material can generate many modifications, which are at the basis of the formation of several types of devices. Although the fundamentals of this interaction is still a topic under discussion and depends on the material used, a simple and common qualitative model is generally accepted. This model involves generation of a free electron plasma followed by energy relaxation and alteration of the material structure. Here, the basic principles of this interaction are presented; a more detailed analysis can be found elsewhere [10].

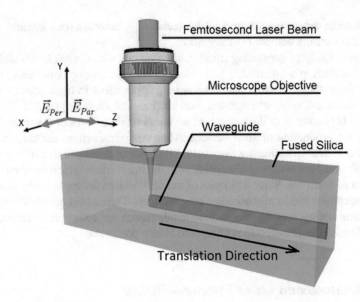

Fig. 6.1 Overview of the laser writing process: optical components are fabricated by writing a single track, while microfluidic channels are usually fabricated by stacking multiple tracks, as described in [17]

A fs-laser pulse, being focused inside a sample, does not have sufficient energy to be linearly absorbed due to the large band gap of the material. However, the high peak intensities of fs-laser pulses (around 10 TW/cm^2) can promote electron transition from the valence band to the conduction band via non-linear absorption. This process is started by multiphoton ionization and/or tunnelling ionization that seed avalanche ionization. Multiphoton ionization corresponds to absorption of multiple photons by an electron at the valence band. The net energy absorbed by the electron is bigger than the band gap energy, thus the electron can transit to the conduction band. Tunneling ionization corresponds to the deformation of the band gap structure due to the strong electric field, which lowers the band gap energy and enables direct transition from the valence band to the conduction band through quantum tunneling. These two processes can occur either simultaneously or separately, depending on the laser intensity and frequency, and their influence can be described by the Keldysh parameter. Both of these processes increase the density of electrons present in the conduction band and act as seed for avalanche ionization: electrons present in the conduction band can excite an electron in the valence band due to collisional ionization. These two electrons, at the conduction band, while exposed to the laser field can absorb laser energy and repeat this process, giving rise to an exponential growth in the number of electrons present in the conduction band. The free electrons form a highly absorbing plasma that oscillates at a frequency that increases with the density of free electrons. In a couple of picoseconds, the plasma frequency matches the laser frequency and the plasma transfers its energy to the lattice. Within the microsecond timescale, heat diffuses away from the irradiated region which melts and resolidifies. If the energy deposited in the lattice is above a threshold level, the material structure is permanently modified. It is possible, however to distinguish between two cases, depending on the pulse repetition rate. At high repetition rates the time between consecutive pulses is shorter than the time for thermal diffusion and heat builds up around the focal volume, while at low repetition rates thermal equilibrium is obtained long before new irradiation. In either case, the modification is confined to the focal volume, and therefore submicrometer resolutions can be obtained.

Different modifications can be produced depending on the pulse characteristics (duration, repetition rate, energy, wavelength) and on the setup used (objective lens and numerical aperture, stages scanning speed, polarisation of the laser beam relative to the scanning direction). This text only focuses on the interactions that enable fabrication of optical circuits, microfluidic systems and polymeric structures. Other induced phenomena can be found elsewhere [10, 18].

6.2.2   Fabrication of Optical Elements

One of the possible modifications resulting from the laser-matter interaction is related to the variation of the refractive index around the focal volume, which can enable fabrication of either passive or active optical components. This modification can be observed in a wide range of materials, but here we focus our attention to fused silica, due to its importance in the optoelectronic industry and its properties, namely chemical and thermal stability and biocompatibility.

Considering the interaction between a fs-laser pulse and fused silica, at low pulse energies slightly above the modification threshold it is observed an increase of the refractive index. The index variation is generally attributed to two mechanisms, densification and formation of colour centres, whose contribution depends on the processing conditions. Densification results from the rapid melting and solidification of the material in the focal volume [19], and has been shown to be linearly related with the refractive index in fused silica [20]. Formation of colour centres, which are defects in the lattice periodicity lead to a change in the refractive index through the Kramers–Kronig mechanism [21]. Analysing the cross-section of the modification pattern, it corresponds to an elongated inverted droplet shape, where the top region is associated to a decrease of the refractive index, while the bottom region is associated with an increase of the refractive index, Fig. 6.2a. By translating the sample in relation to the beam focus, as shown in Fig. 6.1, it is then possible to produce type I optical waveguides embedded in the silica slab, where light is confined to the bottom region. The quality of the waveguides, namely insertion losses and mode field diameter, strongly depends on the irradiation conditions (orientation of the beam polarisation relative to the scanning direction, writing depth, scanning speed, pulse energy) and on the laser pulse characteristics, thus each laboratory may obtain low-loss waveguides at different processing parameters. Of significant interest is the production of single-mode low-loss waveguides at the telecommunication

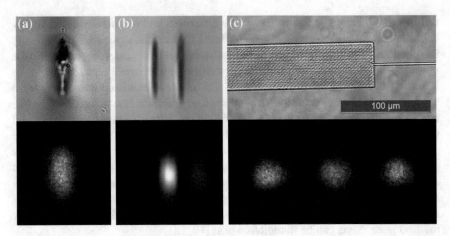

Fig. 6.2 Optical microscope image of the cross-section and mode profile **a** at 1550 nm of type I waveguides made in fused silica and **b** at 633 nm and TM polarization for type II waveguides made as a double line structure with 10 μm separation in *x*-cut LiNbO$_3$. **c** 1 × 3 MMI splitter made in fused silica, with a top view image of the device shown on top (light is incident from the right) and the intensity distribution at 1550 nm shown at the bottom. Image (**b**) was reproduced from [42] (Applied Physics A: Materials Science and Processing, Origins of waveguiding in femtosecond-laser structured LiNbO$_3$, volume 89, 2017, page 132, J. Burghoff, S. Nolte, and A. Tünnermann), © Springer-Verlag 2007, with permission of Springer

wavelength for optical communication devices. By optimizing the irradiation conditions, Fukuda et al. fabricated single-mode waveguides with minimum propagation losses of 0.05 dB/cm in fused silica [22]. The index variation is around 10^{-3} and the mode field diameter of the waveguide matches the mode field diameter of a standard single-mode fiber (~10 μm). Nevertheless, the waveguide mode is asymmetric, and in order to correct this issue several beam shaping techniques have been proposed which consist of slit focusing, astigmatic focusing or multiscan writing [23, 24].

Besides fused silica, other glasses have also shown an increase of the refractive index when exposed to a fs-laser beam. Eaton et al. [25] have fabricated single-mode waveguides in borosilicate glass with propagation losses of 0.2 dB/cm at 1550 nm. They verified that by writing in the high repetition rate regime, the accumulated heat would favour a uniform index distribution and enable production of waveguides with circular cross-section. Ams et al. [26] also wrote waveguides in phosphate glass, having obtained a minimum loss of 0.39 dB/cm at 1550 nm. In this case, a slit was placed in front of the focusing lens in order to shape the beam profile, and obtain symmetric waveguides.

Due to the index modification, several passive elements have already been demonstrated. By making use of the three-dimensional motorized stages, several groups have reported S-bended waveguides with minimum bend losses [22, 27]. This concept was further developed by Thomson et al. [24] where a fan-out device was fabricated in a silica glass chip to couple light from a fiber V-groove array to a multicore optical fiber. Modulation of the fs-laser beam pulse spectrum has also been applied to the production of grating waveguides. By adjusting the stages scanning speed and modulation frequency, first-order Bragg gratings can be produced with tunable Bragg wavelength. Ams et al. [28], for instance, have fabricated Bragg gratings in a silicate glass with a high coupling coefficient of $1457~m^{-1}$, enabling much shorter device lengths. Grenier et al. [29] and Zeil et al. [30] also developed new writing techniques to tailor the grating's coupling strength, which enabled the production of phase-shift and apodized gratings, respectively. Control of the waveguide birefringence for the design of beam splitters and wave retarders has also been demonstrated, by writing stress-tracks next to the optical waveguide [31].

Several optical devices for power splitting have also been reported in a wide range of configurations. Sohn et al. [32] and Liu et al. [33] have extensively studied fabrication of Y-junctions for 1-to-N power division with balanced splitting at 1550 nm. Recently, Amorim et al. [34] proposed a novel writing technique at the junction divergence point, in order to obtain uniform splitting in a broad spectral range (1300 nm to 1600 nm). Fabrication of directional couplers with several coupling ratios has also been reported by a correct adjustment of the interaction length and arm separation [35, 36]. Although, the work developed so far only served as proof-of-concept, multimode interference (MMI) devices for 1-to-N splitting have also been reported in two different configurations, where the multimode region is fabricated by stacking multiple scans close to each other, as shown in Fig. 6.2c [37, 38]. More complex designs, such as channel add-drop filters and arrayed-wave gratings, for channel multiplexing applications have also been reported [39, 40].

Plenty of attention has also been devoted to other materials (rare earth-doped glasses, chalcogenide glasses) and to the production of active elements, namely on-chip laser sources. The induced index variation provides a way to guide light through an optical waveguide, while the medium properties contribute to light amplification. Besides the type I waveguides previously described, two other configurations can be used. In type II waveguides (Fig. 6.2b), an unexposed region surrounded by two damaged regions acts as core: the two damaged regions alter the stress field around the irradiated volume, leading to an index variation through the strain (or stress)-optic effect [41, 42]. The laser-matter interaction can also produce a negative index region, from which depressed cladding waveguides can be obtained where the core is surrounded by irradiated tracks of lower index [43, 44]. These geometries are generally preferred in active waveguides, because the core is not irradiated and, therefore, the medium properties are left unchanged.

Many near infrared continuous-wave laser sources have already been reported in Yb, Nd, Er, Pr doped crystals and glasses [45, 46]. Zhang et al. [47] demonstrated laser oscillation at 1061 nm with a slope efficiency of 25% and maximum output power of 11 mW by fabricating type II waveguides in a Nd:GGG crystal. Palmer et al. [48] reported a depressed cladding waveguide with high index contrast (2.3×10^{-2}) made in Yb-doped glass and demonstrated laser oscillation at 1030 nm in Yb:ZBLAN glass with a slope efficiency of 84% and maximum output power of 170 mW. Della Valle et al. [49] also developed a single longitudinal mode laser emitting at 1550 nm with a slope efficiency of 21% and maximum output power of 50 mW, consisting of a type I waveguide made in Er–Yb-doped phosphate glass. Operation of these devices is based on direct light pumping from a standard optical fiber to a laser written optical waveguide. The cavity feedback elements can be produced by deposition of a reflective coating layer or by integration with external fiber Bragg gratings. The development of a robust monolithic cavity can be achieved by fs-laser writing of Bragg grating waveguides in the sample.

Production of waveguides in chalcogenide glasses has also been explored, with McMillen et al. [50] having reported minimum propagation losses of 0.65 dB/cm in gallium–lanthanum–sulfide chalcogenide glass. The properties of this medium make it highly attractive for the production of nonlinear components, besides laser sources. Hughes et al. [51] have demonstrated spectral broadening of a pulse with duration of 200 fs and wavelength of 1540 nm from an initial width of 50 nm to a final width of 200 nm, in gallium–lanthanum–sulphide glass. Psaila et al. [52] also demonstrated supercontinuum generation in a waveguide written in chalcogenide glass, where the pulse bandwidth increased from 100 to 600 nm.

Type I and type II waveguides can also be fabricated in lithium niobate, where the high electro-optic coefficient makes this glass suitable for the production of on-chip electric modulators [42, 53, 54].

6.2.3 Fabrication of Microfluidic Systems

The interaction between the fs-laser beam and matter can also enable fabrication of microfluidic devices. Production of these structures can be obtained in two different ways depending on the type of interaction (generation of etching selectivity or laser ablation).

Relatively to the former approach, glasses, when immersed in a hydrofluoric acid (HF) solution, are etched isotropically, meaning that all regions are etched at a similar rate. However, if the sample is first subjected to laser writing, then the irradiated regions will be more prone to the etching reaction than the remaining material. In fused silica, when the sample is exposed to a fs-laser beam, besides the refractive index variation, sub-wavelength birefringent structures, called nanogratings, can also be formed. These structures are responsible for the etching selectivity and consist of periodic layers of alternating refractive index, with a period around 100 nm. Formation of the nanogratings is based on the nanoplasmonic model proposed by Taylor et al. [55]. Thus, fabrication of microfluidic channels in fused silica is as follows: (i) the channel pattern is designed following the multiscan writing method (writing of multiple tracks with a separation of a couple of micrometers), (ii) the sample is then, usually, immersed in an HF solution [56]. The presence of the nanogratings leads to an anisotropic etching reaction, where hollow channels are formed inside the silica slab. This fabrication technique is commonly called fs-laser irradiation followed by chemical etching (FLICE).

Microfluidic channels can also be produced in Foturan glass. However, while in fused silica the generation of etching selectivity is based in a photophysical reaction, in Foturan glass it is due to a photochemical one [57, 58]. This photosensitive and amorphous glass is composed by a lithium aluminosilicate matrix doped with trace amounts of a silver and cerium ions. When the laser beam is focused inside the glass, free electrons are generated by inter-band excitation, which reduces the silver ions to silver atoms. By performing a subsequent annealing process, at 500–600 °C, the silver atoms diffuse and agglomerate forming silver nanoclusters. The nanoclusters act as a seed for the formation of a crystalline lithium metasilicate phase that is etched in HF faster than the remaining amorphous material.

Although with limited success, generation of etching selectivity has also been reported in other materials. Matsuo et al. [59] have fabricated a square channel inside a quartz substrate, where selective etching was due to lattice amorphization. The issue was that the etch rate was of 50 μm/h, which resulted in a very long fabrication time. Formation of nanogratings was also verified in sapphire [60], which revealed an enhanced selectivity relatively to fused silica (100:1). The main limitations were the prolonged etching reaction which took several days and the accumulation of unetched material that recrystallized inside the channel. Choudhury et al. [61] also demonstrated that Nd:YAG crystals were selectively etched in an aqueous solution of H_3PO_4, due to the creation of lattice defects and a certain degree of disorder. Though the etching reaction develops slowly, these results can pave the way for production of novel optofluidic devices with integrated laser sources.

The processing conditions affect the properties of the microfluidic channels (shape, surface roughness, aspect ratio) and the fabrication time (etch rate) [62–65]. Considering the formation of channels in fused silica, the orientation of the laser beam polarisation relative to the scanning direction influences the nanograting orientation and, consequently, the etch rate. If the polarisation is set orthogonal to the scanning direction, the etch rate is maximum because the nanogratings align with the channel axis (Fig. 6.3b). Meanwhile, for parallel alignment the HF finds alternate hollow and pristine layers that reduce the cumulative etch rate (Fig. 6.3a). The pulse energy can also affect the channel quality, given that higher pulse energies result in an increase of the stress field around the laser affected zones which can lead to channel deformation and appearance of microcracks [66]. The separation between modification tracks also influences the surface roughness, with lower separations giving rise to stress accumulation and consequent surface deformation, while higher separations result in accumulation of debris inside the channel. Production of channels with very low surface roughness is also possible by performing annealing after the etching reaction [67].

One of the issues inherent to the etching reaction is that the etch rate tends to saturate which leads to the formation of tapered channels with poor aspect ratio. This problem can be corrected by (i) suitably designing the channel so that it can compensate the tapering effect [68]; (ii) fabrication of openings connecting the channel to the glass surface which results in more access points for the acid to diffuse [17] (Fig. 6.3c); (iii) using potassium hydroxide instead of HF which eliminates the saturation effect at the expense of a lower etch rate [69].

In the FLICE method, channels with circular or rectangular cross-section (Fig. 6.3d) can be obtained by carefully patterning the modification tracks geometry [17]. Rectangular cross-sections are preferred for integration with optical waveg-

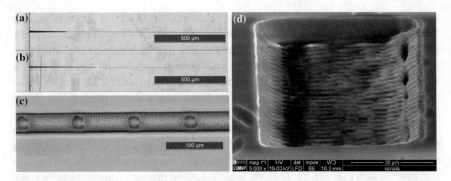

Fig. 6.3 Top view microscope images of microfluidic channels fabricated in fused silica with **a** parallel polarisation and **b** perpendicular polarisation; the remaining processing conditions and the etching time was the same in both cases. The darker regions correspond to hollow channels and, as can be seen, the etched length is bigger in (**b**) than in (**a**), due to the beam polarisation orientation. **c** Top view image of a microfluidic channel made inside fused silica, which shows the access holes, connecting the channel to the silica surface, for acid diffusion. **d** SEM (Scanning Electron Microscope) cross-section image of a rectangular channel made in fused silica by FLICE

uides for cell trapping and stretching applications, while circular channels are better suited for developing on-chip biomimetic applications [46].

Another advantage of FLICE is that the microfluidic channels can be fabricated inside the glass, with the HF acid flowing through reservoirs that are made at the surface. This way it is not necessary to seal the final device and the reservoirs offer a simple way to connect tubes for fluid pumping [70]. Using this approach, Masuda et al. [71] have fabricated a Y-branched microfluidic channel for mixing of two fluids. One of the problems of this configuration is that the fluids only interact at the interface, which leads to an often slow and inefficient mixing process. To solve this issue, Liu et al. [72] developed a helical microchannel in fused silica that promoted a faster mixing reaction, by turning the usual laminar flow into a turbulent one. Due to the selective etching reaction, Sugioka et al. [52] also produced a pressure-actuated microplate inside the microfluidic channel that was used to switch the flow direction of reagents.

Laser ablation can also be used for production of microfluidic devices. At sufficiently high pulse energies, the laser-matter interaction can result in micro-explosions and formation of voids around the irradiated volume, which can be used as microfluidic channels [73]. Although this technique does not involve generation of etching selectivity and can be applied to a wider range of materials, some problems still stand. First, compared to FLICE the channel surface is typically rougher. Second, the channel length is limited to around 100 μm due to formation of debris that deposits along the channel. These issues can be minimized by performing the laser irradiation with the sample immersed in water, which flows through the channel removing the debris and smoothes the surface [74].

6.2.4   Fabrication of Polymeric Structures

Fs-laser micromachining can also be used in liquid resins or solid photoresists (that contain photoinitiators) to assist in two-photon polymerization. After fabricating the microfluidic channel, the channel is filled with the polymer and the fs-laser beam is scanned across it. The irradiation process leads to a two-photon absorption process, where two photons induce an electron transition from the ground state to an excited state. The excited electron then decays to the fundamental state emitting light in the ultraviolet to visible range. The photoinitiators present in the polymer absorb the fluorescent light and generate free radicals. The free radicals react with monomers, producing monomer radicals which react again with other radicals, thereby forming a polymeric chain. A detailed description of this process can be found in [75].

The polymerization reaction only occurs around the irradiated volume, while the unexposed zones can be rinsed in an appropriate solvent. This way, by scanning the laser beam across a determined pattern, 3D polymeric structures can be fabricated inside the microfluidic channel. The spatial resolution is around 100–200 nm, well beneath the diffraction limit and high aspect ratio structures can be fabricated. Additionally, the structures can be fabricated on top of any material, as long as the polymer can adhere strongly to it.

This technique has already been successfully applied to the fabrication of hybrid devices, which combine microfluidic channels and polymeric structures, for fluid mixing or particle filtering. Wang et al. [76] developed several filters consisting of a porous sheet, with variable shape and size (Fig. 6.4a). Therefore, only particles with a size smaller than the opening dimension could pass through the channel. Amato et al. [77] expanded on this concept and fabricated a filter to separate 3 μm polystyrene spheres in a Rhodamine 6G solution. Additionally, in a separate experiment they demonstrated filtering of red blood cells. However, after some time the particles would start to clog the channel.

Regarding mixers, Lim et al. [78], fabricated a polymeric structure after a Y-branched channel for mixing of two separate fluids (Fig. 6.4b). The structure con-

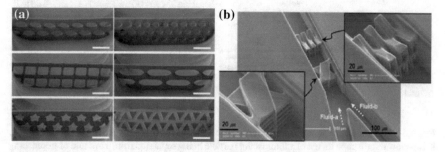

Fig. 6.4   SEM images of **a** microsieves with different openings sizes and shapes for particle filtering and **b** a crossing manifold micromixer embedded in a Y-branched channel. Images (**a**) and (**b**) were reproduced from [76, 78], respectively, with permission of ©The Royal Society of Chemistry

sisted of multiple layers, stacked vertically, which crossed each other in order to turn the flow chaotic. With this geometry, they predicted a mixing efficiency of 90%. Wu et al. [79] developed a similar device that could simultaneously filter and enhance mixing of different solutions, and thus significantly reduce the channel length. The structure consists of a mixer, with an equal configuration as in [78], which is placed in the middle of two sheets with periodic openings of constant dimension, which serves as filter. A mixing efficiency of 87% was obtained.

6.3   Integrated Optofluidic Devices

Fs-laser micromachining offers a unique way to integrate optical layers with microfluidic channels for cell manipulation and flow control applications or for on-chip optical sensing. Applegate et al. [80] were one of the first groups to show this integration for cell trapping and sorting. They fabricated a hybrid device where a microfluidic channel made by soft lithography in a PDMS slab was sealed against a fused silica substrate where optical waveguides were inscribed by fs-laser writing. The channel corresponds to a 1×4 splitter, and the system was assembled in such a way that the waveguides were set orthogonally to the channel and crossed it at the junction. Device operation relied on two laser beams: one beam was focused in the channel to excite the cell's fluorescence, and the other beam ($\lambda = 980$ nm) was coupled to the optical waveguide and was used to trap cells flowing through the stream (Fig. 6.5a). The latter beam was also used to manipulate the cell's position and direct it to a specific output branch, depending on the measured fluorescence. The fluorescence radiation was collected by a focusing lens and was detected by a CCD (charge coupled device) camera. This device was employed to successfully separate dyed-polystyrene particles with a diameter of 10 μm from non-fluorescent particles (Fig. 6.5b).

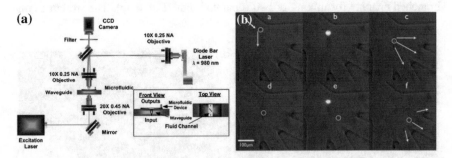

Fig. 6.5   a Schematic of an integrated microfluidic sorting system based on fluorescence detection. **b** Top view microscope image of a fluorescing particle being separated from a mixture of fluorescing and non-fluorescing particles. Reproduced from [80] with permission of ©The Royal Society of Chemistry

Using a similar principle, Hwang et al. [81, 82] developed an optofluidic device for cell detection and counting. The device was made in a fused silica chip entirely by fs-laser micromachining. The circular cross-section channel, made following the FLICE technique, exhibited a variable diameter that decreased from 100 μm at the entrance to 1–2 μm at the channel middle, which was smaller than the diameter of the particles used. This way, the particles would self-align and single cell detection was possible. An optical waveguide was inscribed perpendicular to the channel and crossing it at the smaller-sized segment. To detect particles, two distinct configurations were used as demonstrated in Fig. 6.6a, b. The first relied on measurement of the waveguide transmitted power, where the output power would vary whenever a particle crossed the waveguide. The second approach relied on detection of the cell fluorescence, which was induced by light propagating through the waveguide. In both cases, the number of peaks present in the fluorescence/intensity spectrum corresponds to the number of cells that crossed the waveguide (Fig. 6.6c). In particular, it was possible to detect red blood cells up to a flow rate of 0.5 μL/min.

Optical tweezers were also used to study the deformability properties of cells, given that the exerted optical force on the particle can cause it to stretch. In [15, 83] an optofluidic device was fabricated in fused silica, where optical waveguides

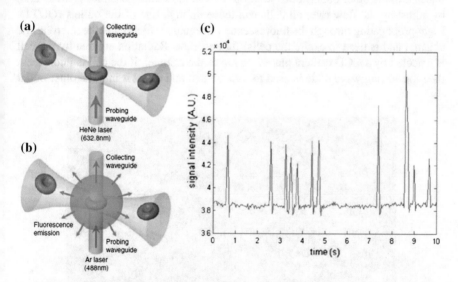

Fig. 6.6 Schematic of the integrated cell detection configuration: **a** measurement of transmitted power and **b** measurement of fluorescence radiation. **c** Output intensity spectrum as a function of time in a red blood cell counting experiment. The peaks present in the spectrum correspond to the passage of a cell through the waveguide. Images (**a**) and (**b**) were reproduced from [82] with permission of ©The Royal Society of Chemistry. Image (**c**) was reproduced from [81] (Applied Physics A: Materials Science and Processing, Three-dimensional opto-fluidic devices fabricated by ultrashort laser pulses for high throughput single cell detection and processing, volume 96, 2009, page 389, D.J. Hwang, M. Kim, K. Hiromatsu, H. Jeon, and C.P. Grigoropoulos), © The Author(s) 2009, with permission of Springer

were inscribed perpendicular to a microfluidic channel as described previously. A schematic of the proposed device is shown in Fig. 6.7. Light coupled to both waveguides was used to trap red blood cells that were flowing in a diluted blood solution. By adjusting the optical power in both waveguides, the cell position in the stream could be manipulated. In addition, by increasing the optical power it was observed that the cell would start to stretch along the optical trap axis. Due to the fused silica transparency, cell deformation could be observed in real time and its size could be easily measured. Study of these properties can be important in biomedical applications, given that the cell's deformability is related to its health. However, this approach is still limited given that the channel curvature and surface roughness cause deformation of the cell contour, which limits measurement resolution.

Using laser-induced fluorescence and optical stretching, other cell sorting configurations were also reported. Bragheri et al. [84] developed an optofluidic device formed by an X-shaped channel, where a single straight channel bifurcates at both ends into two 45° angled branches. The straight waveguide is interrogated by two optical waveguides (fluorescence waveguide and sorting waveguide), as shown in Fig. 6.8. The device was made in fused silica by fs-laser micromachining. In this system, a liquid containing the cells is injected in one inlet and a buffer solution is injected in the other outlet. Both solutions mix in the straight channel segment and, by adjusting the flow rate, all cells can leave through one of the outlets (OUT1). Light propagating through the fluorescence waveguide is focused in the microfluidic channel and is used to excite the cells' fluorescence. Radiation emitted by the cell is detected by a CCD camera placed on top of the channel; if the cell is fluorescent then the sorting waveguide is used to trap the cell and direct it into the other outlet

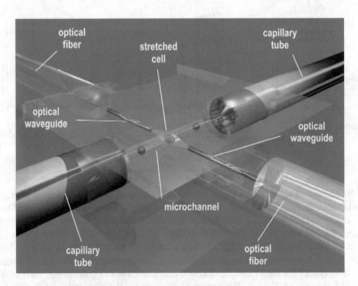

Fig. 6.7 Schematic of the monolithic optical stretcher for study of cell's deformability properties. Reproduced from [15] with permission of The Optical Society (© 2010 OSA)

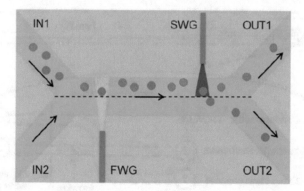

Fig. 6.8 Schematic of the device for cell sorting based on particle fluorescence (green and red particles correspond to fluorescent and non-fluorescent particles, respectively; FWG and SWG correspond to the fluorescence and sorting waveguide, respectively). Reproduced from [84] with permission of ©The Royal Society of Chemistry

branch (OUT2). This device was successfully used to sort, in real time, fluorescent polystyrene beads with a diameter of 7 μm. Yang et al. [85] used a similar device to sort melanoma cells based on their deformability. In this case, light propagating through the waveguide was used to first trap the cell, then to induce cell stretching by increasing the optical power, and finally to direct the cell to a specific branch. Cell deformation was evaluated in real time by imaging the cell and analyzing its shape as a function of the optical force exerted on it.

In microfluidic channels, electrokinetics is one of the main driving mechanisms of fluid flow due to the low surface-to-volume ratio of these systems. Therefore, capillary electrophoresis (separation of particles based on their electrical mobility) is highly attractive in microfluidic devices for the analysis of biomolecules in clinical applications. Chips for capillary electrophoresis are already commercially available and typically consist of two channels (one for injection and one for separation) crossing at 90° (Fig. 6.9). Each channel is terminated by two reservoirs with deposited electrodes, for application of a voltage difference across the channel. Operation of these chips is as follows: (i) a fluid is drawn into the injection channel and fills it due to electrokinetic flow, by applying a voltage difference across the channel; (ii) the injection channel is then kept at the same potential, while an electric field is applied across the separation channel; (iii) the fluid starts to flow along the separation channel and different species present in the stream flow at different speeds, due to the different electrophoretic mobility. By placing a detector at the separation channel end, the different species can be identified based on their arrival times. Vazquez et al. [86] used laser-induced fluorescence for particle detection. They used a commercial capillary electrophoresis chip (model D8-LIF from LioniX BV) and wrote an optical waveguide perpendicular to the separation channel, as shown in Fig. 6.9. Light at 532 nm was coupled to the waveguide via an optical fiber to excite fluorescence and an optical fiber, placed on top of the channel, was used to

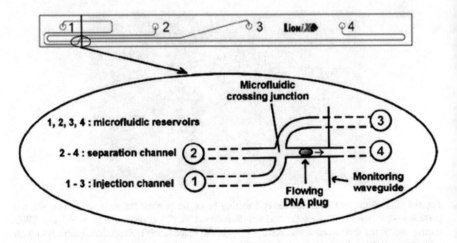

Fig. 6.9 Schematic of a capillary electrophoresis chip with an optical waveguide at the end of the separation channel for particle detection. Reproduced from [87] with permission of The Optical Society (© 2008 Optical Society of America)

collect the emitted radiation. The device was tested by using a solution of Rhodamine 6G (R6G) dissolved in a phosphate buffer. By increasing the R6G concentration, the intensity of the fluorescent signal detected increased linearly. In particular, low analyte concentrations of 40 pM were detected. Dongre et al. [87] used a similar configuration, but injected different rhodamine solutions (R6G and rhodamine-B) simultaneously in the channel. In the end, they were able to distinguish the different species by observing two different peaks in the fluorescent intensity spectrum, which corresponds to two different arrival times. In a separate experiment, Dongre et al. [88] fabricated a capillary electrophoresis chip in a fused silica slab by photolithography and etching and, after obtaining the chip, inscribed an optical waveguide by fs-laser writing as described previously. In order to maximize the detection resolution: (i) the channel walls were coated with a polymer to suppress electro-osmotic flow and minimize particle adsorption to the channel wall, and (ii) the channels were filled with a sieving gel matrix to induce a size-dependent variation in particles that exhibit a very similar mobility. In this case, they were able to detect double-stranded DNA molecules (labelled with a fluorescent dye) with different sizes in the channel, by observing that the molecules arrived at the detection point at different times.

Fluorescence detection is a very sensitive detection technique that allows measurement of very low analyte concentrations. However, the species must be labelled with a fluorescent molecule which may affect the particle properties and interfere with some chemical reactions. Therefore, label-free detection configurations may be more desirable in certain applications. Maselli et al. [89] fabricated an optofluidic device in a fused silica chip for refractive index sensing, where a Bragg grating waveguide (BGW) was written in between two microfluidic channels. Both structures were written in the same laser writing process and, when immersing the chip

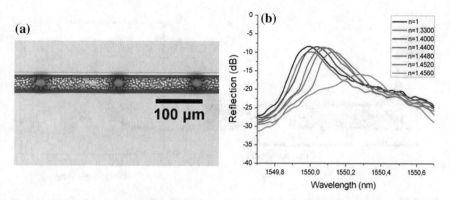

Fig. 6.10 **a** Top view image of a Bragg grating waveguide fabricated 3.8 μm from the microfluidic channel wall. The grating reflection spectrum obtained for this device is displayed in (**b**) for different fluid refractive indexes inside the microfluidic channel

in an HF bath, only the channel was etched while the waveguide remained unaffected. The separation between the BGW and each channel wall was 2 μm. The low separation between both structures enables evanescent interaction; thus light from a broadband source coupled to the waveguide is reflected at the Bragg wavelength. The reflected wavelength depends on the index of the fluid circulating in the channel (Fig. 6.10b). By filling the channel with solutions with different refractive index, a maximum sensitivity of 81 nm/RIU (refractive index unit) was obtained at the index 1.458, which represents a limit of detection of 1.2×10^{-4}. Additionally, compensation for temperature variation was achieved by fabricating a grating, with a different Bragg wavelength, outside the channel range. To obtain higher sensitivities, several solutions have been proposed: fabrication of phase-shifted gratings to engineer spectral defects or fabrication of a channel surrounding the entire channel. Recently, Maia et al. [90] monitored the etching reaction in situ, in order to precisely control the separation between the grating and the channel, which can also increase device sensitivity.

Higher sensitivities can be obtained in sensing schemes where there is a direct interaction between light and analyte. Hanada et al. [91] developed a sensor for detection of low chemical concentration based on absorbance measurements. The device was made in Foturan glass and the fabrication procedure encompasses the following steps: (i) the channel was inscribed by fs-laser writing followed by annealing and chemical etching in HF; (ii) a post-annealing process was realized to smooth the channel surface; (iii) the channel walls were coated with a polymer that had a lower index than water, in order to have light being guided along the channel; (iv) an optical waveguide was fs-laser written between the channel edge and the edge of the chip. The polymer coating process was essential to obtain a longer interaction length. The transmission spectrum was measured in a spectrometer: white light from a halogen lamp was coupled to the waveguide and transmitted through the microfluidic channel; the channel end was slightly tilted to direct light to the spectrometer,

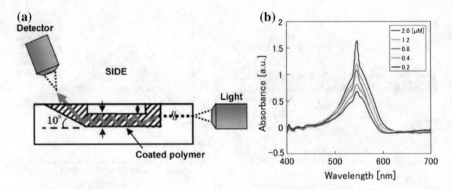

Fig. 6.11 **a** Schematic of the optofluidic chip for liquid concentration detection based on absorbance measurements. Figure **b** displays the corresponding absorbance spectra of glucose-D solutions with different concentrations. Reproduced from [91] with permission of ©The Royal Society of Chemistry

as depicted in Fig. 6.11a. The absorbance spectrum was obtained by subtracting the transmission spectrum of the liquid sample from the spectrum obtained for a hollow channel. The sensor was used, in two separate experiments, to detect trace amounts of protein (bovine serum albumin) and glucose-D. The measurements revealed that an increase of the species concentration resulted in a stronger absorbance (Fig. 6.11b); in particular a limit of detection of 7.5 mM and 200 nM was obtained for the protein and glucose-D, respectively.

Crespi et al. [16] also reported on a Mach–Zehnder interferometer integrated with a microfluidic system for refractive index detection. The entire device was made in a fused silica chip by fs-laser micromachining and a schematic is shown in Fig. 6.12a. In this device, light, from a broadband source, is coupled from an optical fiber to the optical waveguide and is splitted when it reaches the first Y-junction (because the interferometer is unbalanced, power splitting is not uniform). Then, one of the sensing arms of the interferometer crosses the channel (orthogonally to it), while the reference arm passes over the channel. The two arms then join in a second Y-junction, which leads to light interference. The output spectrum corresponds to periodically-spaced fringes. By changing the index of the fluid inside the channel, it is observed a shift in the fringe position which allows refractive index detection (Fig. 6.12b). The index detection can be related to other properties of the fluid. In this case, the channel was filled with different concentrations of aqueous glucose-D solutions, and a linear sensitivity of 1500 nm/RIU was obtained along the entire concentration range, which corresponds to a limit of detection of 1.5×10^{-4} RIU. In a separate experiment, peptide molecules, used in the pharmaceutical industry, were also detected with a limit of detection of 9 mM.

Microresonators can also be fabricated for generation of whispering gallery modes, in which specific resonances are confined to the cavity. These structures have unique properties, suitable for chemical and biosensing applications, such as high cavity quality factor and small resonator size. Lin et al. [92] used the selective

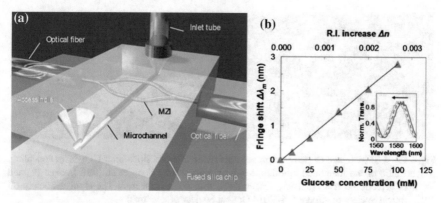

Fig. 6.12 **a** Schematic of the microfluidic channel integrated with a Mach-Zehnder interferometer for refractive index measurements. Figure **b** displays the fringe shift as a function of glucose-D concentration for the device shown in (**a**). Reproduced from [16] with permission of ©The Royal Society of Chemistry

etching effect to produce microresonators in fused silica. These were fabricated by laser-writing the negative image of the final structure and by etching it, leaving a microdisk supported by thin pillars on top of the silica surface. The structures were then annealed with the aid of a CO_2 laser to improve the cavities surface smoothness. However, during this process, surface tension induced a collapse of the disk leading to a toroid-shaped boundary (Fig. 6.13a). Using the optical fiber taper coupling method, they generated whispering gallery modes in the 1530–1565 nm range with a quality factor of 1.07×10^6. One of the limitations is the long time necessary to fabricate the microresonators (6 h in this case). Therefore, two-photon polymerization, assisted by fs-laser writing, has been proposed for the production of these structures. Liu et al. [93] fabricated microcavities in sol–gel deposited on a zirconium/silicon substrate and obtained a quality factor of 1.48×10^5 in the 1550 nm band range. Grossmann et al. [94] also fabricated a microcavity in Ormocomp, a negative-tone photoresist, deposited on a silicon wafer with a quality factor of 1.2×10^6 in the 1300 nm wavelength region.

Song et al. [95] used these microresonators to produce an optofluidic sensor in a fused silica chip. A Y-branched channel, with two inlets for fluid injection, was fabricated using the FLICE technique. The channel outlet port was fabricated closely to a reservoir, where the microresonator was produced in order for light to interact with the fluid; the microresonator was also fabricated using the FLICE technique. A design of the proposed sensor is shown in Fig. 6.13b. To excite whispering gallery modes, a fiber taper was brought to the vicinity of the microresonator, and both were welded by CO_2 laser irradiation. In the final structure, a quality factor of 3.21×10^5 was measured in air. In order to demonstrate refractive index sensing, the channel was filled with purified water mixed with salt. Variations of the salt concentration induced a shift on the wavelength of the mode coupled to the microresonator from

Fig. 6.13 a SEM image of two microresonators made in fused silica after CO_2 laser annealing. **b** Schematic of the integrated optofluidic sensor proposed by Song et al. [95]. Images (**a**) and (**b**) were reproduced from [92, 95], respectively, with permission from The Optical Society (© 2012 Optical Society of America, © 2014 Optical Society of America)

the fiber, and a linear sensitivity of 220 nm/RIU was obtained, yielding a detection limit of 1.2×10^{-4} RIU.

The high quality factor provided by these structures makes them also attractive for the production of active whispering gallery mode microcavities, with lasing sources having already been produced in dye-doped resins with emission at 639 nm and in doped Nd:YAG glass with emission at 1550 nm [96, 97].

SERS (Surface-enhanced Raman Spectroscopy) micromonitors for detection of biomolecules have also been demonstrated by Xu et al. [98]. Production of this device involved the following steps: (i) a microfluidic channel in a glass substrate was produced by photolithography and etching, (ii) then a silver nitrate solution was injected in the channel, (ii) fs-laser writing induced photoreduction of the irradiated silver particles that deposited in the channel surface (Fig. 6.14a). The device was used to detect p-aminothiophenol (p-ATP) and flavin adenine dinucleotide (FAD) molecules, in two separate experiments. The molecules were injected in the channel and adsorbed to the metal surface. An ion laser, emitting at 514 nm, was then focused inside the channel through an objective lens, which was also used to collect scattering light (Fig. 6.14b). Both molecules were detected with an enhancement factor around 10^8 and a detection limit around the nanomolar range. The high surface roughness of the silver layer was shown to enhance the detection limit.

Recently, fs-laser micromachining has also been applied for the formation of 3D photonic crystals. This structure consists of a three-dimensional periodic modulation of the refractive index, and is characterized by a photonic band gap that inhibits light propagation at a certain wavelength. This wavelength depends on the index contrast between the two materials that compose the photonic crystal, which makes these structures attractive for sensing applications. Haque et al. [99] fabricated a woodpile structure in SU-8 photoresist by two-photon polymerization. The photonic crystal was produced inside a microfluidic channel and was interrogated by an optical waveguide (Fig. 6.15a). Both structures were made in a fused silica chip by fs-laser micromachining. Light propagating through the waveguide could transverse the photonic crystal, where the existence of the photonic band gap led to dips in

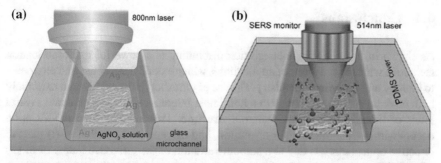

Fig. 6.14 Schematics of **a** silver layer patterning by fs-laser writing and of **b** target molecule detection through SERS. Reproduced from [98] with permission of ©The Royal Society of Chemistry

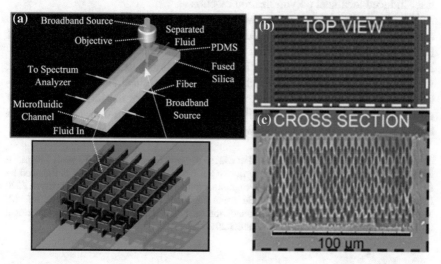

Fig. 6.15 Schematic of the integrated optofluidic device proposed by Haque et al. [99], with a zoom-in image of the proposed woodpile structure. SEM **b** top view image and **c** cross-section image of the inverted woodpile structure fabricated in fused silica by Ho et al. [100]. Images (**a**) and (**b**) were reproduced from [99, 100], respectively, with permission from The Optical Society (© 2013 OSA, © 2012 Optical Society of America)

the transmission spectrum. Theoretical results show that a sensitivity as high as 725 nm/RIU can be obtained. Some limitations of this technique are the low index contrast provided by the SU-8 photoresist, which contributes to a weaker band gap, and shrinkage of the polymer, which disturbs the lattice periodicity. Ho et al. [100] also fabricated an inverted-woodpile structure embedded in a microfluidic channel made in a fused silica chip (Fig. 6.15b). The photonic crystal structure was made based on the etching selectivity modification induced by fs-laser writing. The crystal was written layer-by-layer, which revealed that the processing conditions (scanning speed and pulse energy) had to be changed in order to maintain the hole width and height constant.

6.4 Conclusions

As shown in this chapter, fs-laser micromachining is a powerful microfabrication technique with a broad range of applications, which extend from optics and photonics to biology and analytical chemistry. The use of ultrashort pulse widths and extremely high peak intensities, gives rise to a non-linear interaction that can generate several kinds of modifications in the material structure. This laser-induced modification occurs within a submicrometer resolution, and can enable formation of (i) passive or active optical circuits due to a smooth variation of the refractive index, or of (ii) microfluidic systems combined with polymeric structures to mix or filter particles or to control the flow rate, due to the generation of etching selectivity and to a laser-induced localized polymerization reaction.

The possibility to fabricate three-dimensional components and integrate them in a single glass block has also extended the applications of this technique to the field of lab-on-a-chip devices. As demonstrated, several optofluidic devices for cell manipulation and behaviour analysis or for label-free detection have already been reported, which show the potential of this technique in the biomedical field [101]. Nevertheless, improvements on these devices can still be made. For instance, integration of microfluidics with biochemistry skills for formation of organ-on-a-chip devices is highly regarded for the study of biological micro- and nano-environments [102, 103].

Acknowledgements Project "NanoSTIMA: Macro-to-Nano Human Sensing Towards Integrated Multimodal Health Monitoring and Analytics/NORTE-01-0145-FEDER-000016" is financed by the North Portugal Regional Operational Programme (NORTE 2020), under the PORTUGAL 2020 Partnership Agreement, and through the European Regional Development Fund (ERDF). João M. Maia acknowledges the support of the Foundation for Science and Technology (FCT), Portugal through the Investigation Grant PD/BI/128995/2017.

References

1. G.M. Whitesides, Nature **442**(7101), 368–373 (2006)
2. P.S. Dittrich, A. Manz, Nat. Rev. Drug Discov. **5**(3), 210–218 (2006)
3. T.M. Squires, S.R. Quake, Rev. Mod. Phys. **77**(3), 977–1026 (2005)
4. P. Yager, T. Edwards, E. Fu, K. Helton, K. Nelson, M.R. Tam, B.H. Weigl, Nature **442**(7101), 412–418 (2006)
5. J. Wu, M. Gu, J. Biomed. Opt 16(8), (2011)
6. K. De Vos, I. Bartolozzi, E. Schacht, P. Bienstman, R. Baets, Opt. Express **15**(12), 7610–7615 (2007)
7. S. Suzuki, A. Sugita, NTT Tech. Rev. **3**(7), 12–16 (2005)
8. D.C. Duffy, J.C. McDonald, O.J.A. Schueller, G.M. Whitesides, Anal. Chem. **70**(23), 4974–4984 (1998)
9. H. Becker, L.E. Locascio, Talanta **56**(2), 267–287 (2002)
10. D. Tan, K.N. Sharafudeen, Y. Yue, J. Qiu, Prog. Mater Sci. **75**, 154–228 (2016)

11. K.M. Davis, K. Miura, N. Sugimoto, K. Hirao, Opt. Lett. **21**(21), 1729–1731 (1996)
12. J. Xu, D. Wu, Y. Hanada, C. Chen, S. Wu, Y. Cheng, K. Sugioka, K. Midorikawa, Lab Chip **13**(23), 4608–4616 (2013)
13. A. Marcinkevičius, S. Juodkazis, M. Watanabe, M. Miwa, S. Matsuo, H. Misawa, J. Nishii, Opt. Lett. **26**(5), 277–279 (2001)
14. K. Sugioka, J. Xu, D. Wu, Y. Hanada, Z. Wang, Y. Cheng, K. Midorikawa, Lab Chip **14**(18), 3447–3458 (2014)
15. N. Bellini, K.C. Vishnubhatla, F. Bragheri, L. Ferrara, P. Minzioni, R. Ramponi, I. Cristiani, R. Osellame, Opt. Express **18**(5), 4679–4688 (2010). https://doi.org/10.1364/OE.18.004679
16. A. Crespi, Y. Gu, B. Ngamsom, H.J.W.M. Hoekstra, C. Dongre, M. Pollnau, R. Ramponi, H.H. Van Den Vlekkert, P. Watts, G. Cerullo, R. Osellame, Lab Chip **10**(9), 1167–1173 (2010). https://doi.org/10.1039/B920062B
17. S. Ho, P.R. Herman, J.S. Aitchison, Appl. Phys. A **106**(1), 5–13 (2012)
18. K. Itoh, W. Watanabe, S. Nolte, C.B. Schaffer, MRS Bull. **31**(8), 620–625 (2006)
19. J.W. Chan, T. Huser, S. Risbud, D.M. Krol, Opt. Lett. **26**(21), 1726–1728 (2001)
20. A. Agarwal, M. Tomozawa, J. Non-Cryst. Solids **209**(1–2), 166–174 (1997)
21. K. Hirao, K. Miura, J. Non-Cryst. Solids **239**(1–3), 91–95 (1998)
22. T. Fukuda, S. Ishikawa, T. Fujii, K. Sakuma, H. Hosoya, Photon processing in microelectronics and photonics III, in *Proceedings of SPIE*, vol. 5339, 15 July 2004
23. Y. Nasu, M. Kohtoku, Y. Hibino, Opt. Lett. **30**(7), 723–725 (2005)
24. R.R. Thomson, H.T. Bookey, N.D. Psaila, A. Fender, S. Campbell, W.N. MacPherson, J.S. Barton, D.T. Reid, A.K. Kar, Opt. Express **15**(18), 11691–11697 (2007)
25. S.M. Eaton, H. Zhang, P.R. Herman, Opt. Express **13**, 4708–4716 (2005)
26. M. Ams, G.D. Marshall, D.J. Spence, M.J. Withford, Opt. Express **13**(15), 5676–5681 (2005)
27. S.M. Eaton, M.L. Ng, R. Osellame, P.R. Herman, J. Non-Cryst. Solids **357**(11–13), 2387–2391 (2011)
28. M. Ams, P. Dekker, S. Gross, M.J. Withford, Nanophotonics (to be published) (2017)
29. J.R. Grenier, L.A. Fernandes, J.S. Aitchison, P.V.S. Marques, P.R. Herman, Opt. Lett. **37**(12), 2289–2291 (2012)
30. P. Zeil, C. Voigtländer, J. Thomas, D. Richter, S. Nolte, Opt. Lett. **38**(13), 2354–2356 (2013)
31. L.A. Fernandes, J.R. Grenier, P.R. Herman, J.S. Aitchison, P.V.S. Marques, Opt. Express **19**(19), 18294–18301 (2011)
32. I.-B. Sohn, M.-S. Lee, J. Chung, IEEE Photon. Technol. Lett. **17**(11), 2349–2351 (2005)
33. J. Liu, Z. Zhang, S. Chang, C. Flueraru, C.P. Grover, Opt. Commun. **253**(4–6), 315–319 (2005)
34. V.A. Amorim, J.M. Maia, D. Alexandre, P.V.S. Marques, I.E.E.E. Photon, Technol. Lett **29**(7), 619–622 (2017)
35. K. Suzuki, V. Sharma, J.G. Fujimoto, E.P. Ippen, Y. Nasu, Opt. Express **14**(6), 2335–2343 (2006)
36. W.-J. Chen, S.M. Eaton, H. Zhang, P.R. Herman, Opt. Express **16**(15), 11470–11480 (2008)
37. W. Watanabe, Y. Note, K. Itoh, Opt. Lett. **30**(21), 2888–2890 (2005)
38. D.-Y. Liu, Y. Li, Y.-P. Dou, H.-C. Guo, H. Yang, Q.-H. Gong, Chin. Phys. Lett. **25**(7), 2500–2503 (2008)
39. V.A. Amorim, J.M. Maia, D. Alexandre, P.V.S. Marques, J. Lightwave Technol. **35**(17), 3615–3621 (2017)
40. G. Douglass, F. Dreisow, S. Gross, S. Nolte, M.J. Withford, Rapid prototyping of arrayed waveguide gratings, *Paper presented at Photonic and Fiber Technology 2016*, Sydney, Australia, 5–8 Sept 2016
41. S.-L. Li, P. Han, M. Shi, M. Yao, B. Hu, M. Wang, X. Zhu, Opt. Express **19**(24), 23958–23964 (2011)
42. J. Burghoff, S. Nolte, A. Tünnermann, Appl. Phys. A **89**(1), 127–132 (2007). https://doi.org/10.1007/s00339-007-4152-0
43. A.G. Okhrimchuk, A.V. Shestakov, I. Khrushchev, J. Mitchell, Opt. Lett. **30**(17), 2248–2250 (2005)

44. M.-M. Dong, C.-W. Wang, Z.-X. Wu, Y. Zhang, H,-H. Pan, Q.-Z. Zhao, Opt. Express **21**(13), 15522–15529 (2013)
45. R. Osellame, G. Della Valle, N. Chiodo, S. Taccheo, P. Laporta, O. Svelto, G. Cerullo, Appl. Phys. A **93**(1), 17–26 (2008)
46. D. Choudhury, J.R. Macdonald, A.K. Kar, Laser Photon. Rev. **8**(6), 827–846 (2014)
47. C. Zhang, N. Dong, J. Yang, F. Chen, J.R.V. De Aldana, Q. Lu, Opt. Express **19**(13), 12503–12508 (2011)
48. G. Palmer, S. Gross, A. Fuerbach, D.G. Lancaster, M.J. Withford, Opt. Express **21**(14), 17413–17420 (2013)
49. G. Della Valle, S. Taccheo, R. Osellame, A. Festa, G. Cerullo, P. Laporta, Opt. Express **15**(6), 3190–3194 (2007)
50. B. McMillen, B. Zhang, K.P. Chen, A. Benayas, D. Jaque, Opt. Lett. **37**(9), 1418–1420 (2012)
51. M.A. Hughes, W. Yang, D.W. Hewak, J. Opt. Soc. Am. B **26**(7), 1370–1378 (2009)
52. N.D. Psaila, R.R. Thomson, H.T. Bookey, S. Shen, N. Chiodo, R. Osellame, G. Cerullo, A. Jha, A.K. Kar, Opt. Express **15**(24), 15776–15781 (2007)
53. H.T. Bookey, R.R. Thomson, N.D. Psaila, A.K. Kar, N. Chiodo, R. Osellame, G. Cerullo, I.E.E.E. Photon, Technol. Lett. **19**(12), 892–894 (2007)
54. W. Horn, S. Kroesen, J. Herrmann, J. Imbrock, C. Denz, Opt. Express **20**(24), 26922–26928 (2012)
55. R. Taylor, C. Hnatovsky, E. Simova, Laser Photon. Rev **2**(1–2), 26–46 (2008)
56. A.A. Said, M. Dugan, P. Bado, Y. Bellouard, A. Scott, J.R. Mabesa Jr., Photon processing in microelectronics and photonics III, in *Proceedings of SPIE*, vol. 5339, 15 July 2004
57. B. Fisette, M. Meunier, Photonics north 2004: photonic applications in astronomy, biomedicine, imaging, materials processing, and education, in *Proceedings of SPIE*, vol. 5578, 9 Dec 2004
58. K. Sugioka, Y. Cheng, K. Midorikawa, Appl. Phys. A **81**(1), 1–10 (2005)
59. S. Matsuo, Y. Tabuchi, T. Okada, S. Juodkazis, H. Misawa, Appl. Phys. A **84**(1–2), 99–102 (2006)
60. D. Wortmann, J. Gottmann, N. Brandt, H. Horn-Solle, Opt. Express **16**(3), 1517–1522 (2008)
61. D. Choudhury, A. Rodenas, L. Paterson, F. Díaz, D. Jaque, A.K. Kar, Appl. Phys. Lett **103**(4) (2013)
62. Y. Bellouard, A. Said, M. Dugan, P. Bado, Opt. Express **12**(10), 2120–2129 (2004)
63. Y. Bellouard, A.A. Said, M. Dugan, P. Bado, Commercial and biomedical applications of ultrafast lasers VI, in *Proceedings of SPIE*, vol. 6108, 28 Feb 2006
64. C. Hnatovsky, R.S. Taylor, E. Simova, V.R. Bhardwaj, D.M. Rayner, P.B. Corkum, Opt. Lett. **30**(14), 1867–1869 (2005)
65. J.M. Maia, V.A. Amorim, D. Alexandre, P.V.S. Marques, in *Proceedings of the 5th International Conference on Photonics, Optics and Laser Technology (PHOTOPTICS 2017)*, 27 Feb–1 Mar 2017
66. A. Champion, M. Beresna, P. Kazansky, Y. Bellouard, Opt. Express **21**(21), 24942–24951 (2013)
67. F. He, J. Lin, Y. Cheng, Appl. Phys. B **105**(2), 379–384 (2011)
68. K.C. Vishnubhatla, N. Bellini, R. Ramponi, G. Cerullo, R. Osellame, Opt. Express **17**(10), 8685–8695 (2009)
69. S. Kiyama, S. Matsuo, S. Hashimoto, Y. Morihira, J. Phys. Chem. C **113**(27), 11560–11566 (2009)
70. Y. Temiz, R.D. Lovchik, G.V. Kaigala, E. Delamarche, Microelectron. Eng. **132**, 156–175 (2015)
71. M. Masuda, K. Sugioka, Y. Cheng, N. Aoki, M. Kawachi, K. Shihoyama, K. Toyoda, H. Helvajian, K. Midorikawa, Appl. Phys. A **76**(5), 857–860 (2003)
72. K. Liu, Q. Yang, S. He, F. Chen, Y. Zhao, X. Fan, L. Li, C. Shan, H. Bian, Microsyst. Technol. **19**(7), 1033–1040 (2013)
73. Y. Li, K. Itoh, W. Watanabe, K. Yamada, D. Kuroda, J. Nishii, Y. Jiang, Opt. Lett. **26**(23), 1912–1914 (2001)

74. Y. Li, S. Qu, Curr. Appl. Phys. **13**(7), 1292–1295 (2013)
75. K.-S. Lee, R.H. Kim, D.-Y. Yang, S.H. Park, Prog. Polym. Sci. **33**(6), 631–681 (2008)
76. J. Wang, Y. He, H. Xia, L.-G. Niu, R. Zhang, Q.-D. Chen, Y. Zhang, Y.-F. Li, S.-J. Zeng, J.-H. Qin, B.-C. Lin, H.-B. Sun, Lab Chip **10**(15), 1993–1996 (2010). https://doi.org/10.1039/C003264F
77. L. Amato, Y. Gu, N. Bellini, S.M. Eaton, G. Cerullo, R. Osellame, Lab Chip **12**(6), 1135–1142 (2012)
78. T.W. Lim, Y. Son, Y.J. Jeong, D.-Y. Yang, H.-J. Kong, K.-S. Lee, D.-P. Kim, Lab Chip **11**(1), 100–103 (2011). https://doi.org/10.1039/C005325M
79. D. Wu, S.-Z. Wu, J. Xu, L.-G. Niu, K. Midorikawa, K. Sugioka, Laser Photon. Rev. **8**(3), 458–467 (2014)
80. R.W. Applegate Jr., J. Squier, T. Vestad, J. Oakey, D.W.M. Marr, P. Bado, M.A. Dugan, A.A. Said, Lab Chip **6**(3), 422–426 (2006). https://doi.org/10.1039/B512576F
81. D.J. Hwang, M. Kim, K. Hiromatsu, H. Jeon, C.P. Grigoropoulos, Appl. Phys. A **96**(2), 385–390 (2009). https://doi.org/10.1007/s00339-009-5210-6
82. M. Kim, D.J. Hwang, H. Jeon, K. Hiromatsu, C.P. Grigoropoulos, Lab Chip **9**(2), 311–318 (2009). https://doi.org/10.1039/B808366E
83. F. Bragheri, L. Ferrara, N. Bellini, K.C. Vishnubhatla, P. Minzioni, R. Ramponi, R. Osellame, I. Cristiani, J. Biophotonics **3**(4), 234–243 (2010)
84. F. Bragheri, P. Minzioni, R. Martinez Vazquez, N. Bellini, P. Paiè, C. Mondello, R. Ramponi, I. Cristiani, R. Osellame, Lab Chip **12**(19), 3779–3784 (2012). https://doi.org/10.1039/c2lc40705a
85. T. Yang, P. Paiè, G. Nava, F. Bragheri, R.M. Vazquez, P. Minzioni, M. Veglione, M. Di Tano, C. Mondello, R. Osellame, I. Cristiani, Lab Chip **15**(5), 1262–1266 (2015)
86. R.M. Vazquez, R. Osellame, D. Nolli, C. Dongre, H. Van Den Vlekkert, R. Ramponi, M. Pollnau, G. Cerullo, Lab Chip **9**(1), 91–96 (2009)
87. C. Dongre, R. Dekker, H.J.W.M. Hoekstra, M. Pollnau, R. Martinez-Vazquez, R. Osellame, G. Cerullo, R. Ramponi, R. Van Weeghel, G.A.J. Besselink, H.H. Van Den Vlekkert, Opt. Lett. **33**(21), 2503–2505 (2008). https://doi.org/10.1364/OL.33.002503
88. C. Dongre, J. Van Weerd, G.A.J. Besselink, R. Van Weeghel, R.M. Vazquez, R. Osellame, G. Cerullo, M. Cretich, M. Chiari, H.J.W.M. Hoekstra, M. Pollnau, Electrophoresis **31**(15), 2584–2588 (2010)
89. V. Maselli, J.R. Grenier, S. Ho, P.R. Herman, Opt. Express **17**(14), 11719–11729 (2009)
90. J.M. Maia, V.A. Amorim, D. Alexandre, P.V.S. Marques, J. Lightwave Technol. **35**(11), 2291–2298 (2017)
91. Y. Hanada, K. Sugioka, K. Midorikawa, Lab Chip **12**(19), 3688–3693 (2012). https://doi.org/10.1039/C2LC40377C
92. J. Lin, S. Yu, Y. Ma, W. Fang, F. He, L. Qiao, L. Tong, Y. Cheng, Z. Xu, Opt. Express **20**(9), 10212–10217 (2012). https://doi.org/10.1364/OE.20.010212
93. Z.-P. Liu, Y. Li, Y.-F. Xiao, B.-B. Li, X.-F. Jiang, Y. Qin, X.-B. Feng, H. Yang, Q. Gong, Appl. Phys. Lett **97**(21) (2010)
94. T. Grossmann, S. Schleede, M. Hauser, T. Beck, M. Thiel, G. Von Freymann, T. Mappes, H. Kalt, Opt. Express **19**(12), 11451–11456 (2011)
95. J. Song, J. Lin, J. Tang, Y. Liao, F. He, Z. Wang, L. Qiao, K. Sugioka, Y. Cheng, Opt. Express **22**(12), 14792–14802 (2014). https://doi.org/10.1364/OE.22.014792
96. J.-F. Ku, Q.-D. Chen, R. Zhang, H.-B. Sun, Opt. Lett. **36**(15), 2871–2873 (2011)
97. J. Lin, Y. Xu, J. Song, B. Zeng, F. He, H. Xu, K. Sugioka, W. Fang, Y. Cheng, Opt. Lett. **38**(9), 1458–1460 (2013)
98. B.-B. Xu, Z.-C. Ma, L. Wang, R. Zhang, L.-G. Niu, Z. Yang, Y.-L. Zhang, W.-H. Zheng, B. Zhao, Y. Xu, Q.-D. Chen, H. Xia, H.-B. Sun, Lab Chip **11**(19), 3347–3351 (2011). https://doi.org/10.1039/C1LC20397E
99. M. Haque, N.S. Zacharia, S. Ho, P.R. Herman, Biomed. Opt. Express **4**(8), 1472–1485 (2013). https://doi.org/10.1364/BOE.4.001472

100. S. Ho, M. Haque, P.R. Herman, J.S. Aitchison, Opt. Lett. **37**(10), 1682–1684 (2012). https://doi.org/10.1364/OL.37.001682
101. F. Sima, J. Xu, D. Wu, K. Sugioka, Micromachines 8(2), (2017)
102. D. Huh, G.A. Hamilton, D.E. Ingber, Trends Cell Biol. **21**(12), 745–754 (2011)
103. D.J. Beebe, D.E. Ingber, J. Den Toonder, Lab Chip **13**(18), 3447–3448 (2013)

Chapter 7
Microfiber Knot Resonators for Sensing Applications

A. D. Gomes and O. Frazão

Abstract Microfiber knot resonators are widely applied in many different fields of action, of which an important one is the optical sensing. Microfiber knot resonators can easily be used to sense the external medium. The large evanescent field of light increase the interaction of light with the surrounding medium, tuning the resonance conditions of the structure. In some cases, the ability of light to give several turns in the microfiber knot resonator allows for greater interaction with deposited materials, providing an enhancement in the detection capability. So far a wide variety of physical and chemical parameters have been possible to measure using microfiber knot resonators. However, new developments and improvements are still being done in this field. In this chapter, a review on sensing with microfiber knot resonators is presented, with particular emphasis on the application of these structures as temperature and refractive index sensors. A detailed analysis on the properties of these structures and different assembling configurations is presented. An important discussion regarding the sensor stability is presented, as well as alternatives to increase the device robustness. An overview on the recent developments in coated microfiber knot resonators is also addressed. In the end, other microfiber knot configurations are explored and discussed.

7.1 Introduction

The beginning of optical fibers life history take us back to the 20th century. The first optical fibers were fabricated in the 20s, although they only became more common during the 50s for guiding light in short distances, mainly for medical application.

In 1966, Prof. Charles Kao suggested their application to communication systems, reason why he received the Nobel Prize in 2009. Henceforth, followed by the advances of the telecommunication industry, optical fibers have become a topic of

A. D. Gomes (✉) · O. Frazão
Faculty of Sciences, INESC TEC and Department of Physics and Astronomy, University of Porto, Rua do Campo Alegre 687, 4169-007 Porto, Portugal
e-mail: ardcgomes@gmail.com

© Springer Nature Switzerland AG 2019
P. Ribeiro et al. (eds.), *Optics, Photonics and Laser Technology 2017*,
Springer Series in Optical Sciences 222,
https://doi.org/10.1007/978-3-030-12692-6_7

145

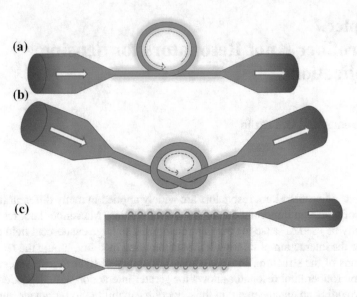

Fig. 7.1 Illustration of resonator-type MNF structures. Adapted from Chen and Ding [3]

extensive research. From the fruits of this research new areas of investigation and application of optical fibers began to grow. A good example is the application of optical fibers to sense physical, chemical and biochemical parameters.

With the technological advances, the fabrication of micro and nanofibers by tapering an optical fiber to micrometric or even nanometric size [1] opened new doors to the study and development of new and enhanced sensors.

Until nowadays, new sensing configurations based on micro and nanofibers were studied and developed. These type of sensors come in many different forms and normally the measured effects are sensed via intensity or phase change of the transmitted light. One type of configuration are the resonator-type micro and nanofiber sensors that involves all the sensors which use resonant structures. The mode of operation of these structures consists in tying a knot or coiling a micro or nanofiber onto itself allowing coupling and evanescent overlap between modes propagating in adjacent turns [2].

This type of structure can be divided into three groups (see Fig. 7.1):

(a) Microfiber loop resonator (MLR): it is a single turn ("loop") of an optical microfiber in the taper waist region [4];
(b) Microfiber knot resonator (MKR): it is made by tying a knot in an optical microfiber creating a ring with micrometer dimensions [5];
(c) Micro-coil resonator (MCR): it can be seen as a high-order MLR, which consists of several independent loop resonators. They present great potential ability of micro-assemblage in 3-D, for example, wrapping a microfiber onto optical rods [4, 6].

Microfiber knot resonators are one of the configurations which had a huge impact due to their potential application to different fields in addition to sensing, such as in ultrafast optics [7]. This chapter will focus the recent progress in microfiber knot resonators as sensing structures, giving special attention to temperature and refractive index monitoring. An important discussion about the sensor stability will also be addressed. In the end, new configurations involving this structure will be presented.

7.2   Microfiber Knot Resonators

A microfiber knot resonator, as previously explained, is performed by tying a knot in the taper waist region of a micro or nanofiber. Light that enters the microfiber knot resonator will be divided in the knot region between the ring and the output (Fig. 7.2).

New light that enters the knot will combine with light going to the output of the structure, while feeding at the same time the ring. For certain wavelengths, light that enters the knot is in phase with the one traveling in the ring, resulting in accumulation of light in the ring. Hence, a reduction in the transmission spectrum is observed at those wavelengths, as shown in Fig. 7.3.

To produce this structure, first, the freestanding end of a microfiber is assembled into a large ring with few millimeters in diameter. At that point, the diameter of the ring is continuously reduced by pulling the free end of the fiber until a microknot with the desired dimensions is obtained (Fig. 7.2). This overlap of the fiber with itself allows no need for a precise alignment [5], revealing to be a great advantage compared with microfiber loop resonators. The low dimension and low fabrication cost of these devices have attracted much attention to fabricate them for sensing purposes [3].

A more detailed analysis regarding the properties of microfiber knot resonator will now be made. Different assembling configurations will also be addressed, as well as an important discussion regarding some methods to increase the sensor stability.

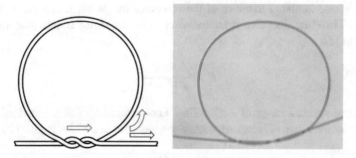

Fig. 7.2 Schematic and picture of a microfiber knot resonator. The structure is produced using a microfiber

Fig. 7.3 Schematic of a
microfiber knot resonator
transmission spectrum [8]

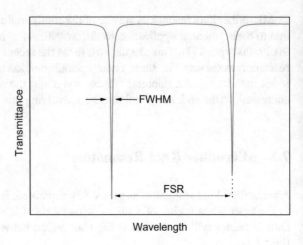

7.2.1 Properties

Microfiber knot resonators present interesting properties: high quality factor and
finesse, and a free spectral range (FSR) that depends on the diameter of the ring. A
schematic of a high-Q microfiber knot resonator transmission spectrum is presented
in Fig. 7.3.

The quality factor (Q) is a useful quantity to evaluate the losses of the resonator.
The higher the Q-factor, the longer the light stays in the ring. This factor can be
expressed as [2, 8]:

$$Q = \frac{\lambda_{\text{res}}}{FWHM} \tag{7.1}$$

where λ_{res} is the resonance wavelength and FWHM is the full width at half-maximum,
also known as the bandwidth of the resonances. Microfiber knot resonators can
achieve Q factors up to 10^5 [5, 9, 10]. In other words, these structures can present
very narrow resonant peaks in the transmission spectrum.

The finesse is also a measure of the resonator losses but it is a dimensionless
quantity. Therefore, its value is independent of the resonator length. The finesse is
defined as [8]:

$$F = \frac{FSR}{FWHM} \tag{7.2}$$

The free spectral range of a microfiber knot resonator, which is defined as the
distance between two adjacent resonance wavelengths λ_1 and λ_2, is given by [8]:

$$FSR \approx \frac{\lambda_1 \lambda_2}{n_{\text{eff}} L} \tag{7.3}$$

where n_{eff} is the effective refractive index of the microfiber and L is the cavity length. In a microfiber knot resonator the cavity length is given by the perimeter of the ring. This property is very useful since one can tune the sensor's free spectral range by adjusting the diameter of the microfiber knot resonator.

For sensing purposes the free spectral range is an important quantity, along with the sensitivity to the measured parameter, as it imposes the limits of the measurement range. As an example, when measuring the wavelength shift of a resonance peak as a function of a given physical or chemical parameter, the maximum displacement of a resonance peak must be lower than the free spectral range. Otherwise, the new resonance peak position will overlap the position of other resonance peak for a previous value of the measured parameter.

The transmission spectrum of a microfiber knot resonator is, from another point of view, similar to a Fabry-Perot cavity: the cavity is the ring and the mirrors are the knot structure that recouples light back into the ring. A Fabry-Perot cavity with high reflectance mirrors means that light stays more time in the cavity before going out. For a microfiber knot resonator this is translated as a greater coupling of light between adjacent fibers in the knot structure (which implies small fiber diameters to have large evanescent fields).

7.2.2 Assembling Configurations

In the beginning, the assembling of microfiber knot resonators was performed using a micro or nanofiber with a freestanding end (like a pigtail) and a second microfiber (output fiber) to collect the light transmitted out of the knot, as depicted in Fig. 7.4. The second microfiber is evanescently coupled to the output of the microfiber knot resonator, connected through Van der Wall attraction. For practical applications of microfiber knot resonators such solution presents some limitations. Despite some authors state that microfiber knot resonators assembled from double-ended micro or nanofibers can easily break when knotted [2, 10, 11], the first microfiber knot resonator made from double-ended tapered fibers was demonstrated by Xiao and Birks back in 2011 [5]. Using a 128 μm-diameter microfiber knot resonator assembled from a 1 μm-diameter microfiber, the authors obtained microfiber knot resonators with high finesse (104). A high Q-factor of 97,260 was also obtained using a 570 μm−diameter microfiber knot resonator assembled from the same fiber taper. The authors reported a minimum knot diameter achieved of 46 μm using the same microfiber.

Still in 2011, the first all-fiber and optically tunable microfiber knot resonator has been proposed [12]. The microfiber knot resonator was coated with a photoresponsive liquid crystal mixture. The mixture presents an easily variable refractive index that can be changed/tuned upon irradiation with ultraviolet (UV) light. Hence, the change in the microfiber knot resonator effective refractive index caused by the liquid crystal induces a change in the resonance wavelengths. A spectral shift of 0.15 nm was obtained when irradiating a 468 μm−diameter knot with 50 mW/cm^3 of UV light.

Fig. 7.4 Schematic of a microfiber knot resonator using a second microfiber as collecting fiber [8]

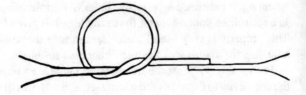

In this case the microfiber knot resonator was assembled from a 2.5 μm−diameter microfiber. A great advantage of this technique is the capability to reverse the process.

7.2.3 Stability

Even though microfiber knot resonators present outstanding mechanical flexibility, many applications require them to be as stable and protected as possible. Therefore it is important to discuss some critical aspects and approaches regarding the stability of these structures.

The stability of the microfiber knot resonators can be increased using different methods and techniques. The first method is to use a polished solid MgF_2 crystal substrate that helps to support the microfiber knot resonator. Since MgF_2 substrates present low refractive index ($n = 1.37$) the resonance is preserved [10, 13, 14]. However a small change in the resonance wavelengths is observed due to a change in the effective refractive index. The substrate also provides good thermal conductivity. This property can be very helpful in temperature measurements for a good thermalization of the sensor [14].

Coating microfiber knot resonators with low refractive index polymers [13, 15] or simply use microfiber knot resonators made out of polymer microfibers [14, 16, 17] is another common way to accomplish long-term stability. With respect to this method, Vienne et al. performed an analysis on the effect of host polymers in microfiber knot resonators [18]. They showed that the free spectral range and the maximum Q-factor and extinction ratio did not change significantly after coating the microfiber knot resonator with low-index polymer. On the other hand, the change in the effective refractive index causes the wavelength region of high Q-factor and extinction ratio to shift. Polymer coatings allow to fix the knot structure, reducing the probability of changing the knot diameter.

Polymer coatings protect microfiber knot resonators against degradation over time. The effect of signal degradation over time was explored by Li and Ding [13]. In fact, the transmission losses of a bare knot resonator increased at the speed of around 0.24 dB/h, reaching 18 dB after 3 days. On the other hand, a Teflon ($n \sim 1.31$) coated microfiber knot resonator presented no changes for half a month.

More recently, the use of hydrophobic aerogel ($n \sim 1.05$) to embed microfiber knot resonators was studied [5]. The losses of the structure off-resonance (off-resonance is defined as the region of wavelengths far from the resonance wavelengths) proved to be much lower than others using different encapsulants. Compared with low-index polymers, the aerogel maintains light confinement and avoids change of dispersion due to its low refractive index.

The use of large knots made out of larger diameter fibers can provide robustness and resistance [19]. However, the resonance property can be lost due to the existence of small evanescent field of light.

7.3 Microfiber Knot Resonators as Sensors

Microfiber knot resonators can be used to sense a wide variety of physical, chemical, and biochemical parameters [8]. As seen previously, the measured effects are sensed via intensity or more usually by phase change of the transmitted light. This last method relies on the change of the microfiber effective refractive index or on the change of the physical length of the structure, for example, by thermal effects. Such effects modify the resonance conditions causing a modification in the resonance wavelengths, as well as in the free spectral range. The sensor characterization is performed by monitoring resonance wavelength shift as a function of the measured parameter. From the sensor characterization a sensitivity to the measured parameter is obtained, together with the measurement range.

A more detailed analysis of some published results will now be made with special focus on the use of microfiber knot resonators for temperature and refractive index sensing.

7.3.1 Temperature Sensing

Exposing a microfiber knot resonator to temperature variations will change its length through thermal expansion. Adding to this effect there is also a change in the fiber refractive index due to thermo-optic effects (temperature dependence of the refractive index). The combination of these effects change the resonance conditions of the knot structure, resulting in a shift of the resonance wavelengths [15]. By monitoring the resonance wavelength variations one can have information about the temperature variations around the sensor. Until now many microfiber knot resonators have been proposed and demonstrated for temperature sensing using this principle of working.

In 2009, two temperature sensors based on silica/polymer microfiber knot resonator were reported by Wu et al. [14]. Both sensors were placed in an MgF_2 crystal plate and covered with an MgF_2 slab to make them more robust and immune to environmental fluctuations, as depicted in Fig. 7.5. Note that MgF_2 substrates ensures good thermal conductivity and presents low refractive index, as discussed previously in Sect. 7.2.3.

The silica microfiber knot resonator was produced from a 1.7 μm-diameter silica microfiber creating a 190 μm-diameter ring. A Q-factor of 12,000 was reported for

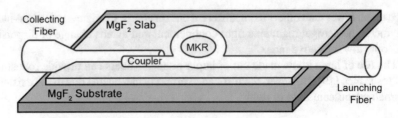

Fig. 7.5 Schematic of the microfiber knot resonator temperature sensor presented by Wu et al. [8, 14]

the structure. For temperature sensing a sensitivity of around 52 pm/°C was obtained between 30 and 700 °C.

In relation to the polymer microfiber knot resonator, the structure was produced using a polymethyl methacrylate (PMMA) microfiber ($n \sim 1.49$) with 2.1 μm-diameter. The ring had 98 μm-diameter. Polymer microfibers have the advantage of bending more easily and with smaller radius of curvature than standard silica microfibers. A lower Q-factor (8000) compared with the previous structure was obtained. On the other hand, a temperature sensitivity of 266 pm/°C was achieved between 20 and 80 °C. In fact, PMMA presents higher thermal expansion and thermo-optic coefficient than silica. Therefore polymer microfiber knot resonators exhibit more temperature sensitivity than silica ones. Despite having higher sensitivity, microfiber knot resonators made from polymers have lower measuring range since they are limited to the low melting point of the polymer.

In the same year, Zeng et al. demonstrated a polymer coated microfiber knot resonator for temperature sensing [15]. The assembling of the sensor is identical to the one presented in Fig. 7.5. Instead of an MgF_2 substrate a glass plate was adopted as substrate due to its good adhesion to the polymer. The polymer coating consists of a 20 μm thin layer of low refractive index polymer (EFIRON UVF PC-373, $n = 1.3759$ @ 852 nm) deposited on the surface of the substrate to isolate the microfiber knot resonator from the high refractive index of the glass plate. The sensor was fabricated from a 1 μm-diameter microfiber creating a 55 μm−diameter knot resonator. A second polymer layer was applied over the structure to make it immune to environmental changes. The authors reported a temperature sensitivity of 270 pm/°C in the heating process from 28 to 140 °C, and 280 pm/°C in the cooling process from 135 to 25 °C. These sensitivity values are similar to the PMMA microfiber knot resonator sensitivity demonstrated by Wu et al. [14]. The sensor resolution is 0.5 °C, but it can be increased if a higher resolution spectrometer is used.

A simple silica microfiber knot resonator for seawater temperature sensing was theoretically an experimentally analyzed by Yang et al. in 2014 [20]. It is important to refer that measuring seawater temperature can be complex and problematic since silica and seawater show opposite thermo-optic coefficients. When the thermo-optic coefficient of silica is predominant the resonant wavelengths with suffer a red shift. However, if there is a predominance of the seawater thermo-optic coefficient, a blue shift is observed. Therefore the sensor behavior depends from case to case. Theoret-

Fig. 7.6 Schematic of the microfiber double-knot resonator in series [21]

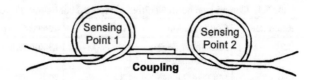

ically, there is a fiber diameter for which the effect of silica thermo-optic coefficient and seawater thermo-optic coefficient cancel each other. According to the author's calculations, this happens for a microfiber diameter of 1.27 μm. Bellow that diameter the temperature sensitivity will be negative and above that value the temperature sensitivity will be positive. For microfibers with diameters above 4 μm the temperature sensitivity saturates. Furthermore, the sensitivity will also increase if the probing wavelength is increased. The proposed microfiber knot resonator presented a diameter of 473 μm and it was produced using a 3.91 μm−diameter microfiber. A maximum temperature sensitivity of 22.81 pm/°C was achieved for the sensing structure in the range from 23 to 33 °C. This value of sensitivity was obtained at a probing wavelength of 1599.6 nm. The structure presents a Q-factor of 3000 and a finesse of 11.69.

Microfiber knot resonators can also be applied for multi-point temperature sensing. For such purpose a configuration based on a microfiber double-knot resonator was developed by Wu et al. [21]. The sensing structure consists of two microfiber knot resonators coupled using the free-end of each microfiber knot resonator. A schematic of the sensor is depicted in Fig. 7.6. The two microfiber knot resonators, one with 506 μm−diameter and the other with 500 μm−diameter, were fabricated using a 2.3 μm−diameter microfiber. The length of the coupling region can be tuned by adjusting the distance between the two knots.

In terms of transmission spectrum, if the diameter of both microfiber knot resonators is the same (same loop length), the interference spectrum shows a regular resonance like a single microfiber knot resonator. In other words, the spectrum of each microfiber knot resonator will add. On the other hand, if now the diameter of one ring changes, a new phase matching condition will arise. Hence, a second-order resonant peak will appear between the primary resonant peaks. The authors demonstrate that if the temperature changes locally on the first microfiber knot resonator, only the primary resonant peak will shift while the second-order resonant peak is constant. Contrariwise, if the temperature of the second microfiber knot resonator changes, the second-order resonant peak will shift while the primary resonant peak is constant. So, multi-point temperature sensing can be achieved in a very small scale using this sensor.

A similar sensor was demonstrated by Yang et al. in 2017, for dual-point seawater temperature simultaneous sensing [22]. The two microfiber knot resonators, one with 858 μm-diameter and other with 883 μm-diameter, made from a 2.8 μm-diameter microfiber are connected by a single mode fiber. The microfiber double knot resonator obtained temperature sensitivities of −13.86 pm/°C and −5.6 pm/°C, respectively.

Table 7.1 Comparison between different reported configurations for temperature sensing [8]

Reference	Configuration	Q-factor	Sensitivity (pm/°C)	Range (°C)
[22]	Microfiber double knot resonator	X	13.86/−5.6	25.8–33.8/26.1–34.4
[23]	MKR in Sagnac loop	X	20.6	30–130
[20]	Simple Silica MKR (for seawater)	3000	22.81	23–33
[14]	Silica MKR with MgF$_2$	12,000	52	30–700
[15]	Polymer (PMMA) MKR	8000	266	20–80
[15]	Polymer (EFIRON) MKR	X	270 (heating)	28–140
–	–	X	−280 (cooling)	135–25

Independent and simultaneous measurement of seawater temperature is achieved using this structure.

A comparison between the different configurations discussed previously is summarized in Table 7.1.

7.3.2   Refractive Index Sensing

The use of microfiber knot resonators as refractive index sensors consists mainly of taking advantage of the large evanescent field of light to interact with the external environment. Alterations in the external medium refractive index will change the properties of light traveling in the microfiber, which leads to a change in the knot resonance conditions. As discussed before, this effect is manifested as a shift in the resonance wavelengths and as a change in the free spectral range that depends on the external refractive index variations.

It is worth mentioning that the control of temperature can be crucial for refractive index sensing. In fact, if temperature changes during refractive index measurements, the refractive index of silica and the measured solution will change due to thermo-optic effects. Such effect may introduce small errors in the determination of refractive index. Usually a second sensor sensitive to temperature but not to refractive index is used to detect temperature fluctuations in order to further correct the refractive index measurements.

A Teflon coated microfiber knot resonator ($n_{\text{Teflon}} \sim 1.31$) for refractive index sensing was demonstrated by Li and Ding [13]. The polymer was applied using a dip-coating technique, whose process is outlined in Fig. 7.7.

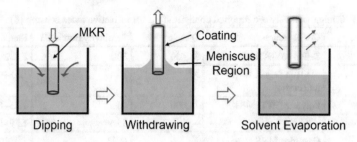

Fig. 7.7 Illustration of the microfiber knot resonator dip-coat process in Teflon [13]

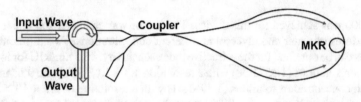

Fig. 7.8 Schematic diagram of a microfiber knot resonator in a Sagnac loop reflector [23]

The sensor was produced from a microfiber with an initial diameter of 2.92 μm, creating a 1.02 mm-diameter microfiber knot resonator. After the coating process the microfiber diameter increased to 3.95 μm, allowing to estimate a value of 0.5 μm for the Teflon thickness. Tensions suffered during the dip-coating process reduced the microfiber knot resonator diameter to around 450 μm. Before coating, the structure presented a Q-factor of 20,000, increasing to around 31,000 after the Teflon coating. A refractive index sensitivity of approximately 30.5 nm/RIU was reported for this sensor in the refractive index range from 1.3322 to 1.3412.

Before that, Lim et al. proposed a microfiber knot resonator with around 0.5 mm-diameter in a Sagnac loop reflector configuration [23]. The Sagnac loop reflector configuration allows the signal collection through the incident path. A low refractive index Teflon ($n \sim 1.31$) was used to embed the structure, except the microfiber knot resonator sensing region, to obtain a balance between responsiveness and robustness. An illustration of the experimental setup is depicted in Fig. 7.8.

The sensor achieved a sensitivity of 30.49 nm/RIU in a refractive index range from 1.334 to 1.348. Such value of sensitivity is similar to the previously explored Teflon coated microfiber knot resonator proposed by Li and Ding [13]. In terms of temperature response, the sensor showed a sensitivity of 20.6 pm/°C between 30 and 130 °C.

Also in 2014, Yu et al. demonstrated a polymer microfiber knot resonator for refractive index sensing [17]. A microfiber of Poly(trimethylene terephthalate) (PTT) was produced with 1.3 μm-diameter. PPT presents strong flexibility, large refractive index ($n = 1.638$) and also good transparency in the visible and near-infrared spectrum. The PPT microfiber was used to create an 85 μm−diameter microfiber knot resonator. The structure was placed on an MgF_2−coated glass substrate. A Q-factor

Table 7.2 Comparison between reported configurations for refractive index sensing [8]

Reference	Configuration	Q-factor	Sensitivity (nm/RIU)	Range
[13]	Teflon coated MKR	31,000	30.5	1.3322–1.3412
[23]	MKR in Sagnac loop reflector	X	30.49	1.334–1.348
[17]	Polymer (PTT) MKR	11,000	46.5	1.35–1.38
–	–	–	95.5	1.39–1.41
[24]	Cascaded MKR	X	6523	1.3320–1.3350

of 11,000 was achieved for this device. The sensor was characterized under different mixtures of water and glycerol at different concentrations. Two different linear regimes can be obtained for the refractive index sensitivity: 46.5 nm/RIU for low concentration solutions (1.35–1.38) with a resolution 4.3×10^{-5} RIU, and 95.5 nm/RIU for high concentration solutions (1.39–1.41) with a resolution of 2.1×10^{-5} RIU.

A comparison between the different configurations discussed previously is summarized in Table 7.2.

7.3.3 Other Sensing Applications

Microfiber knot resonators can also be applied for sensing many other types of parameters in addition to temperature and refractive index. Examples of other applications will now be explored.

The first microfiber knot resonator accelerometer by optically interrogating the vibration of a microoptoelectromechanical system (MOEMS) was proposed by Wu et al. [25]. The microfiber knot resonator accelerometer used a 1.1 μm-diameter microfiber to create a 386 μm diameter knot. The structure achieved a Q-factor of 8500. The mode of operation of the sensor is very simple. The microfiber knot resonator is placed at the top surface of a MEMS cantilever and covered with 6 μm of low refractive index polymer (Efron PC-373). A schematic of the sensing structure is presented in Fig. 7.9. When the system is submitted to acceleration, the microfiber knot resonator will experience strain producing a shift in the resonant wavelength. Measuring the detected intensity with a photodetector, a sensitivities of 654 mV/g was achieved in a dynamic range of 20 g (gravitational force). Using a spectral method, like monitoring the wavelength shift, a sensitivity of 29 pm/g (at 856 nm) was obtained. The resolution of this system is 80 μg/$\sqrt{\text{Hz}}$ at 300 Hz.

In 2015, the first underwater acoustic sensor using a microfiber knot resonator was reported by Freitas et al. [9]. A 4 μm-diameter microfiber was used to produce an 800 μm-diameter microfiber knot resonator. The sensor was encapsulated with polymer by placing the microfiber knot resonator in a polytetrafluorethylene (PTFE) tray and embedded in silicone rubber. The structure achieved a Q-factor of 41,100, being the highest Q-factor obtained until that point for an encapsulated microfiber knot resonator. Acoustic wavefields will change the external pressure, causing vari-

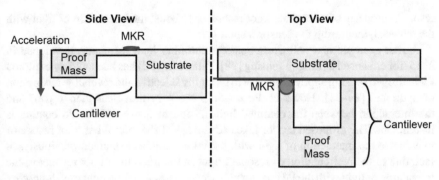

Fig. 7.9 Schematic of the MOEMS accelerometer based on microfiber knot resonator [25]

ations in the microfiber knot resonator optical path length, as well as changes in the fiber refractive index due to elasto-optic effects. The sensor achieved a normalized sensitivity, $(\delta\lambda/\delta p)/\lambda$, of 288 dB re/$\mu$Pa using acoustic frequencies from 25 to 300 Hz. Converting to different units, the sensitivity can also be expressed as 5.83×10^{-3} pm/Pa. However, to use such sensing structure a fast spectrometer is needed to pick the transmission spectrum as it changes in time.

Salinity sensing can also be performed with microfiber knot resonators, which is somehow similar to refractive index sensing. For this purpose Liao et al. demonstrated a microfiber knot resonator for monitoring sodium chloride (NaCl) concentration in solution [26]. Microfibers with diameters from 2.5 to 3.5 μm were used to create 855 μm-diameter microfiber knot resonators. A maximum sensitivity of 21.18 pm/% was obtained in the NaCl concentration range from 20.494 to 37.178%, at a probing wavelength of 1600 nm.

Microfiber knot resonators for relative humidity sensing were proposed in 2014 [27]. A 3 μm-diameter microfiber was used to produce a 150 μm-diameter microfiber knot resonator. The whole structure was coated with Nafion. Nafion is a perfluorosulfonated-based polymer which presents high hydrophilicity, chemical and thermal stability, high conductivity, high adherence to silica, and also mechanical toughness. Two regimes were observed, one for low humidity (30–45% relative humidity (RH)) obtaining a maximum sensitivity of (0.11 ± 0.02) nm/%, and other for higher-mid humidity (45–75% RH) where a maximum sensitivity of (0.29 ± 0.01) nm/% was achieved.

7.4 Recent Developments in Coated Microfiber Knot Resonators

Nowadays, the target of the researchers has been to incorporate other materials in microfiber knot resonators in order to obtain new sensing devices with enhanced sensitivity to the measured parameters. The main idea is to take advantage of the resonant property of microfiber knot resonator to boost the interaction of light with the material, increasing the sensor response.

A first example is a palladium-coated microfiber knot resonator, developed in 2015, for enhanced hydrogen sensing [28]. Palladium (Pd) has been widely explored for hydrogen sensing applications due to its highly selective and reversible absorption of hydrogen [29–31]. However, Pd coatings are very thin (less than 1 μm) and the interaction between the confined light in optical fibers and the Pd coating is insufficient. The proposed sensor takes advantage of the microfiber knot resonator to enhance the interaction of light with the Pd coating, as explained previously. In fact, due to the recirculation of resonant light in the microfiber knot resonator, the interaction of light with the Pd coating will accumulate each time light travels another turn in the ring. Such effect enhances the sensor sensitivity with the advantage of using just small interaction lengths and a thin Pd film (~13 nm-thick), compared with other existing optical fiber hydrogen sensors. For this sensor, the Pd coating was created using a plasma sputtering device in a vacuum chamber. The hydrogen concentration was monitored by measuring the wavelength shift and absorption at a resonant peak.

More recently, graphene oxide (GO) coated microfiber knot resonators were studied for gas sensing (carbon monoxide (CO) and ammonia (NH_3)) [32]. When gas molecules are absorbed on the surface of GO, the refractive index of GO is modified. Hence, the refractive index of the GO film will change according to the concentration of gas molecules. A 5.1 μm-diameter microfiber was used to perform a 1.85 mm-diameter microfiber knot resonator. The structure was placed over an MgF_2 substrate and covered with a GO sheet, as depicted in Fig. 7.10.

To produce the GO film, a droplet of a 100 mg/L GO solution was dropped onto the knot and heated up (~40 °C) until it was dried. The sensor achieved a Q−factor of 78,000 before the introduction of the GO sheet. Such value was reduced to 49,000 after depositing the GO film. The device was used to monitor CO and NH_3 concentrations. A CO sensitivity of ~0.17 pm/ppm and a NH_3 sensitivity of ~0.35 pm/ppm were achieved for concentrations of these molecules bellow 150 ppm.

Fig. 7.10 Schematic of a graphene oxide-coated microfiber knot resonator [32]

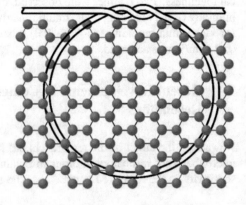

7.5   Other Microfiber Knot Resonator Configurations

Over the last years, in addition to the normal microfiber knot resonator structures, new ways of creating different microfiber knot resonators were also studied and developed. One example of those configurations is the reef knot microfiber resonator demonstrated by Vienne et al. [33]. The reef knot was produced using 2 microfibers: a 1.2 μm-diameter biconical microfiber and a 1.5 μm-diameter microfiber with a free end, as shown in Fig. 7.11. The structure is produced by bending the biconical microfiber in a "U" form and the free standing end of the second microfiber is guided through the first taper, also in a "U" shape, forming the reef knot. A non-circular reef knot was created with 340 μm-diameter in the short axis and 450 μm-diameter in the long axis. The device can be used as an add-drop filter. Q-factors of 10,000 and 3500 were obtained for the "through" port and the "drop" port, respectively.

A microfiber double-knot resonator with a Sagnac loop reflector was demonstrated in 2015 [34]. This double-knot device is different in structure and behavior compared to the one explored back in Sect. 7.3.1. In this case, only one microfiber is used to create two knots, being the second knot inside the first one (parallel configuration), as shown in Fig. 7.12. A Sagnac loop is then created in the end of the device so that light can experience the double-knot structure twice, enhancing the sensor response, and also returning from the input port allowing to monitor the sensor in a reflection configuration. The whole structure is placed in an MgF_2 substrate. The transmission spectrum of a double knot in parallel is not the overlap of the transmission spectrum of each knot independently, just as it happens with the double-knot in series [35].

Playing with different microfiber knot resonator configurations may lead in some cases to interesting results. For example Xu et al. demonstrated a small-size refrac-

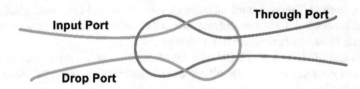

Fig. 7.11 Schematic of a reef knot microfiber resonator [33]

Fig. 7.12 Schematic of a microfiber double-knot resonator in parallel [34]

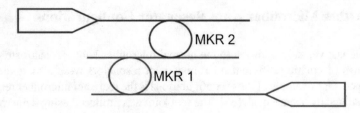

Fig. 7.13 Schematic of a cascaded microfiber resonator [24]

tometer for detecting very small refractive index variations based on cascaded microfiber knot resonators (CMKR) with Vernier effect [24]. The Vernier effect is generally used in calipers and barometers to increase the measurement accuracy by overlapping lines on two scales with different periods. Translating this method to microfiber knot resonators, the setup uses two microfiber knot resonators with millimeters of diameter (1.178 and 1.230 mm) assembled from 1.9 μm-diameter microfibers. An illustration of the configuration is presented in Fig. 7.13. To perform the experiment, the ambient refractive index around the first microfiber knot resonator was kept constant (1.3315), serving as a reference, while the ambient refractive index around the second microfiber knot resonator was slightly changed. A sensor sensitivity to the external refractive index change of 6523 nm/RIU was achieved. The configuration allows a resolution of 1.533×10^{-7} RIU because the measured wavelength shift is an absolute parameter dependent on the relative optical intensity variation. Hence, relative intensity noise in the light source, thermal noise and shot noise in the photo-detector of the spectrum analyzer do not affect the refractive index measurements.

Furthermore, embedding the first microfiber knot resonator in low refractive index polymer, such as Teflon, would increase its robustness and long-term stability, as explored in Sect. 7.2.3, but it would also ensure that the first microfiber knot resonator is immune to ambient refractive index changes.

In the end of 2016, a Mach-Zehnder interferometer with a microfiber knot resonator was demonstrated for simultaneous measurement of seawater temperature and salinity [36].

Two microfibers, one with 2.8 μm diameter and other with 3.3 μm diameter, were used to create the sensor. The thinner microfiber was used to create the microfiber knot resonator. To form the second arm of the interferometer, the thicker fiber was used creating the structure depicted in Fig. 7.14.

The transmission spectrum of the structure is the combined response of the Mach-Zehnder interferometer and the microfiber knot resonator. The spectral characteristics of the device were already theoretically studied [37, 38]. Each of these components can be monitored as a function of the measured parameters. Temperature sensitivities of -112.33 pm/°C and 13.96 pm/°C were achieved for the interference peak and the resonant peak respectively, from 13.7 °C to 25 °C. As for salinity, sensitivities of 208.63 pm/‰ and 16.21 pm/‰ were achieved for the interference peak and the

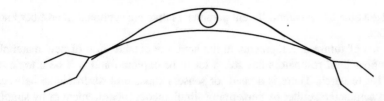

Fig. 7.14 Schematic of a Mach-Zehnder interferometer with a microfiber knot resonator [36]

Fig. 7.15 On the left—schematic of the sensor: milling of an air cavity in the taper region next to the microfiber knot resonator. On the right—focused ion beam image of the cavity [39]

resonant peak, respectively, between 25‰ and 37‰. With these values of sensitivity, a simple matrix method can be used to discriminate between temperature and salinity.

A similar sensor but with a completely different approach was presented in 2017 [39]. In this structure the Mach-Zehnder interferometer is created by focused ion beam milling a cavity on the microfiber next to the microfiber knot resonator, as shown in Fig. 7.15.

Part of light is able to go through the air cavity and the other through the silica microfiber. The phase difference between the two optical paths depends on the refractive index of the cavity. The authors reported a preliminary refractive index sensitivity of −8935 nm/RIU for the Mach-Zehnder interferometer between 1.333 and 1.341 RIU. The microfiber knot resonator can be used to allow discrimination between temperature and refractive index. Moreover, if the microfiber knot resonator is coated, becoming immune to ambient refractive index variations, it can be used as a reference to extract temperature effects from refractive index measurements.

Final Remark

From the creation of the first microfiber knot resonator until nowadays these structures have been studied and applied in different fields, particularly in sensing applications. The multiple characteristics addressed in this chapter show the great potential of microfiber knot resonator structures to be used as sensors, in some cases in place of conventional ones. Different techniques were explored to overcome fragility and stability problems, increasing therefore the sensors durability and its range of applications. It was also shown that microfiber knot resonators were object of several studies for application in temperature and refractive index sensing.

In the last couple of years, researchers have put their effort to develop new configurations of microfiber knot resonators. The incorporation of different materials in the structure allows to increase the scope of these sensors. As a result we have new alternative configurations that are getting more and more complex but, on the other hand, the new sensors present increased sensitivity to the measured parameters and are able to measure different parameters than conventional microfiber knot resonators.

In terms of future developments in the area, the combination of new materials of microfiber knot resonators has still a lot to be explored and will be a topic of interest for research. There is a need for sensors capable of discriminate between different parameters, either by performing simultaneous measurement or by simply correcting the influence of cross-sensitivity when measuring a single parameter. To match these needs, the combination of microfiber knot resonators with other well-known structures, such as interferometers, Bragg gratings, Fabry-Perot cavities, and others, can be a good solution.

References

1. J. Lou, Y. Wang, L. Tong, Microfiber optical sensors: a review. Sensors (Switzerland) **14**, 5823–5844 (2014)
2. G. Brambilla, F. Xu, P. Horak, Y. Jung, F. Koizumi, N.P. Sessions, E. Koukharenko, X. Feng, G.S. Murugan, J.S. Wilkinson, D.J. Richardson, Optical fiber nanowires and microwires: fabrication and applications. Adv. Opt. Photonics **1**, 107–161 (2009)
3. G. Chen, M. Ding, A review of microfiber and nanofiber based optical sensors. Open Opt. J. 32–57 (2013)
4. M. Sumetsky, Y. Dulashko, J.M. Fini, A. Hale, D.J. Digiovanni, The microfiber loop resonator: theory. Exp. Appl. **24**, 242–250 (2006)
5. L. Xiao, T.A. Birks, High finesse microfiber knot resonators made from double-ended tapered fibers. Opt. Lett. **36**, 1098–1100 (2011)
6. M. Sumetsky, Basic elements for microfiber photonics: micro/nanofibers and microfiber coil resonators. J. Light. Technol. **26**, 21–27 (2008)
7. Y. Xu, L. Ren, J. Liang, C. Ma, Y. Wang, X. Kong, X. Lin, Wideband slow light in microfiber double-knot resonator with a parallel structure. J. Appl. Phys. **118**, 73105 (2015)
8. A.D. Gomes, O. Frazão, Microfiber knot resonators as sensors: a review, in *Proceedings of the 5th International Conference on Photonics, Optics and Laser Technology (PHOTOPTICS 2017)* (2017), pp. 356–364
9. J.M. De Freitas, T.A. Birks, M. Rollings, Optical micro-knot resonator hydrophone. Opt. Express **23**, 5850–5860 (2015)
10. X. Jiang, L. Tong, G. Vienne, X. Guo, A. Tsao, Q. Yang, D. Yang, Demonstration of optical microfiber knot resonators. Appl. Phys. Lett. **88**, 2004–2007 (2006)
11. X.D. Jiang, Y. Chen, G. Vienne, L.M. Tong, All-fiber add-drop filters based on microfiber knot resonators. Opt. Lett. **32**, 1710–1712 (2007)
12. Z. Chen, V.K.S. Hsiao, X. Li, Z. Li, J. Yu, J. Zhang, Optically tunable microfiber-knot resonator. Opt. Express **19**, 14217 (2011)
13. X. Li, H. Ding, A stable evanescent field-based microfiber knot resonator refractive index sensor. IEEE Photonics Technol. Lett. **26**, 1625–1628 (2014)
14. Y. Wu, Y.-J. Rao, Y. Chen, Y. Gong, Miniature fiber-optic temperature sensors based on silica/polymer microfiber knot resonators. Opt. Express **17**, 18142–18147 (2009)
15. X. Zeng, Y. Wu, C. Hou, J. Bai, G. Yang, A temperature sensor based on optical microfiber knot resonator. Opt. Commun. **282**, 3817–3819 (2009)
16. Y. Wu, T. Zhang, Y. Rao, Y. Gong, Miniature interferometric humidity sensors based on silica/polymer microfiber knot resonators. Sens. Actuators B Chem. **155**, 258–263 (2011)
17. H. Yu, L. Xiong, Z. Chen, Q. Li, X. Yi, Y. Ding, F. Wang, H. Lv, Y. Ding, Solution concentration and refractive index sensing based on polymer microfiber knot resonator. Appl. Phys. Express **7**, 3–7 (2014)

18. G. Vienne, Y. Li, L. Tong, Effect of host polymer on microfiber resonator. IEEE Photonics Technol. Lett. **19**, 1386–1388 (2007)
19. A.D. Gomes, O. Frazão, Mach – Zehnder based on large knot fiber resonator for refractive index measurement. IEEE Photonics Technol. Lett. **28**, 1279–1281 (2016)
20. H. Yang, S. Wang, X. Liao, J. Wang, X. Wang, Y. Liao, J. Wang, Temperature sensing in seawater based on microfiber knot resonator. Sensors **14**, 18515–18525 (2014)
21. Y. Wu, L. Jia, T. Zhang, Y. Rao, Y. Gong, Microscopic multi-point temperature sensing based on microfiber double-knot resonators. Opt. Commun. **285**, 2218–2222 (2012)
22. H. Yang, J. Wang, Y. Liao, S. Wang, X. Wang, Dual-point seawater temperature simultaneous sensing based on microfiber double knot resonators. IEEE Sens. J. **17**, 2398–2403 (2017)
23. K.-S.S. Lim, I. Aryanfar, W.-Y.Y. Chong, Y.-K.K. Cheong, S.W. Harun, H. Ahmad, Integrated microfibre device for refractive index and temperature sensing. Sensors (Basel) **12**, 11782–11789 (2012)
24. Z. Xu, Q. Sun, B. Li, Y. Luo, W. Lu, D. Liu, P.P. Shum, L. Zhan, Highly sensitive refractive index sensor based on two cascaded microfiber knots with vernier effect. Opt. Express **23**, 6662–6672 (2015)
25. Y. Wu, X. Zeng, Y.J. Rao, Y. Gong, C.L. Hou, G.G. Yang, MOEMS accelerometer based on microfiber knot resonator. IEEE Photonics Technol. Lett. **21**, 1547–1549 (2009)
26. Y. Liao, J. Wang, H. Yang, X. Wang, S. Wang, Salinity sensing based on microfiber knot resonator. Sens. Actuators A Phys. **233**, 22–25 (2015)
27. M.A. Gouveia, P.E.S. Pellegrini, J.S. dos Santos, I.M. Raimundo, C.M.B. Cordeiro, Analysis of immersed silica optical microfiber knot resonator and its application as a moisture sensor. Appl. Opt. **53**, 7454–7461 (2014)
28. X. Wu, F. Gu, H. Zeng, Palladium-coated silica microfiber knots for enhanced hydrogen sensing. IEEE Photonics Technol. Lett. **27**, 1228–1231 (2015)
29. T. Hübert, L. Boon-Brett, G. Black, U. Banach, T. Hübert, L. Boon-Brett, G. Black, U. Banach, Hydrogen sensors: a review. Sens. Actuators B Chem. **157**, 329–352 (2011)
30. S.F. Silva, L. Coelho, O. Frazão, J.L. Santos, F.X. Malcata, A review of palladium-based fiber-optic sensors for molecular hydrogen detection. IEEE Sens. J. **12**, 93–102 (2012)
31. C. Wadell, S. Syrenova, C. Langhammer, Plasmonic hydrogen sensing with nanostructured metal hydrides. ACS Nano **8**, 11925–11940 (2014)
32. C.-B. Yu, Y. Wu, X.-L. Liu, B.-C. Yao, F. Fu, Y. Gong, Y.-J. Rao, Y.-F. Chen, Graphene oxide deposited microfiber knot resonator for gas sensing. Opt. Mater. Express **6**, 727–733 (2016)
33. G. Vienne, A. Coillet, P. Grelu, M. El Amraoui, J.-C. Jules, F. Smektala, L. Tong, Demonstration of a reef knot microfiber resonator. Opt. Express **17**, 6224–6229 (2009)
34. Y. Xu, L. Ren, Y. Wang, X. Kong, J. Liang, K. Ren, X. Lin, Enhanced slow light in microfiber double-knot resonator with a Sagnac loop reflector. Opt. Commun. **350**, 148–153 (2015)
35. Y. Xu, L. Ren, J. Liang, C. Ma, Y. Wang, N. Chen, E. Qu, A simple, polymer-microfiber-assisted approach to fabricating the silica microfiber knot resonator. Opt. Commun. **321**, 157–161 (2014)
36. Y. Liao, J. Wang, S. Wang, H. Yang, X. Wang, Simultaneous measurement of seawater temperature and salinity based on microfiber MZ interferometer with a knot resonator. J. Light. Technol. **34**, 5378–5384 (2016)
37. Y. Liao, J. Wang, S. Wang, H. Yang, X. Wang, Spectral characteristics of the microfiber MZ interferometer with a knot resonator. Opt. Commun. **389**, 253–257 (2017)
38. Y.H. Chen, Y. Wu, Y.J. Rao, Q. Deng, Y. Gong, Hybrid Mach-Zehnder interferometer and knot resonator based on silica microfibers. Opt. Commun. **283**, 2953–2956 (2010)
39. A.D. Gomes, R.M. André, S.C. Warren-Smith, J. Dellith, M. Becker, M. Rothhardt, O. Frazão, Combined microfiber knot resonator and focused ion beam-milled Mach-Zehnder interferometer for refractive index measurement. Presented at the (2017)

Chapter 8
High Power Continuous-Wave Er-doped Fiber Lasers

Leonid V. Kotov and Mikhail E. Likhachev

Abstract Recent years, high power fiber lasers at the spectral region near 1550 nm attract a lot of attention as promising sources for numerous civil and military applications. There are four main type of lasers allowing to generate high power radiation at this spectral range: Er/Yb-co doped fiber lasers, Yb-free Er-doped fiber lasers cladding-pumped at 980 nm, Er-doped fiber lasers core-pumped by Raman lasers at 1480 nm, and Er-doped fiber laser in-band pumped at 1532 nm. In this chapter we review current states, limiting factors and future perspectives of all these approaches.

8.1 Introduction

An emission band of erbium ions near 1.5 μm which coincides with the silica fibers minimum loss spectral range made Er-doped fibers (EDFs) indispensable for the telecom industry. These fibers were actively studied during 1990s in order to develop efficient high-gain optical amplifiers. The output power of such amplifiers required by telecom applications does not exceed 1 W. At the same time, there is large variety of applications with a great demand of the high power (1 W–1 kW) laser sources operating at 1.5 μm wavelength. There are three main reasons for that. First, this spectral range falls into the transparency window of the atmosphere, so a laser beam can be transmitted over a long distance with low loss. Second, a radiation at wavelengths longer than ~1400 nm is not focused by an eye lens on a retina. For this reason, 1.5 μm lasers are concerned to be so-called "eye-safe" type of light sources. Thus, for many applications high power Er-doped fiber lasers are much more preferable than e.g. Yb-doped counterparts operating near 1 μm due to safety reasons. Third,

L. V. Kotov (✉)
College of Optical Sciences, University of Arizona, 1630 E. University Blvd., Tucson, AZ 85721, USA
e-mail: alterlk@yandex.ru

L. V. Kotov · M. E. Likhachev
Fiber Optics Research Center of the Russian Academy of Sciences, 38 Vavilova str., Moscow 119333, Russia

© Springer Nature Switzerland AG 2019
P. Ribeiro et al. (eds.), *Optics, Photonics and Laser Technology 2017*,
Springer Series in Optical Sciences 222,
https://doi.org/10.1007/978-3-030-12692-6_8

the active development of the optical communication industry at 1.55 μm during last decades resulted in the large amount of available high-quality low-cost optical components (isolators, wavelength division multiplexers (WDMs), circulators, couplers etc) operating at this wavelength range. This frequently makes the Er-doped fiber lasers (EDFLs) more attractive than Yb- and especially Tm-fiber sources from the commercial point of view. Therefore, high-power EDFLs are quite interesting for such applications as light detection and ranging, power beaming, illumination, free-space optical communications, remote sensing and bio-medical applications.

Despite the great demand of the high-power Er-doped fiber lasers for civil and military applications, demonstrated output power of such lasers are lower than that from any other commonly used rare-Earth-doped silica-based fiber laser. Thus, to date, 10 kW [1], 1 kW [2], and 400 W [3] power levels were demonstrated from Yb-, Tm- and Ho-doped fiber lasers respectively. At the same time, the highest power from a fiber laser at 1.55 μm is only 300 W [4]. The reasons for that are low Er^{3+} ions absorption and emission cross-sections and concentration effects caused by erbium ions clustering.

In this chapter we consider the main factors limiting manufacturing of the high power fiber lasers near 1.55 μm, present state of the art solutions, and discuss prospects of the power scaling of such systems.

8.2 Properties of Erbium Ions in Silica Glass

8.2.1 Spectroscopic Properties

There are several absorption bands of the Er-doped silica glass in the visible and near IR spectral ranges. Figure 8.1 demonstrates energy diagram of erbium ions in silica glass [5]. Transitions between the ground state $4I_{15/2}$ and higher-lying energy levels might be observed by absorption measurements. Figure 8.2 represents typical absorption spectrum of the Er-doped silica fiber measured by a cut-back technique [5].

The energy level $4I_{13/2}$ has a rather long lifetime of ~10 ms (it slightly varies with the glass composition). Electron transition from this level to $4I_{15/2}$ provides lasing in ~1530–1610 nm spectral range. Due to a large energy gap between $4I_{13/2}$ and the ground state a population of this level is almost equal to zero when not pumped. Population inversion might be achieved by an optical pumping into any upper energy levels (then ions quickly, mostly nonradiatively relax to the $4I_{13/2}$ level) or directly into short-wave edge of the absorption band $4I_{13/2}$. As the lower laser level is the ground state rather high population inversion (~50% depending on wavelength) is required to have a positive gain of Er-doped fiber.

Absorption and emission spectra of EDFs as well as the lifetime slightly depend on the glass host. This is a result of different positions of Stark levels, intensities of transitions between them and their homogeneous and inhomogeneous broadening

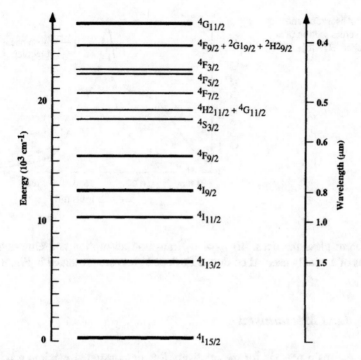

Fig. 8.1 Energy diagram of erbium ions in silica glass [5]

Fig. 8.2 Experimentally measured loss spectrum of an Er-doped silica glass. The absorption at 400–600 nm region was divided by the factor of 10; the small oscillatory structure near 1100 nm corresponds to the cutoff of the second-order mode of the fiber [5]

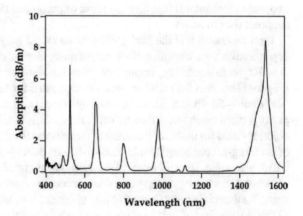

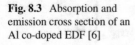

Fig. 8.3 Absorption and emission cross section of an Al co-doped EDF [6]

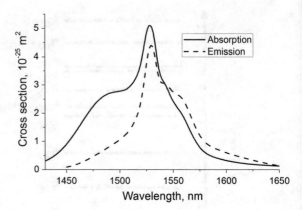

in different glass matrices. $4I_{13/2}$–$4I_{15/2}$ transition absorption and emission cross sections of a widely used Al co-doped Er-doped glass are presented in Fig. 8.3 [6].

8.2.2 Loss Mechanisms

One of the main reasons for the relatively low demonstrated maximum power of Er-doped fiber lasers is their low pump-to-signal conversion efficiencies. There are two main mechanisms that results in loss of pump and signal powers and, therefore, suppress the efficiency.

First mechanism is the background or so-called grey loss. This loss are weakly dependent on a wavelength and are sum of many factors: electron and photons absorption, Rayleigh scattering, impurities, glass non-uniformities, stresses etc Although grey loss less than 0.17 dB/km were demonstrated for telecom fibers typical loss of EDFs are 1–50 dB/km. This is caused by erbium and co-dopants precursors whose purity is often much worse than that of commercially available germanium and silicon chlorides used for the manufacturing of the telecom fibers. It should be noted, that loss of cladding-propagating light in state-of-the-art double-clad fibers is ~10–50 dB/km, which also affect the efficiency of the cladding-pumped lasers.

Second mechanism is up-conversion processes caused by a clustering of erbium ions. The thing is in rather low solubility of erbium ions in silica glass. It was shown [7, 8] that distribution of Er^{3+} in silica glass is not fully uniform and a noticeable amount of ions are much closer to each other than it can be expected from the statistics. In other words, erbium ions form clusters in the glass matrix. Short distance between ions within a cluster leads to a fast ion-ion energy exchange and cause an effect called up-conversion illustrated below.

Let's consider cluster that consists from only two erbium ions. When both of ions are excited one of them transfers its energy to another and goes into a ground state. The other one simultaneously will go to the upper $4I_{9/2}$ state and then fast

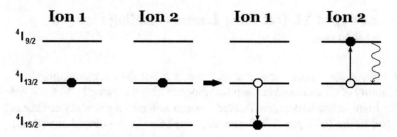

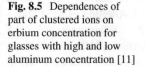

Fig. 8.4 Illustration of the up-conversion process

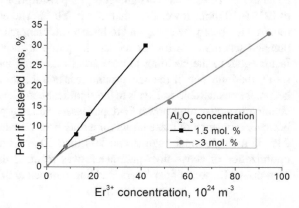

Fig. 8.5 Dependences of part of clustered ions on erbium concentration for glasses with high and low aluminum concentration [11]

non-radiatively relaxes back to the $4I_{13/2}$ state (Fig. 8.4). Thus, up-conversion results in the loss of a signal or pump photon. For clusters of bigger size the up-conversion process proceeds in a similar manner.

Co-doping of the silica glass with aluminum or phosphorus oxides allows significantly decrease a number of clustered ions of rare earth elements [9]. For Er-doped silica glass it was shown that namely aluminum co-doped fibers demonstrates the best performances [10]. Thus, to date, Al_2O_3 is the most usable co-dopant for Er-doped fibers. Figure 8.5 shows dependences of part of clustered ions for fibers with low (<1.5 mol%) and high (>3 mol%) concentrations of the aluminum oxide [11]. It can be seen, that an addition of aluminum allows significantly decrease a number of the erbium clusters in the silica glass. However, it should be noted, that the clustering suppression saturates with the aluminum concentration at doping level near 3 mol%. Further increase of the aluminum content does not affect the clustered ions number [10].

Thus, up-conversion might be considered as a source of unbleachable loss at signal and pump wavelengths. The value of this loss depends on the erbium concentration, the population inversion and the glass composition.

8.3 Er-doped Yb-free Fiber Lasers Cladding Pumped at 980 nm

The erbium absorption band near 980 nm is often used for pumping due to the availability of efficient and cheap pump diodes at this wavelength. Worth to note that rather high—close to quantum limited—pump conversion efficiency could easily be reached in the Er-doped fiber lasers and amplifiers core-pumped at 980 nm [12]. However, the development of efficient cladding-pumped EDF lasers is much more challenging. The reason is in rather low pump absorption and emission cross-sections of Er^{3+} (~10 times lower than that of e.g. Yb^{3+}). Thereby a relatively long fiber is required to absorb the pump and to have a sufficient gain. Exploiting of a cladding-pumped scheme even further reduce the pump absorption (approximately by the factor equal to the cladding to core areas ratio) and increase a fiber length. As a result, the influence of the background loss and the loss associated with clustering becomes considerable and leads to the significant efficiency drop. There were several attempts to realize high-power Er-doped lasers cladding-pumped fiber lasers [13–17], however efficiency of these lasers for a long time remained relatively low: less than 24% in a single-mode regime and less than 30% for multimode lasers. Accurate optimization of active fiber parameters was required in order to develop efficient high-power Er-doped fiber lasers cladding-pumped at 980 nm.

8.3.1 Optimization of a Double-Clad Er-doped Fiber Design

In 2013 an influence of double-clad Er-doped fibers parameters on their efficiencies was studied through a numerical modeling [11, 18]. It can be seen from Fig. 8.5 that fibers with high Al_2O_3 concentration exhibit a better Er^{3+} ions solubility. However, it worth to note that the high aluminum refractivity leads to a high core-to-clad index difference Δn, limiting the core diameter to 6–10 μm for a single mode regime. A simulation result for amplifiers based on the double-clad EDFs with high Al_2O_3 concentration and core diameter of 10 μm operating at 1550 nm are shown in Fig. 8.6 (curve (1)). It can be seen that the efficiency of the amplifiers based on this type of fibers does not exceed 5%. Reduction of the aluminum content in the core reduces Δn and allows increasing the core diameter. There are commercially available EDFs with 20 μm core ($\Delta n \sim 0.002$) which are operated singlemode by properly bending the fiber [15, 19]. Despite a poor solubility of erbium in such a glass (see Fig. 8.5), increasing the core-to-clad ratio enables the higher clad absorption and therefore the shorter active fiber. This reduces the influence of the up-conversion and grey loss per amplifier length. Calculations show significant increase of the slope efficiency for the 20 μm core fiber design (see curve (2) in Fig. 8.6). The maximum slope efficiency of 16% might be achieved with such fibers.

In order to clearer demonstrate the influence of the background loss and the loss caused by the clustering calculations taking into account each effect separately were

Fig. 8.6 Calculated slope efficiencies of cladding pumped at 980 nm Er-doped fiber amplifiers with various Al_2O_3 concentration, core diameter and operating wavelength (clad diameter is 125 μm). 1: >3 mol%, 10 μm, 1550 nm; 2: >1.5 mol%, 20 μm, 1550 nm; 3: 1.5 mol%, 20 μm, 1585 nm; 4: 1.5 mol%, 35 μm, 1585 nm [11, 18]

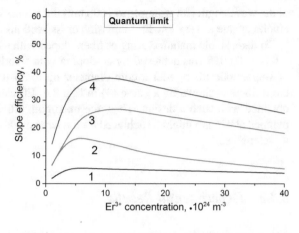

Fig. 8.7 Calculated slope efficiency of 20/125 μm EDF with taking into account only clustering (dotted), only background loss (dashed), and both (solid)

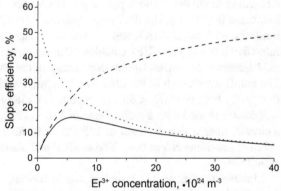

performed in [18]. The simulation was made for the same fiber parameters as for the curve (2) in Fig. 8.6. Results are presented in Fig. 8.7. As follows from it the up-conversion causes a significant degradation of the efficiency in heavily doped fibers. It is caused by the growth of the number of Er-ion pairs with erbium concentration. On the contrary, for fibers with low Er^{3+} content, their length reaches hundreds of meters and the efficiency decreases due to the high net value of background loss. Thereby, there is an optimal concentration of Er^{3+} where sum of clustering and background loss is minimized.

At longer wavelengths Er^{3+} ions and therefore clusters have a lower absorption cross-section. This decreases the number of ions in the excited state and therefore suppresses the up-conversion. So, growth of the efficiency can be obtained by operating at longer wavelengths. The curve (3) in Fig. 8.6 shows the efficiency for the same fibers as in (2) however operated at 1585 nm. The highest efficiency in this case is ~27% at the Er^{3+} content ~10^{25} m^{-3}. This result is in a good agreement with [15] where 24% slope efficiency was obtained in an EDF with core and cladding diameters of 20 and 125 μm and slightly higher erbium concentration. Further increase of the wavelength (for example to 1600 nm) did not show more improvement of the efficiency due to very low Er^{3+} emission cross-sections at long wavelengths.

To date, stable manufacturing of fibers doped with ~1.5 mol% of Al_2O_3 and Δn ~ 0.001–0.0015 was achieved by co-doping with fluorine enabling the fabrication of singlemode fibers with a core diameter up to 35 μm [11]. The efficiency for these fibers is shown by a curve (4) in Fig. 8.6. Efficiency as high as 40% can be obtained with such a design. Thus, the maximum efficiency of an EDF cladding-pumped at 976 nm might be achieved for the large mode area fibers operating at long wavelengths.

8.3.2 Experimental Results

According to the calculations presented above a fiber with optimal parameters was fabricated in [11, 18]. The fiber core contained ~0.017 mol% Er_2O_3 (~7 · 10^{24} Er^{3+} ions/m^3), ~1.5 mol% Al_2O_3 and ~1 wt% of F. The core diameter was 34 μm. The refractive index profile (RIP), calculated fundamental mode field distribution (mode field diameter: 24 μm) and microscope image of the fiber facet are shown in Fig. 8.8. The cutoff wavelength of the fiber was 1.7 μm, ensuring the singlemode operation in a gently bent fiber (R_b < 35 cm) operated at 1585 nm. The cladding of the fiber was square-shaped with a surface area of 110 $\times$ 110 μm^2 (corresponding to that of a circular fiber with a diameter of 125 μm). The fiber was coated with a low index polymer providing NA = 0.46. The small-signal cladding absorption was 0.6 dB/m at 980 nm.

Manufactured fiber was tested as a gain medium in high power amplifier and oscillator laser schemes. Sketches of experimental setups are presented in Fig. 8.9. In both cases a pump radiation from pigtailed 980 nm multimode diodes (105/125 μm, 0.22 NA) was launched into the active fiber through a commercially available 6+1 to 1 pump combiner. A fiber laser with output power of 4 W at 1585 nm was used as a seed source for the amplifier. A cavity of the oscillator was formed by highly reflective fiber Bragg grating (FBG) and a straight cleaved output fiber facet providing 4% Fresnel reflection. The FBG was written in a passive fiber with 20/125 μm core/clad diameters (NA = 0.08/0.46). Output end of the amplifier was angle-cleaved. In both schemes a high-power cladding light stripper similar to that described in [20] was made near the output end of the active fiber. The combined available pump power was 275 W and 190 W in amplifier and oscillator experiments respectively. Dependences of output powers on launched signal powers for the amplifier and the oscillator are presented in Fig. 8.10. In both cases maximum output powers were limited only by available pump powers.

Maximum output powers of 100 and 75 W, and slope efficiencies of 37 and 40% were achieved. The output spectrum didn't show any amplified spontaneous emission (ASE) over the emission band and the beam quality was close to the diffraction limit as presented in the insets of Fig. 8.10. Taking into account high electrical to optical

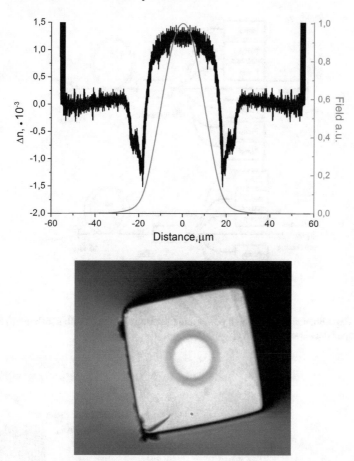

Fig. 8.8 Top: Refractive index profile and calculated mode field distribution in the manufactured fiber (top) [11]; Bottom: and microscope image of the fiber facet [18]

(E–O) efficiency of 980 nm diodes (~50%), the E–O efficiency of the developed lasers was ~20%.

To date, the output power of 100 W and the slope efficiency of 40% are the highest demonstrated power and efficiency for the cladding-pumped at 980 nm Yb-free Er-doped fiber lasers. Concerning power scaling of such type of lasers, to date, 980 nm pump diode with 105/125 output fiber (0.22 NA) and power of 100 W are commercially available now. Thus, using a standard 7 to 1 pump combiner one can build 280 W Er-doped all-fiber laser. Moreover, recently 270 W diodes were reported as experimental samples [21]. Therefore power scaling of such Er-doped lasers to the ~850 W level might be performed in the near future.

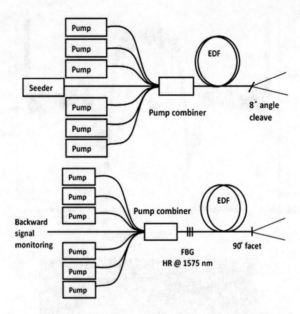

Fig. 8.9 Experimental set up of high power amplifier (top) [18] and oscillator (bottom) based on the developed double-clad EDF [11]

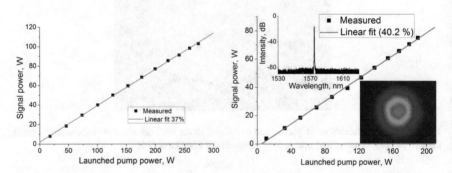

Fig. 8.10 Output power of the amplifier (left) [18] and oscillator (right). Insets: output spectrum of the oscillator and its beam profile [11]

8.4 Er-Yb Co-doped Fiber Lasers

8.4.1 Yb Sensitization of Er-doped Fibers

As it was shown above, the low value of Er^{3+} ions cross-sections results in the increase of the active fiber length and corresponding raise of net loss in laser/amplifier. One solution to this problem is to sensitize fiber by co-doping the EDF with ytterbium. The Yb^{3+} then absorbs most of the pump light and cross relaxation between adjacent

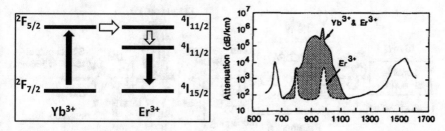

Fig. 8.11 Left: The energy levels for the Er-Yb co-doped system. Right: Attenuation spectrum of a co-doped fiber with Er:Yb concentration ratio 1:30

ions of Er^{3+} and Yb^{3+} allows the absorbed energy to be transferred to the Er^{3+}, thus providing another route for pumping the laser (see Fig. 8.11).

It should be noted that the absorption cross-section of ytterbium is order of magnitude higher than that of erbium. In addition, Yb^{3+} ions have only one energy transition in the near infrared region, thus clustering of ytterbium ions does not lead to up-conversion processes. Therefore, much higher ytterbium concentrations (up to several wt%) might be used comparing to that of erbium (~0.1 wt%) without impact on the efficiency. As a result, co-doping of an Er-doped fiber with Yb^{3+} allows increasing pump absorption by several orders of magnitude (see Fig. 8.11). It is also important that Yb co-doping increase solubility of the Er ions [22], which also allows one to increase pump-to-signal conversion efficiency.

Several groups successfully demonstrated low power (<1 W) Er-Yb lasers in the early 1990s [23, 24]. However, when more powerful pump diodes became available it turned out that power scaling of Er-Yb lasers to levels higher than several Watts is quite challenging. The main limiting factor for that is a parasitic emission (ASE or even lasing) of ytterbium ions near 1 μm. It results in an efficiency rollover at high powers and a self-pulsing near 1 μm causing the catastrophic damage of the fiber and optical components.

Reasons for the parasitic lasing are the back energy transfer from Er to Yb ions and the fact that there are some amount of isolated Yb^{3+} ions (i.e. not coupled with Er ions) that might emit light near 1 μm. Thus, careful core composition optimization is required to increase a threshold of the parasitic emission. Particularly, heavily phosphorous doping of the active core is usually used. The addition of phosphorous to the silica glass increase the phonon energy and ensure rapid multiphonon decay from the $4I_{11/2}$ level of Er^{3+} so minimizing back transfer of energy to the donor Yb^{3+} ion. Another way to avoid self pulsing near 1 μm is simultaneous operation at 1 and 1.55 μm. This might be realized by building additional cavity or seeding at 1 μm [25, 26].

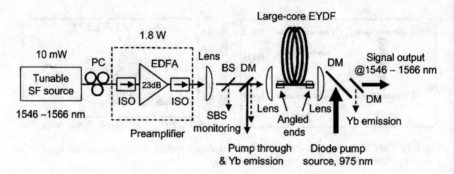

Fig. 8.12 MOPA setup from [29]

8.4.2 High-Power Er-Yb Lasers

To date, the maximum power demonstrated with standard singlemode fibers with 5–10 μm core diameters is only ~10 W [24–27]. This level is limited by parasitic ASE and self-pulsing near 1 μm.

The best results in power scaling of Er-Yb fiber lasers were achieved by the group from Optoelectronic Research Center. The authors proposed exploiting of large-mode area active fibers. Increase of the core diameter allows reduction of the power density and, therefore, increasing the Yb lasing threshold. At the same time, the requirement of a high phosphorous concentration which increase refractive index of the glass leads to the fiber multimodeness.

In [29] an Er-Yb fiber with core/cladding diameters of 30/650 μm and the core NA of 0.2/0.45 was developed. The Er-Yb co-doped core provided a core absorption of 67 dB/m at 1535 nm (from the Er doping) and an inner cladding absorption for the pump light at 975 nm of 1.4 dB/m (from the Yb doping). The fiber was tested in a MOPA scheme presented in Fig. 8.12. The fiber was end-pumped from the signal output end by a free-space coupled diode stack at 975 nm with up to 470 W of launched pump power. Both fiber ends were angle cleaved to eliminate signal feedback. The signal from the preamplifier was launched via a free-space coupling arrangement, through a beam splitter and a dichroic mirror that were inserted to monitor the backscattered signal and to separate the residual pump and any spurious Yb emission at 1 μm from the signal beam path.

The dependences of the signal power near 1.55 μm and the spurious lasing at 1 μm, and their spectra are presented in Figs. 8.13 and 8.14.

The highest output power of 151 W at 1563 nm was achieved at a launched pump power of 457 W. The slope efficiency of 35% at low powers dropped to 29% at higher powers because of the onset of Yb emission. The maximum power of this emission exceeded 70 W at the maximum pump power. Despite the high value of V factor ($V = 12$) singlemode operation was achieved through the fiber bending and beam quality parameter $M^2 = 1.1$ was measured.

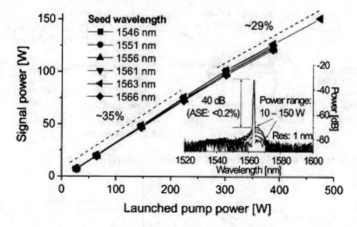

Fig. 8.13 Signal power versus final-stage pump power. Inset: output spectra at 1563 nm [29]

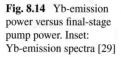

Fig. 8.14 Yb-emission power versus final-stage pump power. Inset: Yb-emission spectra [29]

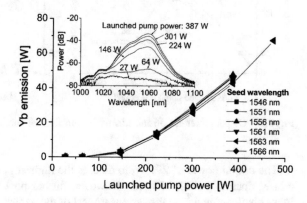

In their following work the authors built a high-power oscillator schematically presented in Fig. 8.15 [4]. The fiber used in the experiment had a 30 μm core and a 600 μm cladding with NAs of 0.21 and 0.48. The absorption for light in the inner cladding from the Er and Yb ions are 0.11 dB/m at 1535 nm and 3.8 dB/m at 976 nm. The EYDF used for the laser is 6-m long, and both ends were cleaved perpendicularly to the fiber axis. The laser cavity was formed between the bare fiber facet at the pump launch end of the fiber and a dichroic mirror butt-coupled to the fiber at the rear end. The dichroic mirror had high reflection at 1.5–1.6 μm and high transmission for the pump and Yb emission wavelengths at ~1.07 μm.

Figure 8.16 shows the laser performance at 1.56 μm (left) and 1.07 μm (right). As well as in the MOPA setup discussed above a significant rollover of efficiency (from 40 to 19%) associated with an Yb parasitic emission can be observed. 297 W of power at 1.56 μm and total 340 W of power at 1 μm were measured at the maximum pump power. The beam quality parameter for the output at 1.56 and 1.07 μm were 3.9 and 12.

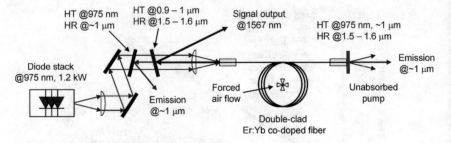

Fig. 8.15 Schematic of the Er:Yb codoped fiber laser system [4]

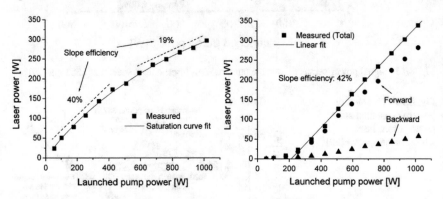

Fig. 8.16 Output power at 1.56 and 1.07 μm as functions of pump power [4]

The output power of 297 W, to date, is the highest power ever demonstrated for a laser operating near 1.55 μm. However, further power scaling of such systems looks challenging due to the serious impact of the parasitic Yb emission on the laser performance. The requirement of the 1 μm lasing control prevents building of all-fiber system and makes Er-Yb lasers rather complex and impractical. Significant improvement of a fiber design is required to overcome this issue.

8.5 Er-doped Fiber Lasers Cladding-Pumped at 1530 nm

In recent years in-band (or resonant) pumping has been the focus of attention as a promising approach for building high average power fiber lasers. For these lasers low quantum defect has allowed the demonstration of much higher optical-to-optical (O–O) efficiency and thus lower heat loads compare to their off-band pumped counterparts. For Er-doped fibers absorption peak near 1530 nm is often used for in-band pumping.

8.5.1   Pump Sources Near 1530 nm

There are commercially available InGaAsP/InP diode systems operating near 1535 nm that can be used for Er-doped fibers pumping. At the same time, to date, their parameters are much inferior to that of well-developed diodes at 915/980 nm. Available power of these diodes typically doesn't exceed ~30 W (for 105/125 μm fiber), and efficiency is ~25%. To compare, 980 nm diodes available power and efficiency are ~100 W and 50%. In addition, a price of 1530 nm diodes is rather high. Another problem of diode pump sources is a thermal drift of the central wavelength. The point is in the rather small (several nanometers) spectral width of the erbium absorption peak near 1530 nm. Therefore, a pump wavelength variation with power or temperature can lead to a decrease of the pump absorption and, thus, efficiency reduction. Moreover, it is known that even diodes with wavelengths locked by a volume Bragg grating frequently generate significant amounts of out-of-band radiation when operated near the maximum current and/or at high heat sink temperatures. The highest demonstrated to date pump power from diode sources near 1535 nm in 125 μm fiber is 170 W [19]. It was obtained by combining of 6 multimode diodes through a commercially available 6+1 to 1 pump combiner.

In order to overcome disadvantages of diode sources fiber pump lasers were developed by several groups [28, 30–32]. For instance, in [30] single-mode 1535 nm an Er-Yb laser with 20 W of output power was used to core- and clad-pump Er-doped fiber. At the same time, as it was shown in Sect. 8.4 power scaling of Er-Yb lasers to 100 W and above power level is quite challenging due to the parasitic Yb lasing. To avoid this limitation Jebali et al. [28] proposed to combine several single-mode Er-Yb laser. The setup of the laser is schematically shown in Fig. 8.17. A radiation of 36 fiber lasers at 1535 nm was coupled into the cladding of an active double-clad fiber through specialty-developed combiner. Each Er-Yb pump laser was, in turn, pumped by 976 nm fiber coupled multimode diodes (see dashed box in Fig. 8.17). A commercially available (CoractiveDCF-EY-10/128) Er-Yb fiber was used as an active medium for these lasers. The fiber had relatively high Yb concentration that caused high heat load. As a result, the output power of each pump laser was thermal limited at level of ~10 W. The average efficiency of pump lasers was 35.6%.

The outputs of these pump lasers were combined using a custom-made pump combiner that merges 37 standard single-mode fibers inputs (SMF28) into one core-less 125 μm fiber with a NA of 0.45. The manufacturing technique of the 37 $\rightarrow$ 1 fused fiber bundle combiner is described in [33] and [34]. The pump combiner losses was evaluated to be ~0.3 dB. As a result, ~370 W of multimode pump radiation at 1535 nm in 125 μm fiber was generated.

Another approach to build a high power pump laser at 1535 nm was demonstrated in [31]. It was proposed to use multimode Yb-free Er-doped fibers cladding-pumped at 980 nm. As it was shown in Sect. 8.3.1, increase of the core diameter of cladding-pumped Er-doped fibers results in significant improvement of their efficiency. It worth to note, that there is no need for a high beam quality of the 1535 nm pump source for cladding-pumped lasers. The parameters of the EDF core are limited only

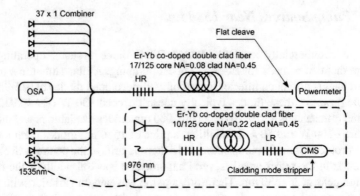

Fig. 8.17 Experimental setup from [28]

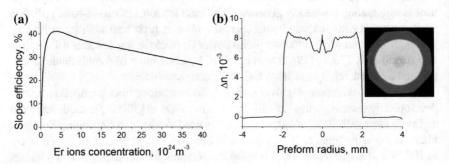

Fig. 8.18 **a** Computed slope efficiency of the amplifier at 1535 nm. **b** Refractive index profile and of the Er-doped preform and microscope image of the fiber facet 32

by the requirement to match them to the pump ports of standard pump combiners: a core/cladding diameter of 105/125 μm and a numerical aperture of 0.15. Therefore, the active fiber core could be multimode and has diameter up to 105 μm.

A numerical analysis similar to that described in Sect. 8.3.1 was performed in [31, 32] to define the optimum erbium concentration for a multimode Er-doped fiber. The slope efficiency of a co-pumped power amplifier based on 100/125 μm double-clad Er-doped fiber operating at 1535 nm was computed (Fig. 8.18a). It can be seen that the efficiency as high as 40% could be achieved.

Based on the simulation results, a fiber preform with optimum parameters was produced [31, 32]. The refractive index profile of the preform core is presented in Fig. 8.18b. The preform was polished to an octagonal shape and a double clad fiber was drawn down from it. The resulting fiber had core/cladding diameters of 95/125 μm and was coated with a polymer providing a pump NA of 0.46. A microscope image of the fiber facet is presented in Fig. 8.18b.

The manufactured fiber was tested in an original laser scheme described below. It should be noted that, in addition to a high E–O efficiency, there are other important demands for the pump source: compactness, high long-term stability, small size and

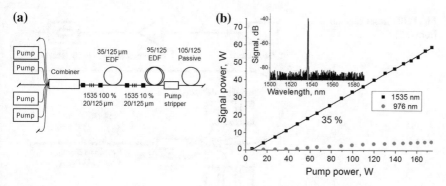

Fig. 8.19 **a** Scheme of the multimode laser. **b** Output power of the multimode laser. Inset-spectrum at full power [32]

low cost. The simplest and therefore the cheapest scheme for a fiber laser consists of a gain fiber spliced between two FBGs. However, it is known that different modes have different reflection spectra from an FBG written in a multimode fiber. As a result, the spectrum of such a multimode laser has several peaks [13], leading to the effective spectral broadening. In addition, this effect could result in unstable operation because of the mode competition. To ensure a narrow spectrum of a multimode laser, a MOPA scheme could be used. However, the requirement of a seed laser that is powerful enough to saturate the amplifier increases the cost and footprint of the system and makes it more cumbersome.

In [31, 32] the new, simple multimode laser scheme was proposed (Fig. 8.19). A pump at 980 nm was launched into the cavity formed by 1.5 m of the single-mode double clad EDF developed in [11] and a pair of FBGs. Four multimode pump diodes at 980 nm and overall maximum power of 173 W were coupled through a commercially available 7 × 1 pump combiner into the laser resonator. The FBGs were written in a 20/125 passive fiber (NA~0.08/0.45) and had reflections of ~100 and 10% at 1535 nm with bandwidths <0.8 nm. The small-signal absorption from the cladding near 980 nm of the single-mode EDF was ~0.6 dB/m [11], so only ~5% of the overall pump power was absorbed in the single-mode laser cavity. It generated light at 1535 nm with a slope efficiency of ~45% with respect to the absorbed power. This signal was used as the seed radiation for the 12 m piece of the multimode EDF described above that was spliced to the 10% FBG of the single-mode cavity. A commercially available 105/125 μm multimode fiber was spliced at the output of the laser, and a cladding pump stripper similar to that described in [20] was built at the splice point. Therefore, the singlemode seed laser and the multimode amplifier were both pumped by the same pump diodes at 976 nm, and the spectral width of the laser was locked by the FBGs written in the single mode fiber, resulting in a relatively narrow output spectrum.

Figure 8.19b depicts the output power of the developed multimode laser. 60 W of output power, which was limited by the available pump power, was achieved. The slope efficiency of the laser was estimated to be 35%. The output spectrum measured

Fig. 8.20 Laser scheme
from [35]

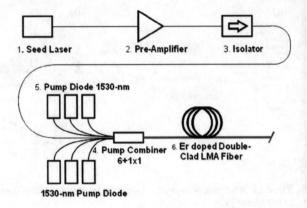

near 1535 nm with 0.1 nm resolution is presented in the inset of Fig. 8.19b. A small part (~8% relative to the output power of the laser) of unabsorbed pump at 976 nm was propagated in the multimode laser core together with the signal near 1535 nm.

8.5.2 Er-doped Fiber Lasers Pumped at 1535 nm

The first high-power Er-doped fiber laser resonantly cladding-pumped near 1535 nm was demonstrated in [35]. Experimental setup from this work is presented in Fig. 8.20. It is comprised of a diode laser seeder, a preamplifier and a booster amplifier. The preamplifier provided a maximum output power of 0.4 W in order to maintain a signal level sufficiently close to saturating the booster gain. The Yb-free fiber booster amplifier was based on a 9.5 m Liekki Er 60-20/125 double-clad large mode area fiber with a core diameter and a numerical aperture of 20 and 0.07 respectively. The booster amplifier was cladding co-pumped by six 5–6 W fiber-coupled 105/125 μm, NA of 0.15 InGaAsP/InP laser diode modules coupled into the active fiber via a 6+1 to 1 pump combiner. The maximum combined launched power was 30.2 W.

The maximum output signal power of 9.3 W at 1570 nm was obtained. The slope efficiency with respect to the absorbed pump power and O–O efficiency were 45 and 33%. In their following works the authors built high-power all-fiber oscillators with similar pump configuration and improved wavelength-stabilized diode sources [19]. As a result, the maximum output power of 88 W was achieved. The slope efficiency relatively to the absorbed pump power and O–O efficiency were increased to 69 and 52%.

It worth to note, that the highest achieved by the authors optical-to-optical of 52% efficiency is quite far from the quantum limit (96%). Thus, exploiting of a specialty optimized fiber instead of a commercially available one was required in order to increase the output power. Such optimization was done in [28]. As it was discussed above, first of all, erbium ions clustering should be suppressed. An alumino-phosphosilicate glass matrix with addition of Yb and Er was used as a material of the core. In contrast to Er-Yb lasers discussed above, Yb ions do not participate in the laser transition of in-band pumped EDFs. It was shown that addition of other rare earth elements such as lanthanum or ytterbium allows to increase efficiency of the Er-doped fiber lasers [22]. Apparently, a portion of the clusters that are consist of different rare earth ions, whose energy levels do not coincide, and Er^{3+} ions are separated inside the clusters by a second type of rare earth ions (e.g. Yb^{3+} or La^{3+}). This effect may prevent up-conversion processes in these clusters. At the same time, aluminum and phosphorus oxides forms $AlPO_4$ joints in silica glass. These structure units act as solvation shells for Er^{3+} ions and by this way also improve erbium solubility in the glass [10]. Both these approaches were applied by the authors of [28] to reduce up-conversion loss.

The manufactured in [28] active fiber had core and cladding diameters of 17 and 125 μm. Small signal absorption of the fiber from the cladding was 1.1 dB/m at 1535 nm. Er:Yb concentrations ratio was 1:3.3. The laser setup depicted in Fig. 8.17 was used. The cavity consisted of a FBG acting as the high reflector and 18 m of Er–Yb co-doped double-clad fiber with the cleaved end acting as the output coupler. The 1585 nm FBG, with a 3 dB bandwidth of 2 nm and a reflectivity of 30 dB, was UV written in a single-mode double-clad fiber having an 11 μm core diameter and a NA of 0.08. 264 W of output power was achieved with the slope efficiency of 75% with respect to the launched pump power in this experiment.

To date, 264 W [28] is a record demonstrated output power for the singlemode Er-doped fiber lasers. However, available pump power for the scheme demonstrated at Fig. 8.17 is quite limited by thermal effects and a parasitic emission of the single-mode Er-Yb fibers. Further power scaling might be achieved by combination of the fiber from [28] and multimode pump sources discussed above (Fig. 8.19). Indeed, the power of the described multimode fiber laser could be easily scaled. As it was already mentioned, 100 W multimode pump diode sources at 980 nm with a 105/125 μm fiber pigtail are commercially available now. Using these diodes together with 7-to-1 pump combiners ~250 W of power at 1535–1590 nm from 100 μm core output fiber (NA = 0.15) at the output of multimode laser. Finally, even this output power level from multimode fiber laser is not limited. It is shown in [31] that decrease of the pump conversion efficiency of multimode fibers is not very significant when core-to-cladding diameters ratio is decreased to 0.5. The efficiency of 100/200 μm core/cladding diameter with optimized erbium concentration and decreased background will have pump conversion efficiency of ~35% at 1535 nm. Currently pump combiners with 19 pump ports (105/125 μm, NA = 0.15) and output fiber with an outer diameter of 200 μm and a NA = 0.45 are commercially available. This means that there are no fundamental limitation to scale output power of multimode Er-doped lasers (output 105/125 μm fiber with NA = 0.15) to the level of 600 W of output power.

Utilization of such powerful sources instead of semiconductor pump diodes or Er-Yb fiber lasers could be very promising. In particular using of 7+1-to-1 pump combiner and 7 such pump sources with a highly efficient in-band pumped Er-doped fiber laser from [28] could allow one to scale output power of the single-mode laser to the unprecedented level of more than 3 kW. Therefore, to date tandem pumping of single-mode Er-doped fiber by the multimode fiber lasers is the most promising approach to build a multi-kW laser source at 1.55 μm.

8.6 Er-doped Fiber Lasers Core-Pumped at 1480 nm

As it was mentioned above Er-doped fibers demonstrated close to the quantum limited efficiency when core-pumped at 980 [12] or 1480 nm [36]. In this case, required length of the fiber is much smaller (approximately by factor of cladding-to-core areas ratio) compare to the cladding-pumped schemes. This significantly reduces the impact of clustering-induced loss on the efficiency. However, to date, the output power of the available singlemode pump diode sources at these wavelengths is less than few Watts. Similar to the pumping near 1530 nm good alternative to diode sources is a usage of fiber lasers.

There were some attempts to manufacture a high-power Yb-doped fiber laser operating at 980 nm [37–39]. However, efficient operation of Yb-doped fiber lasers at this wavelength is greatly obstructed by a competitive lasing near 1030 nm. As a result, proposed single-mode Yb-doped fiber lasers at 980 nm either demonstrated low efficiencies and powers [38, 39] or were too cumbersome and expensive for mass-production as a pump source [37]. As a result, no high-power Er-doped fiber lasers core-pumped by fiber lasers at 980 nm have been reported yet to the best of our knowledge. The development of fiber Raman lasers operating near 1480 nm was more successive and are discussed in the following section.

8.6.1 Raman Lasers at 1480 nm

Stimulated Raman scattering is an effective method to generate new wavelengths in optical fibers. Typical scheme of a Raman fiber laser at 1480 nm is presented in Fig. 8.21 [40]. Radiation of a single-mode high-power Yb-doped fiber laser is coupling into a cascaded Raman resonator. The resonator is formed by sets of Raman input gratings (RIG), Raman output gratings (ROG), and a Raman fiber. Wavelengths of grating sets are defined by a Stokes shift of the Raman fiber. For standard Ge-doped silica fibers the shift is ~440 cm^{-1}, so five FBGs is required for each set to reach wavelength of 1480 nm. For instance for Yb-doped fiber laser operating at 1117 nm grating sets are consisting from FBGs at 1175, 1240, 1310, 1390 and 1480 nm [40]. All FBGs except output one at 1480 nm should be highly reflective. At the same time it was shown that the P-doped fibers have a Stokes shift of ~1300 cm^{-1}. In this case only two FBGs per set (at 1240 and 1480 nm) are required.

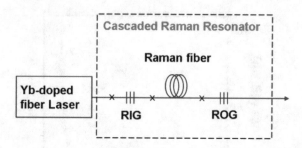

Fig. 8.21 Schematic of a typical Raman fiber laser [40]

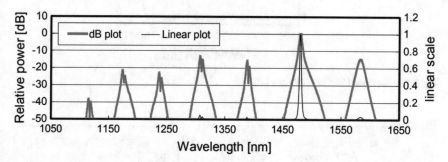

Fig. 8.22 Output spectrum of Raman laser at 41 W of output power [41]

Using setup described above and Ge-doped fiber as an active media the maximum of 41 W of output power at 1480 nm was achieved with 1117-to-1480 nm O–O efficiency of ~40% [41]. The power limitation for such a setup came from an unwanted scattering into the next Stokes order near 1580 nm. For 125 m of Raman fiber, the 1580 nm peak was only 15 dB below the 1480 nm peak at 41 W of output power (Fig. 8.22). The 1580 nm peak could be reduced to 30 dB below the 1480 nm peak by shortening the fiber length to 65 m, but with a negative impact on conversion efficiency. Furthermore, as the fiber length was shortened, a smaller fraction of light at the output was contained at 1480 nm with more residual radiation at the intermediate Stokes wavelengths [41].

Thereby, the lasing suppression at 1580 nm is required for further power scaling of Raman lasers at 1480 nm. In order to do that the specialty Raman filter fiber (RFF) was developed by the group from OFS Laboratories [42]. The RFF was a germanosilicate fiber that used a fundamental mode cutoff to achieve distributed loss at wavelengths longer than 1480 nm [42]. The RFF provided cascaded Raman gain up to 1480 nm and distributed loss at longer wavelengths. The measured loss curve is shown in Fig. 8.23.

Figure 8.24 shows the output power from the Raman laser as a function of 976 nm pump power. The maximum total output power measured was 88.1 W, with 81 W at 1480 nm, for a pump power of 250 W. The output of the 1117 nm, Yb fiber laser at this pump power was 162 W. The measured spectrum at full power is shown in

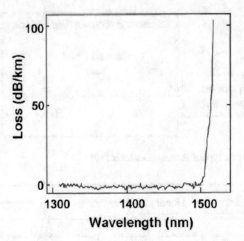

Fig. 8.23 Measured loss of RFF fiber [40]

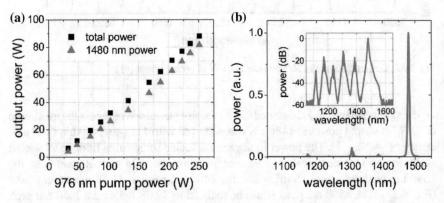

Fig. 8.24 **a** Output power as a function of 976 nm pump power (including both oscillator and amplifier pump power). **b** Output spectrum at maximum pump power on a linear and a log scale (inset) [42]

Fig. 8.24b on a linear scale and on a decibel scale in the inset. At 1552 nm, the signal is 42 dB below the peak at 1480 nm, and more than 60 dB down at 1580 nm. The maximum output power was limited by the available pump power in this experiment.

Further Raman fiber laser construction optimization was done in [40]. The main factors decreasing the efficiency of the Raman lasers were identified as follows [40]:

1. Transmission loss in the Raman input grating set and output grating set.
2. Two intra-cavity splices between low effective area (possibly dissimilar) fibers constituting the grating sets and the Raman gain fiber.
3. Linear loss in the Raman fiber.

4. Enhanced backward and forward light at the intermediate Stokes wavelengths due to their bandwidth being higher than the grating bandwidths.
5. Splice loss between the Yb-doped fiber laser output and the low effective area Raman fiber.

Thus, a number of loss components are associated with the cascaded Raman resonator assembly. To eliminate the cascaded Raman resonator exploiting of a single pass cascaded Raman scheme was developed in [40]. Two key ingredients were necessary to make the single pass cascaded amplifier feasible. Firstly, a simple multi-wavelength source which can simultaneously provide sufficient powers at all the intermediate wavelengths. A lower power conventional cascaded Raman laser lends itself ideally for this purpose. Light present at the output at all the intermediate Stokes wavelengths provides sufficient seed power at the exact required wavelengths. Secondly, scattering of the output wavelength to the next Stokes order can be further enhanced in a single pass configuration. The use of Raman filter fiber eliminates this problem and provides a technique to terminate the cascade of wavelength conversion. The proposed in [40] laser design is presented in Fig. 8.25. A high power Yb-doped fiber laser was combined with a lower power Raman seed laser. The Yb-doped fiber laser was emitting at 1117 nm and the power was combined using an 1117/1480 nm fused fiber wavelength division multiplexer (WDM). This is then sent through a Raman filter fiber.

Figure 8.26 shows the total output power and components at each Stokes wavelength measured as a function of input power at 1117 nm to the cascaded amplifier. A progressive growth and decay of all the intermediate Stokes components with increasing power can clearly be observed. A rapid growth of the final output wavelength is seen beyond a power threshold.

Figure 8.27a shows the total output power and the 1480 nm component as a function of input power at 1117 nm. The maximum output power was ~204 W at 1480 nm. The conversion efficiencies were 65% from 1117 to 1480 nm (for a quantum limited value of 75%) and 43% O–O from 975 nm pump to 1480 nm. Figure 8.27b shows the measured output spectrum at full power (log scale in the inset). More than 95% of the power is in the 1480 nm band indicating the high level of wavelength conversion.

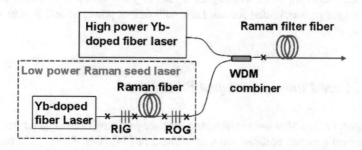

Fig. 8.25 Experimental scheme of single-pass cascaded Raman laser

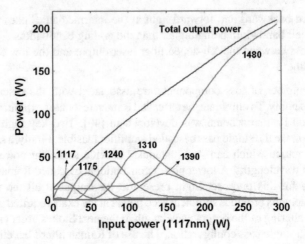

Fig. 8.26 Plot of total output power and components at each Stokes wavelength as a function of input power

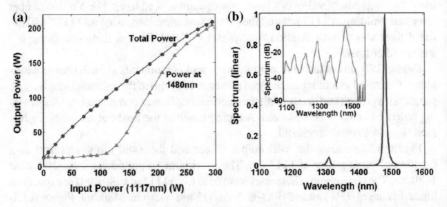

Fig. 8.27 **a** Plot of total output power and output power at 1480 nm as a function of input power at 1117 nm. **b** Spectrum of the output in linear and log scale (inset) at maximum power [40]

In their following work with higher available pump power the authors realized 300 W single-pass cascaded Raman laser with output power of 300 W and similar efficiency [43].

8.6.2 Core-Pumped Er-doped Fiber Laser

Developed Raman fiber lasers were used for core-pumping of Er-doped fiber lasers. A backward pumped architecture was utilized in [44] (Fig. 8.28). 1554 nm FBG with a FWHM of 6 nm served as the high reflector and a flat cleave (which provides

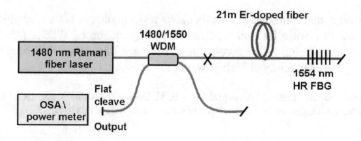

Fig. 8.28   Experimental setup from [44]

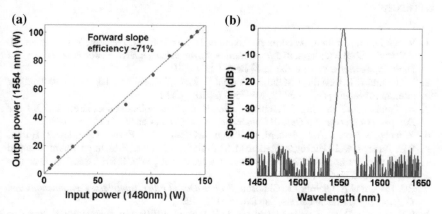

Fig. 8.29   **a** Output forward power at 1554 nm as a function of 1480 nm input. **b** Output spectrum of the laser [44]

a 4% reflection) was used as the output coupler. An advantage of the backward pumped architecture was that no additional filtering was necessary for applications which might be sensitive to the power at the intermediate wavelengths. The output of the FBG was fusion spliced to 21 m of length of MP980 Er-doped fiber from OFS Specialty Photonics Division. The fiber had a peak absorption of approximately 6 dB/m at 1530 nm, 0.23 NA and a mode field diameter of approximately 5.6 μm.

Figure 8.29a shows the forward output power at 1554 nm as a function of input 1480 nm pump power. The maximum power of 101 W was obtained with 142 W of pump power corresponding to a conversion efficiency of 71%. The output power was only limited by the availability of pump power. Figure 8.29b shows the output spectrum at maximum power. The laser FWHM is ~3.5 nm. No ASE was observed even at 50 dB below the emission peak.

It should be noted that the high power WDMs is a critical part of core-pumped high-power oscillators/amplifiers. It was shown above, these WDMs are used to build 1480 nm pump source, and to couple pump radiation into EDF. The intensity of the light in such WDMs are much higher than that in pump combiners used for cladding-pumped scheme. This makes the development of WDMs which are capable to handle multi-hundred of Watts or even kilowatts power level rather challenging. Moreover it was shown, that additional efforts should be taken to filter out intermediate Stokes waves that can cause long-term performance degradation of WDMs [45]. Thus, further power scaling of the core-pumped Er-doped fiber lasers will be probably limited by the availability of high-power multiplexors.

Acknowledgements The authors are grateful to E. M. Dianov, the scientific director of the Fiber Optics Research Center, for his continuous interest in and support of this work.

References

1. V. Fomin, M. Abramov, A. Ferin, A. Abramov, D. Mochalov, N. Platonov, V. Gapontsev, 10 kW single-mode fiber laser, in *5th International Symposium on High-Power Fiber Lasers and Their Applications*, St. Petersburg, 28 June–1 July 2010
2. T. Ehrenreich, R. Leveille, I. Majid, K. Tankala, G. Rines, P.F. Moulton, 1-kW all-glass Tm:fiber laser, in *Proceedings of the SPIE*, vol. 7580 (2010), p. 112
3. N. Simakov, A. Hemming, J. Haub, A. Carter, High power holmium fiber lasers, in *2014 The European Conference on Optical Communication (ECOC)*, Cannes (2014), pp. 1–3
4. Y. Jeong, S. Yoo, C.A. Codemard, J. Nilsson, J.K. Sahu, D.N. Payne, R. Horley, R. Horley, P.W. Turner, L.M.B. Hickey, A. Harker, M.J. Lovelady, A.N. Piper, Erbium:ytterbium co-doped large-core fiber laser with 297 W continuous-wave output power. IEEE J. Sel. Top. Quantum Electron. **13**, 573–579 (2007)
5. P.C. Becker, N.A. Olsson, J.R. Simpson, *Erbium-Doped Fiber Amplifiers: Fundamentals and Technologies* (Academic Press, San Diego, USA, 1999)
6. L.V. Kotov, A.D. Ignat'ev, M.M. Bubnov, M.E. Likhachev, Effect of temperature on the active properties. Quantum Electron. **46**, 271 (2016)
7. P.F. Wysocki, J.L. Wagener, M.J.F. Digonnet, H.J. Shaw, Evidence and modeling of paired ions and other loss mechanisms in erbium-doped silica fibers, in *Proceedings of the SPIE*, vol. 1789 (1993), pp. 66–79
8. P. Myslinski, D. Nguyen, J. Chrostowski, Effects of concentration on the performance of erbium-doped fiber amplifiers. IEEE J. Lightwave Technol. **15**, 112–120 (1997)
9. K. Arai, H. Namikawa, K. Kumata, T. Honda, Aluminium or phosphorous co-doping effects on fluorescence and structural properties of neodimium-doped silica glass. J. Appl. Phys. **59**, 3430–3436 (1986)
10. M.E. Likhachev, M.M. Bubnov, K.V. Zotov, D.S. Lipatov, M.V. Yashkov, A.N. Guryanov, Effect of the $AlPO_4$ join on the pump-to-signal conversion efficiency in heavily Er-doped fibers. Opt. Lett. **34**(21), 3355–3357 (2009)
11. L.V. Kotov, M.E. Likhachev, M.M. Bubnov, O.I. Medvedkov, M.V. Yashkov, A.N. Guryanov, J. Lhermite, S. Février, E. Cormier, 75 W 40% efficiency single-mode all-fiber erbium-doped laser cladding pumped at 976 nm. Opt. Lett. **38**, 2230–2232 (2013)
12. R.I. Laming, J.E. Townsend, D.N. Payne, F. Meli, G. Grasso, E.J. Tarbox, High-power erbium-doped-fiber amplifiers operating in the saturated regime. IEEE Photon. Tech. Lett. **3**(3), 253–255 (1991)
13. A.S. Kurkov, V.M. Paramonov, M.V. Yashkov, S.E. Goncharov, I.D. Zalevskii, Multimode cladding-pumped erbium-doped fibre laser. Quantum Electron. **37**(4), 343 (2007)
14. M. Dubinskii, V. Ter-Mikirtychev, J. Zhang, I. Kudryashov, Yb-free, SLM EDFA: comparison of 980-, 1470- and 1530-nm excitation for the core- and clad-pumping, in *Proceedings of the SPIE*, vol. 6952 (2008), p. 695205
15. V. Kuhn, D. Kracht, J. Neumann, P. Weßels, Yb-free Er-doped 980 nm pumped single-frequency fiber amplifier with output power of 54 W and near-diffraction limited beam quality, in *CLEO/Europe and EQEC 2011 Conference Digest*, CJ7_5 (2011)

16. V. Kuhn, D. Kracht, J. Neumann, P. Weßels, Er-doped photonic crystal fiber amplifier with 70 W of output power. Opt. Lett. **36**, 3030–3032 (2011)
17. V. Kuhn, D. Kracht, J. Neumann, P. Wessels, 67 W of output power from an Yb-free Er-doped fiber amplifier cladding pumped at 976 nm. IEEE Photonics Technol. Lett. **23**(7), 432 (2011)
18. L.V. Kotov, M.E. Likhachev, M.M. Bubnov, O.I. Medvedkov, M.V. Yashkov, A.N. Guryanov, S. Fevrier, J. Lhermite, E. Cormier, Yb-free Er-doped all-fiber amplifier cladding-pumped at 976 nm with output power in excess of 100 W, in *Proceedings of the SPIE*, vol. 8961 (2014), p. 89610X
19. J. Zhang, V. Fromzel, M. Dubinskii, Resonantly cladding-pumped Yb-free Er-doped LMA fiber laser with record high power and efficiency. Opt. Express **19**, 5574–5578 (2011)
20. S. Aleshkina, T.A. Kochergina, K.K. Bobkov, L.V. Kotov, M.M. Bubnov, J. Park, M.E. Likhachev, High-power 125-μm-optical-fiber cladding light stripper, in *Conference on Lasers and Electro-Optics*, OSA Technical Digest (online), paper JTu5A.106 (2016)
21. J. Skidmore, M. Peters, V. Rossin, J. Guo, Y. Xiao, J. Cheng, A. Shieh, R. Srinivasan, J. Singh, C. Wei, R. Duesterberg, J.J. Morehead, E. Zucker, Advances in high-power 9XXnm laser diodes for pumping fiber lasers, in *Proceedings of the SPIE*, vol. 9733 (2016), p. 97330B
22. N.V. Kiritchenko, L.V. Kotov, M.A. Melkumov, M.E. Likhachev, M.M. Bubnov, M.V. Yashkov, A.Y. Laptev, A.N. Guryanov, Effect of ytterbium co-doping on erbium clustering in silica-doped glass. Laser Phys. **25**(2), 025102 (2015)
23. E. Snitzer, H. Po. F. Hakimi, R. Tumminelli, B.C. McCollum, Erbium fiber laser amplifier at 1.55 μm with pump at 1.49 μm and Yb sensitized Er oscillator, in *OFC'88 Optical Fibre Communications Conference*, paper PD2-1 (1988)
24. J. Minelly, W. Barnes, R. Laming, P. Morkel, J. Townsend, S. Grubb, D. Payne, Diode-array pumping of Er/sup 3+//Yb/sup 3+/ Co-doped fiber lasers and amplifiers. IEEE Photon. Technol. Lett. **5**(3), 301–303 (1993)
25. G. Sobon, P. Kaczmarek, A. Antonczak, J. Sotor, K.M. Abramski, Controlling the 1 μm spontaneous emission in Er/Yb co-doped fiber amplifiers. Opt. Express **19**, 19104–19113 (2011)
26. Qun Han, Yunzhi Yao, Yaofei Chen, Fangchao Liu, Tiegen Liu, Hai Xiao, Highly efficient Er/Yb-codoped fiber amplifier with an Yb-band fiber Bragg grating. Opt. Lett. **40**, 2634–2636 (2015)
27. E. Yahel, A. Hardy, Modeling high-power Er3+-Yb3+ codoped fiber lasers. J. Lightwave Technol. **21**(9), 2044–2052 (2003)
28. M.A. Jebali, J.-N. Maran, S. LaRochelle, 264 W output power at 1585 nm in Er–Yb codoped fiber laser using in-band pumping. Opt. Lett. **39**, 3974–3977 (2014)
29. Y. Jeong, J.K. Sahu, D.B.S. Soh, C.A. Codemard, J. Nilsson, High-power tunable single-frequency single-mode erbium:ytterbium codoped large-core fiber master-oscillator power amplifier source. Opt. Lett. **30**(22), 2997–2999 (2005)
30. Ee-Leong Lim, Shaif-ul Alam, David J. Richardson, Optimizing the pumping configuration for the power scaling of in-band pumped erbium doped fiber amplifiers. Opt. Express **20**, 13886–13895 (2012)
31. L. Kotov, O. Medvedkov, M. Bubnov, D. Lipatov, A. Guryanov, M. Likhachev, High brightness multi-mode fiber lasers—a novel sources for in-band cladding pumping of singlemode fiber lasers, in *Proceedings of the 5th International Conference on Photonics, Optics and Laser Technology (PHOTOPTICS)*, vol. 1 (2017), pp. 99–105
32. L.V. Kotov, O.I. Medvedkov, M.M. Bubnov, D.S. Lipatov, N.N. Vechkanov, A.N. Guryanov, M.E. Likhachev, High power pump source at 1535 nm based on multimode Er-doped fiber, in *Proceedings of the SPIE*, vol. 10083 (2017), pp. 100831N-1
33. F. Gonthier, L. Martineau, N. Azami, M. Faucher, F. Seguin, D. Stryckman, A. Villeneuve, High-power all-fiber components: the missing link for high-power fiber lasers, in *Proceedings of the SPIE*, vol. 5335 (2004), p. 266
34. B. Wang, E. Mies, Review of fabrication techniques for fused fiber components for fiber lasers, in *Proceedings of the SPIE*, vol. 7195 (2009), p. 71950A
35. M. Dubinskii, J. Zhang, I. Kudryashov, Single-frequency, Yb-free, resonantly cladding-pumped large mode area Er fiber amplifier for power scaling. Appl. Phys. Lett. **93**, 031111 (2008)

36. M. Dubinskii, J. Zhang, V. Ter-Mikirtychev, Record-efficient resonantly-pumped Er-doped singlemode fibre amplifier. Electron. Lett. **45**(8), 400–401 (2009)
37. J. Boullet, Y. Zaouter, R. Desmarchelier, M. Cazaux, F. Salin, J. Saby, R. Bello-Doua, E. Cormier, High power ytterbium-doped rod-type three-level photonic crystal fiber laser, Opt. Express **16**, 17891–17902 (2008)
38. M. Leich, M. Jäger, S. Grimm, D. Hoh, S. Jetschke, M. Becker, A. Hartung, H. Bartelt, Tapered large-core 976 nm Yb-doped fiber laser with 10 W output power. Laser Phys. Lett. **11**, 045102 (2012)
39. S.S. Aleshkina, M.E. Likhachev, D.S. Lipatov, O.I. Medvedkov, M.M. Bubnov, A.N. Guryanov, All-fiber single-mode laser at 977 nm with 5.5 W output power, in *CLEO/Europe-EQEC 2015*, paper CJ-P.13 (2015)
40. V.R. Supradeepa, J.W. Nichsolson, C.E. Headley, M.F. Yan, B. Palsdottir, D. Jakobsen, A high efficiency architecture for cascaded Raman fiber lasers, Opt. Express **21**, 7148–7155 (2013)
41. Y. Emori, K. Tanaka, C. Headley, A. Fujisaki, High-power cascaded Raman fiber laser with 41-W output power at 1480-nm band, in *Conference on Lasers and Electro-Optics/Quantum Electronics and Laser Science Conference and Photonic Applications Systems Technologies*, paper CFI2 (2007)
42. J.W. Nicholson, M.F. Yan, P. Wisk, J. Fleming, F. DiMarcello, E. Monberg, T. Taunay, C. Headley, D.J. DiGiovanni, Raman fiber laser with 81 W output power at 1480 nm. Opt. Lett. **35**(18), 3069–3071 (2010)
43. V.R. Supradeepa, J.W. Nicholson, Power scaling of high-efficiency 1.5 μm cascaded Raman fiber lasers. Opt. Lett. **38**, 2538–2541 (2013)
44. V.R. Supradeepa, J.W. Nicholson, K. Feder, Continuous wave Erbium-doped fiber laser with output power of >100 W at 1550 nm in-band core-pumped by a 1480 nm Raman fiber laser, in *presented at CLEO*, paper CM2N.8 (2012)
45. X. Peng, K. Kim, X. Gu, M. Mielke, S. Jennings, A. Rider, N. Fisher, T. Woodbridge, R. Dionne, F. Trepanier, Root cause analysis and solution to the degradation of wavelength division multiplexing (WDM) couplers in high power fiber amplifier system. Opt. Express **21**, 20052–20061 (2013)

Chapter 9
Characterization of the Charge Transfer in an Enhanced Pinned Photodiode with a Collection Gate

L. Girgenrath, M. Hofmann, R. Kühnhold and H. Vogt

Abstract An implantation scheme which enhances the readout speed of a silicon pinned photodiode (PPD) with large pixel length is presented. A special type of pinned photodiode which was developed by the Fraunhofer IMS in Duisburg, Germany for Time-of-Flight distance measurement applications is taken as the starting point. The sensor which was fabricated in a standard 0.35 μm CMOS process and the optimized design introduces a second gate, the Collection Gate (CG), to the pinned photodiode which will be analysed. Based on this PPD, a second well implantation is described which improves the electron transfer. Furthermore, the influence of the Collection Gate on the electron transfer is described. The CG can alter the conduction band energy of the PPD. It is shown that the barrier at the interface between well and GC can be reduces by applying a voltage to the CG. The second implantation in combination with the CG creates a designated electron path which introduces the possibility to enlarge the PPD without affecting the performance of the sensor.

L. Girgenrath (✉) · M. Hofmann · R. Kühnhold
ELMOS Semiconductor AG, Heinrich-Hertz-Str. 1, 44227 Dortmund, Germany
e-mail: lutz.girgenrath@elmos.com

M. Hofmann
e-mail: martin.hofmann@elmos.com

R. Kühnhold
e-mail: ralf.kuehnhold@elmos.com

L. Girgenrath · H. Vogt
Faculty of Engineering, EIT, EBS, University of Duisburg-Essen, Bismarkstr. 81,
47057 Duisburg, Germany
e-mail: holger.vogt@ims.fraunhofer.de

H. Vogt
Fraunhofer Institute for Microelectronic Circuits and Systems, Finkenstrasse 61,
47057 Duisburg, Germany

© Springer Nature Switzerland AG 2019
P. Ribeiro et al. (eds.), *Optics, Photonics and Laser Technology 2017*,
Springer Series in Optical Sciences 222,
https://doi.org/10.1007/978-3-030-12692-6_9

9.1 Introduction

In recent years, the usage of optical and contactless distance measurements like phase-modulation or time-of-flight has increased. Only a few years ago, time-of-flight sensors were mainly used in radar applications to measure large distances. The constant miniaturization of the sensor systems introduced these sensors into numerous other applications fields like gesture control for entertainment systems, passenger and pedestrians detection for the automotive industry or for area surveillance in safety applications. All applications have in common that the basic principle of the distance measurement is to determine the time of flight of the light. It is possible to measure the time of flight directly or indirectly. The direct measurement is mainly applicable to measure only one distance at a time like the distance between the earth and the moon. The measurement of a whole scenery requires the distance to be calculated indirectly which is done pixelwise by the internal comparison of the laser pulse with the measured signal. Each sensor has to be designed to satisfy special requirements for each application. As faster the systems has to work, as higher the requirements are especially if operated in unfavourable conditions. If the system is used in the everyday life, the usage of IR-light is mandatory for scanning applications in terms of eye safety and distraction.

Basically, these so called 3D Time-of-Flight (ToF) sensors must meet three different requirements. They have to be fast and must be able to measure small and large signals with the same resolution while maintaining a reasonable frame rate. This means that the internal signal processing and signal acquisition has to be fast. As nearly all relevant sensor parameters are defined by the sensor length [3], the resolution and transfer times can be enhanced by decreasing the sensor length. Simultaneously the sensitivity decreases as well and the noise level increases as well as the cross talk especially if IR-Light is used. Furthermore, to maximize the measurement range all additionally error sources has to be reduces as much as possible. The above mentioned dependences show that the length of a photodetector for 3D ToF measurements cannot simply by reduced to achieve a better performance as it reduces the sensitivity. The best way to enhance the performance of the sensor is to increase the sensor area while keeping the electron transfer time small. In this work, we will analyse the basic structure of a pinned photodiode in terms of electron transfer speed and collection gate voltage.

9.2 Transfer Times and 3D ToF Measurements

The basic principle of the indirect time of flight measurement is shown in Fig. 9.1. A scenery is illuminated by a short light pulse and the time till the reflection can be measured is defined as the time of flight. Since the light travels the distance D twice in t_1 with the velocity c the distance can be calculated as:

Fig. 9.1 Pulse scheme to measure the distance in a 3D-ToF application

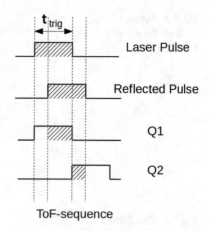

t_{trig}

Laser Pulse

Reflected Pulse

Q1

Q2

ToF-sequence

$$D = \frac{1}{2} \times t_1 \times c. \tag{9.1}$$

The determination of t_1 in the real application is done with two separate measurements with slightly different timings. The time of flight is calculated by comparing the charge informations generated for each measurement. The first measurement is done while the laser is on and the second measurement is done after the laser is shut off after t_{trig} for the same amount of time. The two generated charge informations are called Q_1 and Q_2. The time of flight t_1 can be calculated as:

$$t_1 = t_{trig} \times \frac{Q_2}{Q_1 + Q_2}. \tag{9.2}$$

The smallest measurable distance is defined by the transfer time T_{trans} of the electrons. The distance of 0 m could only be measured if every generated electron is read out instantaneously so that no electrons are generated in Q_2. Since the electron speed is not infinitely high, no generated electrons in Q_2 is physically impossible. Therefore, both of the charge informations will be affected by the transfer time (see Fig. 9.2). The fastest electrons which are read out almost immediately define the rising edge of the of the signal. These electrons originates from the area around the read-out path which is always filled with electrons over the length of the laser pulse. Therefore, the signal response is rather fast. The falling edge of the signal however is defined by the slowest electrons. If the transfer time is high, the amount of electrons that reaches the FD delayed increases. In a real measurement, the charge information Q_1 will always be measured too low and Q_2 too high. The actual measured distance D_{mess} can be calculated as:

$$D_{mess} = \frac{1}{2} \times c \times t_{trig} \times \frac{Q_2 + \Delta_{Q2}}{Q_1 + Q_2 + \Delta_{Q2} - \Delta_{Q1}}. \tag{9.3}$$

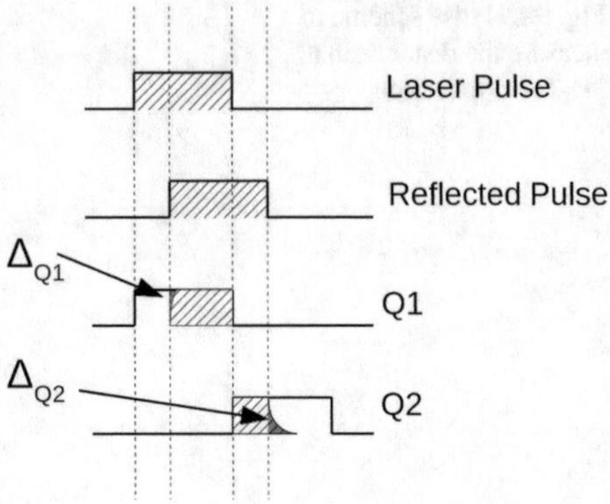

Fig. 9.2 Influence of the
electron transfer time on Q_1
and Q_2

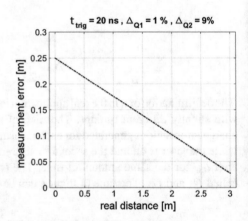

Fig. 9.3 Calculated
measurement error over the
measurable distance of 3 m

Δ_{Q1} and Δ_{Q2} are the errors which are introduced to Q_1 and Q_2. The difference ΔD
between the real distance and the measured distance which is introduced by the finite
electron speed can be calculated as:

$$\Delta D = \frac{c \times t_{\text{trig}}}{2} \frac{Q_1 \Delta_{Q2} + \Delta_{Q1} Q_2}{(Q_1 + Q_2 + \Delta_{Q2} - \Delta_{Q1})(Q_1 + Q_2)} \tag{9.4}$$

As discussed above, Δ_{Q1} is always smaller than Δ_{Q2}. This dependency leads to a
significant influence of the electron transfer time to the measured distance. On the one
hand, it results in a constant offset. The error on the other hand depends also on the
measured distance and is larger for small distances. Figure 9.3 shows the calculation
of ΔD for the whole measurable distance ($t_{\text{trig}} = 20$ ns, $Q_1 = 1\%$ and $Q_2 = 9\%$).
Eventually the delayed electron transfer reduces the measurement range.

9.3 The Enhanced Pinned Photodiode

A standard pinned photodiode is composed of two implantations. These two implantations are a low n-type implantation to form a well inside the p-type substrate and a shallow p+ implantation to separate the well from the semiconductor-oxide interface. The schematic cross-section of the well and the relevant depletion regions is shown in Fig. 9.4. Both implantation define the properties of the well as both of the pn-junctions define the potential inside the well [12]. The pn-junction [9] on top of the well can be approximated as an abrupt junction because the acceptor concentration N_p is much larger than the donator concentration N_n. The pn-junction at the bottom of the well must be considered as continuous. In case of an abrupt junction, the depletion layer will only be formed inside the n-well and the depletion layer width W_{n1} is defined as follows:

$$W_{n1} = \sqrt{\frac{2\epsilon_s (V_{\text{nwell}} + V_{\text{bi1}})}{q N_n}}.$$ (9.5)

ϵ_s describes the permittivity and V_{bi1} the built-in potential of the abrupt junction. q stands for the electrical charge and V_{nwell} describes the potential inside the well. The depletion region of the continuous junction extends in the n-well as well as in the p-epi substrate. The p-epi doping concentration N_{epi} is considered constant and the n-doping concentration is considered to decrease linearly into the substrate. The space charge density $\rho(x)$ can now be calculated in respect to the n-well concentration gradient a.

$$\rho(x) \approx q(ax + N_{\text{epi}}) \quad \text{for} \quad -W_{n2} \leq x \leq 0$$ (9.6)

$$\rho(x) \approx -q N_{\text{epi}} \quad \text{for} \quad 0 \leq x \leq W_{p2}$$ (9.7)

If the boundary conditions in the n-well are considered, the poisson equations gives the electrical field on both sides of the pn-junction.

Fig. 9.4 Schematic drawing of a p+np junction with the resulting depletion layer

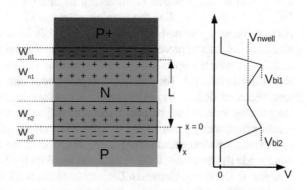

$$E_n(x) = \frac{q}{\epsilon_s} \left(a \times x^2 + N_{epi} \times x + N_{epi} W_{n2} - a W_{n2}^2 \right) \qquad (9.8)$$

$$E_p(x) = -\frac{q N_{epi}}{\epsilon_s} \left(x - W_{p2} \right) \qquad (9.9)$$

The electrical field in the n-region $E_n(x)$ and the electrical field in the p-region $E_p(x)$ will be the same at the junction and the width of the depletion regions can be calculated (9.10).

$$W_{p2} = \left(W_{n2} - \frac{a W_{n2}^2}{N_{epi}} \right) \qquad (9.10)$$

nwell If the pinned photodiode is used as an optical sensor, the substrate is connected to ground and the potential inside W_{p2} is zero. The potential distribution inside the depletion regions are calculated as:

$$\Phi(x) = \frac{q N_{epi}}{\epsilon_s} \left(\frac{x^2}{2} - W_{p2} \times x \right) + \frac{q N_{epi}}{2\epsilon_s} W_{p2}^2 \quad 0 \le x \le W_{p2} \qquad (9.11)$$

$$\Phi(x) = \frac{q}{\epsilon_s} \left(\frac{1}{3} a \times x^3 + \frac{1}{2} N_{epi} \times x^2 + \left(N_{epi} W_{n2} - a W_{n2}^2 \right) \times x \right) + \frac{q N_{epi}}{2\epsilon_s} W_{p2}^2 \quad - W_{n2} \le x \le 0 \qquad (9.12)$$

If $x = 0$, the potential inside the p- and n-regions must be the same. Since the p-epi layer is connected to ground, the built-in potential V_{bi2} of this pn-junctions defines the potential inside the n-well ($x = -W_{n2}$). Since the Potential outside of the depletion regions inside the n-well still influences the width of the depletion region, the potential V_{nwell} can be used to widen or to reduce both of the depletion regions.

$$V_{bi2} + V_{nwell} = \frac{q}{\epsilon_s} \left(\frac{a^2 W_{n2}^4}{N_{epi}} - \frac{4}{3} a W_{n2}^3 + \frac{1}{2} N_{epi} W_{n2}^2 \right). \qquad (9.13)$$

The potential V_{nwell} which is necessary to bring both of the regions in contact ($W_{n1} + W_{n2} \ge L$) is called the pinning potential V_{pin}. If the depletion layer are in contact, the well potential is pinned and it will only be defined by V_{bi2}. Therefore, no electrical field can be present inside the well and the dominating transport mechanism is diffusion which is up to two orders of magnitude slower than the electron transport in an electric field [6]. The pinning voltage depends only on the technology parameters, which are defined by the implantation dose and implantation energy. Therefore, the pinning voltage can be manipulated by adjusting the implantation scheme. In recent years, many groups tried to implement an electrical field inside the n-well to enhance the electron transfer [5, 10, 11]. One of the most recent improvements were made by the group of the Fraunhofer Institute for Microelectronic Circuits ans Systems in Duisburg, Germany. Derived from the work of Durini [2], Spickermann [7], Süss [8] and Driewer [1] further increased the performance of the sensor. This

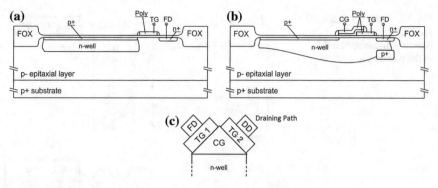

Fig. 9.5 **a** Schematic cross section of a standard pinned photodiode. **b** The enhanced version which was introduced by the Fraunhofer IMS. **c** Top view of the read-out path

already enhanced sensor will be the basic structure for this work. Figure 9.5 shows the main differences between this new sensor architecture and a standard pinned photodiode, which are the connection of the n-well with the Floating Diffusion (FD) and a second gate, the so called Collection Gate (CG). Due to the connection of the n-well and the FD, the charge can not longer be stored in the well but is accumulated in the FD. The increase in n-type doping concentration affects the Collection Gate (CG) and Transfer Gate (TG) to be self-conducting and they can be described as tunable resistors. Therefore, a second draining path (fig. fig:ppdverglc) is mandatory to measure a time-dependent signal. The electron current is switched between the two path by changing the TG voltage V_{TG} of both of the Transfer Gates. If TG1 is opened, TG2 will be closed. Since the electrons which are transferred into the draining path can not be measured, the TG1 will be referred as TG. Every electron which is generated outside the designated time steps should not be able to participate in the signal generation. The connection of FD and n-well are suppresses the formation of a barrier at the interface to the TG so that an adjustment of the well configuration doesn't influence the charge transfer. This design introduces the possibility to manipulate the pinning voltage which limits the performance of a pinned photodiode. The built-in electrical field is realized through an implantation gradient. The doping concentration decreases through the n-well so that the lowest concentration is implemented in the edge regions. The lower doping concentration widens the depletion layer which results in a reduced pinning voltage. The change of the concentration over the entire well results in a potential gradient which introduces an electrical field inside the diode. However, the detection of photons requires the well to be completely covered by the depletion layer. This boundary condition limits the electrical field as the doping concentration needs to be relatively low. Since a gradient depends on the length it is applied on, this effect also vanishes in bigger sensors. The function of the CG is to collect the electrons which are generated inside the well so that they can be transferred to the FD by opening the TG. The CG is set to a constant positive potential which is lower as the highest possible TG potential V_{TG}. The voltage distribution across the diode forms a "stairway" for the electrons from

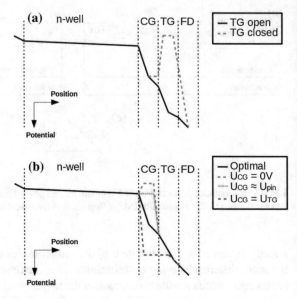

Fig. 9.6 Influence of the TG and CG voltage on the potential distribution inside of the pinned photodiode

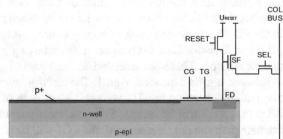

Fig. 9.7 Schematic drawing of the pinned photodiode as a pixel

the well into the FD as it can be seen in Fig. 9.6a. Figure 9.6b shows the schematic influence of the CG potential V_{CG} on the electron transfer. If V_{CG} is lower than the pinning voltage (see red curve) a barrier is formed at the transition between well and CG. If V_{CG} is either set to the pinning potential or the TG potential, an area is formed in which the potential is constant. Eventually, this will lead to a reduce in electron speed beneath the gates.

If the pinned photodiode is used as a sensor, in the open-state the potential of the FD is set to $V_{FD} = 5$ V, V_{TG} is set to 3 V and V_{CG} can be vary between 0 and 2 V. The open-state is the default state as one of the two read-out path are always open. A schematic drawing of the pinned photodiode used as a pixel sensor is shown in Fig. 9.7. Only the reset voltage V_{reset} directly influences the diode as it defines the FD voltage V_{FD}. CG and TG only influences the voltage distribution from the FD over the read-out to the well. Beneath the gates, the potential is only affected by the sheet resistance which is defined by V_{TG} and V_{CG}. The Potential inside the well is constant due to the pinning voltage and the substrate is set to 0 V.

As mentioned above, the absence of a potential gradient inside the well results in a slower electron transfer. A strong gradient is only possible if the pinning voltage can be mitigated. The highest possible electrical field would be achieved if the FD voltage V_{FD} drops across the whole diode. A global increase in doping concentration could enable but it also decreases the optical performance of the sensor. Therefore, another approach must be considered.

9.4 Manipulation of the Pinning Voltage

Figure 9.8a shows a schematic drawing of the depletion layers for a constant doping concentration inside the well. If the depletion regions are separated, the diode resembles a JFET which is only connected from one side. The main difference is the absence of the gate and a source contact. However, opening a designated electron transfer path remains as the basic principle of operation. For a JFET, the electron path is formed by applying a voltage to the gate which affects the width of the depletion layer W_D. Furthermore, the voltage applied at the drain node influences the depletion layer (see 9.14) [9].

$$W_D(x) = \sqrt{\frac{2\epsilon_s V_{bi} + V(x) - V_G}{qN_D}} \qquad (9.14)$$

The depletion layer width of a pinned photodiode is only defined by the potential inside the well. The well is pinned to the pinning potential as long as V_{FD} is bigger than V_{pin}. Otherwise, electrons would be injected into the diode and the two depletion layer are separated. Instead of adding electrons from outside the diode into the well, the same effect can be achieved by simply increasing the doping concentration N_{DPI} (**DPI** = **D**rift **P**ath **I**mplantation) [4] locally in the middle of the well. Figure 9.8b shows the schematic trend of the depletion layers with the doping gradient as well as the DPI. The higher doping concentration near the read-out path leads to clearer separation of the depletion layers. The FD voltage V_{FD} is reduced by the sheet resistance of the

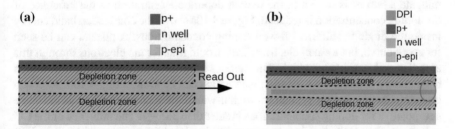

Fig. 9.8 a Schematic drawing of the depletion layers for a constant doping concentration inside the well. **b** Schematic drawing of the depletion layers for a doping gradient inside the well

Fig. 9.9 TCAD simulation of the electrostatic potential inside the diode. Top: without DPI, bottom: with DPI [4]

well. Analogous to the JFET, the voltage inside the well changes the width of the depletion layer. Equation (9.14) changes for a pinned photodiode to:

$$W_D(x) = \sqrt{\frac{2\epsilon_s V_{\mathrm{bi}} + V(x)}{q N_{\mathrm{DPI}}}}. \qquad (9.15)$$

The implantation gradient prevent the cut off of the n-well from the read-out path. Without the implantation gradient the depletion layers would connect in the front of the well as the maximum voltage inside the well can be found at the front of the well. As mentioned above, the voltage inside the n-well is pinned, if the depletion layer are connected. the depletion layers in the rest of the well would never be connected and the electrons are stored inside the well. Since the separation of the depletion layers is big enough at the front of the well, the path closes at the edge of the well first. Eventually, the pinning voltage across the diode resembles the actual voltage drop across the diode of the FD voltage. The total potential difference is only depending on the doping concentration of the DPI. A TCAD simulation of the electrostatic potential inside the diode is shown in Fig. 9.9 as verification for the theoretical discussion. Every change in doping concentration inside a pinned photodiode could lead to the formation of a barrier at the beginning of the read-out path [3]. The best way to mitigate a barrier is to reduce the overall doping concentration as the influence of the doping concentration is reduced. Figure 9.10a shows the conduction band energy inside the diode for different n-well doping concentrations. A plateau can be seen for every curve, but a small electrical field would transfer the electrons through this area. The only parameter which characterises a barrier is the slope of the conduction band energy beneath the collection gate which can be seen in Fig. 9.10b. As it can be seen, a barrier is only present for high n-well doping concentrations as the slope gets positive for doping concentrations greater than $3.0 \frac{1}{\mathrm{cm}^2}$. The influence of the CG voltage on the conduction band energy was simulated for a sensor without the DPI (Reference), and two versions with the DPI from which one was designed with an intentional barrier. The results are presented in Fig. 9.11. The barrier can be observed as the conduction band energy shows a maximum for small V_{GC}. By increasing the

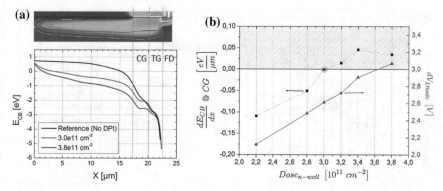

Fig. 9.10 Left: TCAD simulation of the conduction band energy for different doping concentrations. Right: maximum slope of the conduction band energy beneath the collection gate [4]

CG voltage, the barrier is reduced and eventually vanishes. Therefore, the CG can be used to reduce a barrier at the interface of the n-well and the read-out path. The design of a standard pinned photodiode can also be simplified by introducing a CG to the structure. Small barriers can be completely reduced and the diode must not be optimized by numerous simulations and experiments. However, it is not practicable for bigger barriers as the voltage which is applied at the CG reduces the number of electrons which can be detected as the FD can only be filled until the CG Potential is reached.

The conduction band energy of a pinned photodiode with only the doping gradient and without the DPI shows a slightly different behaviour. The collection gate voltage still influences the conduction band. It can be seen that the coupling between the CG, TF and the well changes for different V_{GC}. If V_{GC} is nearly as high as the pinning voltage, then the conduction band energy beneath the CG is coupled with the well. The conduction band energy follows the slope of the well and then bends towards the TG. If V_{GC} increases, the coupling is shifted to the TG as the influence of the well is reduced and the band bending changes the convexity.

The conduction band energy of the sensor without the intentional built-in barrier is only coupled with the TG. Since the pinning voltage follows the voltage drop of the FD voltage across the diode, the potential of the well does not have any influence on the voltage distribution. The influence of the CG voltage on the electron transfer should be reduced as the coupling with the TG is increased.

We calculated the transfer times based on the electron velocity simulation with the assumption that the electrons travel one-dimensionally through the diode. Since the pinning potential for a diode with the DPI isn't pinned any more, the voltage drop across the pn-junction is relatively high which leads to strong electric fields at the well interfaces as it can be observed in Fig. 9.12. The electron speed in these regions is nearly 100 times higher in comparison to a diode without the DPI. Eventually, this leads to a improved electron flow across the diode as all electrons are forced into a designated path. In Fig. 9.13, the Path-Time diagram for one electron can be seen

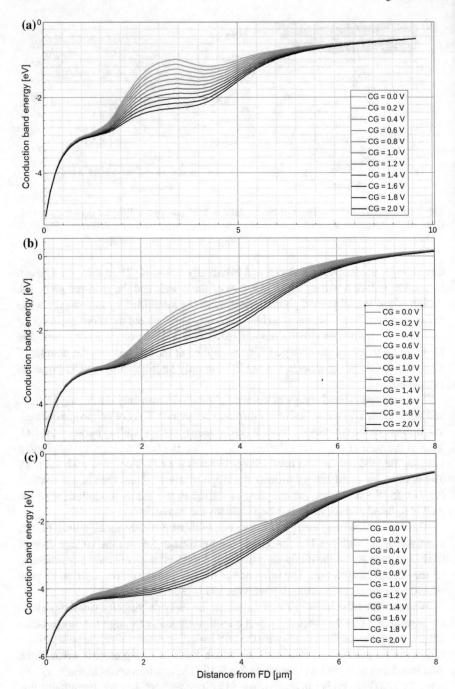

Fig. 9.11 Simulation of the conduction band energy for different sensors. **a** With an intentional barrier, **b** Reference and **c** DPI without barrier

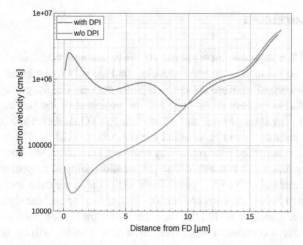

Fig. 9.12 Simulated electron velocity inside the n-well with the DPI (red) and without the DPI (green)

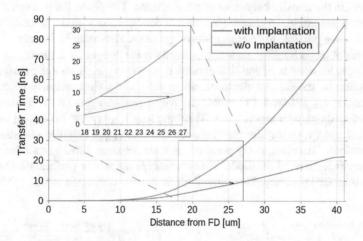

Fig. 9.13 One-dimensionally simulation of the transfer time of one electron for a pinned photodiode without the DPI (green) and with the DPI (red) [4]

which was calculated using (9.16). The transfer time is calculated for diodes length of up to 40 μm with and without the DPI. In this one-dimensionally calculation, a 26 μm long diode can achieve the same performance as a 19 μm lond one.

$$t = \int_0^S \frac{1}{v(x)} dx \qquad (9.16)$$

9.5 Measurement

We measure laser response curves to extract the parameters which describe the electron transfer to achieve comparable results. In Fig. 9.14a, the basic principle of operation is shown to measure the laser response curves. The triggering of the laser and the shutters is performed on-chip and the positions of the pulses refer to this global trigger. The laser pulse is shifted towards the TG shutter which is positioned at 75 ns. The laser delay can be adjusted in time steps of 6.25 ns. The laser response is then plotted against the time remaining between TG shutter and laser pulse (time before shutter: t_{BFS}) as a function of the propagation delay in respect to the time of flight of the laser pulse. An ideal photodiode with infinite electron speed should not show any signal before the laser and the TG shutter overlap and the signal resembles the laser pulse. This means that the time the first signal is recorded is equal to the transfer time for a complete charge transfer. The intensities will be normalized in respect to I_0.

Since the electrons are not drained immediately, the form of the curve is altered in respect to the transfer behaviour of the electrons. Therefore, the measured signal provides an insight in the electron transfer. In combination with the second draining path, it is even possible to determine the origin of the electrons. The electrons which are slower or travel longer distances will reach the FD delayed. The time t_{compl} which is necessary between laser and shutter pulse to ensure complete charge transfer gives information on the slowest electrons. Since the slowest electrons may be affected by a barrier, t_{compl} does not resemble the whole read-out process. To measure the whole electron transfer process, the TG shutter length t_S is set to the lowest value so only electrons with a transfer time between t_{BFS} and $t_{BFS} + t_S$ will contribute to the signal output. Therefore, every point of the laser response curve gives information on different sections of the diode. We also analyse the sensor performance by their charge transfer efficiency. Since the diode is always connected to one of the two

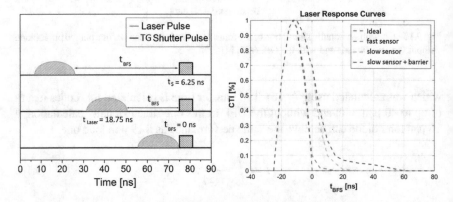

Fig. 9.14 Left: measurement principle to extract laser response curves. Right: schematic drawing of possible laser response curves

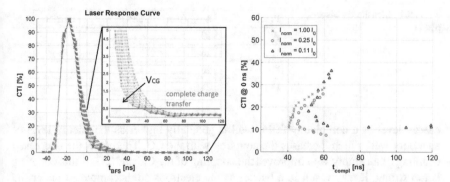

Fig. 9.15 Left: laser response curves of pixel A for different V_{CG}. Right: parametric curves of the laser response curves for different light intensities

draining paths, the charge transfer inefficiency (CTI) is given by the amount of generated electrons which are not effectively drained from the n-well before the TG shutter opens and therefore contributing to the FD discharge. The CTI is equal to the voltage drop if the signal reaches the maximum possible output as the accumulations are set to a constant value. For lower signal levels the CTI is calculated in respect to the highest measured signal. Figure 9.14b shows possible measurement curves based on the discussion above. An ideal diode with infinite electron speed should show the first signal after the laser and TG-shutter overlaps with no CTI at 0 ns. The CTI for $t_{BFS} > 0$ resembles the amount of charges which remain inside the well. Therefore, the CTI resembles the transfer behaviour of the fastest electrons.

If the global electron transfer time rises, the peak is shifted towards longer transfer times and broadens the measured pulse as the electrons are not drained fast enough from the diode. A barrier however will affect the transfer time of the slowest electrons which leads to longer times to ensure a complete charge transfer while the CTI may be low. The presented measurement results originates from different chips. The laser drivers may differ from each other but since the internal oscillators are trimmed, the moment the laser pulse is completely shifted behind the TG shutter pulse is a characteristic point of the curve. All laser response corves are normalized in respect to this point. Based on the discussion above, we measured four different sensor pixels. The reference is pixel A which is the pinned photodiode presented by the Fraunhofer IMS. In pixel B we have enlarged the photo active area and the maximum drift distance L_{max}. Pixel C and D are manufactured with the DPI. In pixel C we intentionally included a barrier. Figure 9.15a shows the laser response curves of the reference (Pixel A) for different CG voltages. They look similar although it seems that there is a optimal voltage. To verify this assumption, the CTI of the curves is plotted against the time t_{compl} to ensure complete charge transfer for different intensities (see Fig. 9.15b). It can be seen that a decrease in intensity shows a significant influence on the charge transfer. An CG voltage optimum is observable and increasing the CG voltage will eventually results in a slower sensor. Since this behaviour can only be seen for lower intensities, the self-induced drift [1] affects the charge transfer. The

Table 9.1 Measured sensor pixels

Pixel	L_{max} (μm)	DPI	Barrier
A	19		
B	26		
C	26	x	x
D	26	x	

energy level of a completely full n-well is generally higher as the energy level of an empty well which increases the coupling with the CG. Therefore, the decreased coupling of an empty well improves the coupling of the CG and TG. If this coupling is too strong, it may result in a barrier as the electrons are accumulated under the TG. This assumptions seems correct as the influence of the intensity on the laser response curves for small CG voltages is relatively low. However, the time to ensure complete charge transfer varies over the measurement rang and it differs by up to 40 ns. For low light intensities and high CG voltages, the transfer time isn't measurable since the internal clock only provides a shift between the laser pulse and the shutter pulse of $T_{BFS} = 120$ ns. If the time to ensure complete charge transfer is longer than 120 ns, we set the corresponding parameter to 120 ns as we wanted to still be able to analyse the CTI (Table. 9.1).

Figure 9.16 shows the laser response curves for all pixels. Pixel B shows a slightly different curve as the optimum CG voltage is 2 V. An optimum below 2 V is only observable if the intensity is reduced. If the complete transfer time can not be measured, the optimal CG voltage will always be 2 V as the higher CG voltage increases the electron velocity in the front section of the diode which leads to fewer electron inside the diode. So the CTI will always be reduced by applying a higher voltage on the CG. However, the read-out path of the pixels is identically while the longer diodes provides more electrons due to the 1.6 times bigger photo active area. The increase in numbers of electrons reduces the coupling between the CG and the TG while the self-induced drift also enhances the electron transfer. Therefore, the influence of the CG on the large diode isn't as strong as on the small reference.

The influence of the barrier in the transfer path is visible for pixel C as the transfer time is significantly increased and a complete transfer time can only be measured for high CG voltages and high intensities. Small CG voltages results in a CTI of nearly 100% which means that nearly no electrons are drained from the well. Only a high CG voltage can reduce the barrier so that a charge transfer is possible as observed in the simulation.

If a barrier-free electron transfer can be achieved, the negative influence of high CG voltages on the electron transfer vanishes. This means that an increase in CG voltage will always result in a faster electron transfer which leads to a significant decreased time for complete charge transfer for DPI enhanced pinned photodiodes. Furthermore, the influence of the intensity on the charge transfer is also reduced. The transfer time doesn't change for high CG voltages. Only the CTI gets bigger if the intensity is increased. For lower CG voltages however, a stronger influence on the charge transfer can be observed. A low intensity results in longer transfer times. This behaviour can be explained using (9.15) and the basic principle of operation of

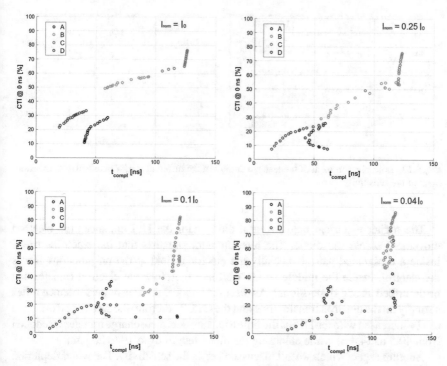

Fig. 9.16 Parametric laser response curves of the pixels for different intensities

a JFET. As mentioned, the voltage which is applied to the CG affects the potential inside the read-out path. If the voltage is set too low, the depletion regions will be separated and a barrier is formed which height depends on the applied voltage. If the depletion layers aren't connected any more, the electron generation will decrease and electrons which are generated in the border area of the diode are drained slower. If enough electrons are present inside the well, then the self-induced drift compensates the small barrier as it can also be observed for pixel B. The laser respond curves of the small reference and the large diode with the DPI are similar for high intensities and low collection gate voltages. If a CG voltage similar to the pinning voltage is applied, both of the diodes are comparable. The higher electrical field at the pn-junction enhances the electron transfer that the long diode performs as good as the small reference as it was already shown in the simulation.

If the CG voltage is high enough, all electrons can be drained faster. The variation in light intensity will only results in a different CTI. This fact can be used to conclude that the electrons in the whole well are accelerated and not just the electrons in the front section. The intensity only varies the number of electrons inside the well. A higher number of electrons therefore results in a higher CTI as more electrons remain in the well. The electrons in Sensors without the DPI remains in the border areas of the diode because of the reduced electrical field. The CTI remains stable over a large intensity range.

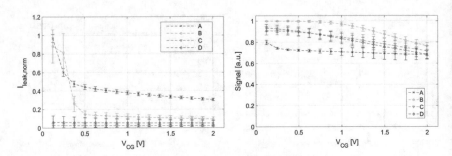

Fig. 9.17 Left: measurement of the leakage current of the different sensor pixels. Right: measurement of the sensitivity

The higher electrical field inside a diode with the DPI enhances the electron movement across the diode. The measurement suggests that the electrons travel inside a designated path. Especially the electrical field at the pn-junctions forces the electron inside the middle of the well and therefore the electron transfer can be described as one-dimensional. As discussed abobe, the sensor performance relies strongly on the electron transfer time and the CTI. Therefore, the sensor performance can be increased with the DPI as the time to ensure complete charge transfer is reduced from 40.5 to 14.4 ns while enlarging the diode length from 19 to 26 μm.

Another aspect which wasn't discussed yet, is the sensitivity. The wider depletion region should enhance the electron detection in the depth which should lead to a higher sensitivity for IR-light. Figure 9.17 shows the measurements with constant light to determine the sensitivity. The sensors are illuminated permanently with a LED (Peak-Wavelength 905 nm) and therefore the n-well is always completely filled with electrons. The electron current is switched every 257 μs (200 accumulations) for 31 ns (2000 accumulations) from the draining path to the FD. This measurement enables the determination of the sensitivity and the leakage current which indicates the shutter effectiveness. Since the the well should always be filled, the number of detected electrons should be proportional to the number of generated electrons. The shutter effectiveness is also important for the real application because every electron which reached the FD outside the designated times contributes to Δ_{Q1} or Δ_{Q2}. A high leakage current will eventually further reduce the measurement range. The sensitivity seems to be dependent from the CG voltage as the sensitivity is reduced for higher CG voltages. The sensitivity of the large diodes is similar the the small reference for high CG voltages. It must be considered that the read-out process of the electrons is still pulsed. The higher coupling between the TG and the CG could affect the switching from the draining path to the FD. The higher sheet capacity beneath the CG could result in a slower switching which eventually reduces the effective pulse length. A decrease in pulse length will logically reduce the number of transferred electrons. Since the small reference is composed of three diodes whose FD are shorted, the effect can be compensated. A small CG voltage will lead to the highest sensitivity in all sensor pixels. This can be explained with the maximum leakage current and the beforehand mentioned dependence of the gates. Furthermore, the leakage current of the DPI enhanced diodes isn't affected by the CG voltage any more. Even for low voltages, the electron current is well separated between the two paths. The number of electrons which reach the FD over the closed TG is low even if the difference in potential between the closed TG and the CG is small. The intentionally implemented barrier reduces the leakage current even more as the total number of electron which reach the read-out path is also reduced. We showed that the electrons travel linearly through the diode if the DPI is used and it can be suggested that the same effects is present in the read-out path. The electrons travel inside of a designated path across the diode so that both of the transfer paths are perfectly separated. The CG voltage for diodes without the DPI has to be adjusted in order to achieve the same effect. Because of the three read-out paths in the small reference, the leakage current of this device is increased.

9.6 Conclusion

In this an enhanced pinned photodiode which was developed by the Fraunhofer IMS, was presented and analysed. Based on the theoretical conclusions we proposed a new type of n-well enhancement. It has been shown that a pinned photodiode can be interpreted as a one-sided JFET and an intended separation of the depletion layers gives the possibility to use the FD voltage to enhance the potential distribution across the diode. The separation of the depletion layers can be achieved by locally increase the doping concentration inside the middle of the n-well. The drift path implantation DPI affects the well and the pinning voltage follows the voltage drop of the FD voltage over the diode. Without the DPI, the pinned voltage inside the well would suppresses the FD voltage. Normally, the slope of the voltage gradient only depends on the implantation dose and implantation energy. By mitigating the pinning behaviour, the higher electrical field inside the diode accelerates the electrons faster and the time to ensure complete charge transfer is reduced. It has also been shown that the electrons will travel on a designated path across the diode and their movement can be described as linear.

Besides the development of the drift path implantation, the influence of the second gate, the Collection Gate, was analysed. A significant dependency of the Collection Gate voltage on the electron transfer was found. Without the drift path implantation, an optimum for the Collection Gate voltage was observed since the coupling between the well, the Collection Gate and the Transfer Gate is altered. Every deviation from this optimum leads to longer transfer times. Sensors which was manufactured with the DPI doesn't show this behaviour and an increase in Collection Gate voltage only increase the sensor performance. The combination of an adjusted Collection Gate voltage and the DPI enhances the electron transfer and the performance of a 26 μm long diode was increased significantly. The time to ensure complete charge transfer of a 19 μm long diode without the DPI of 40.5 ns was reduced to 14.4 ns inside the larger diode with the DPI. The proposed implantation can be used to built larger sensors which performs better than their small equivalent.

References

1. A. Driewer, B.J. Hosticka, A. Spickermann, H. Vogt, Modeling of the charge transfer in a lateral drift field photo detector. Solid State Electron. **126**, 51–58 (2016)
2. D. Durini, A. Spickermann, R. Mahdi, W. Brockherde, H. Vogt, A. Grabmaier, B.J. Hosticka, Lateral drift-field photodiode for low noise, high-speed, large photoactive-area CMOS imaging applications. Nucl. Instrum. Methods Phys. Res. Sect. A: Accel. Spectrom. Detect. Assoc. Equip. **624**, 470–475 (2010)
3. E.R. Fossum, D.B. Hondongwa, A review of the pinned photodiode for CCD and CMOS image sensors. IEEE J-EDS **2**, 33–43 (2014)
4. L. Girgenrath, M. Hofmann, R. Kühnhold, H. Vogt, Optimization of transfer times in pinned photodiodes, in *Proceedings of the 5th International Conference on Photonics, Optics and Laser Technology*, vol. 1 (2017), pp. 312–316
5. Han, A time-of-flight range image sensor with background canceling lock-in pixels based on lateral electric field charge modulation. IEEE J-EDS **3**, 639–641 (2015)
6. C. Jacoboni, C. Canali, G. Ottaviani, A.A. Quaranta, A review of some charge transport properties of silicon. Solid State Electron. **20**, 77–89 (1977)
7. A. Spickermann, D. Durini, A. Süss, W. Ulfig, W. Brockherde, B.J. Hosticka, S. Schwope, A. Grabmaier, CMOS 3D image sensor based on pulse modulated time-of-flight principle and intrinsic lateral drift-field photodiode pixels, in *Proceedings of the ESSCIRC* (2011), pp. 111–114
8. A. Süss, C. Nitta, A. Spickermann, D. Durini, G. Varga, M. Jung, W. Brockherde, B.J. Hosticka, H. Vogt, S. Schwope, Speed considerations for LDPD based time-of-flight CMOS 3D image sensors, in *European Solid State Circuits Conference (ESSCIRC)* (2013), pp. 299–302
9. S.M. Sze, K.K. Ng (2007). Physics of Semiconductor Devices. ISBN: 978-0-47 1-1 4323-9
10. C. Tubert, L. Simony, F. Roy, A. Tournier, L. Pinzelli, P. Magnan, High speed dual port pinned-photodiode for time of flight, *Proceeding of IISW* (2009)
11. Y. Xu, A.J.P. Theuwissen, Image lag analysis and photodiode shape optimization of 4T CMOS pixels, *Proceeding of IISW*, 153 (2013)
12. Y. Xu, Fundamental characteristics of a pinned photodiode CMOS pixel. Ph.D. Thesis, 2015

Chapter 10
Passive Beam Combining for the Development of High Power SOA-Based Tunable Fiber Compound-Ring Lasers Using Low Power Optical Components

Muhammad A. Ummy, Simeon Bikorimana, Abdullah Hossain and Roger Dorsinville

Abstract A simple, stable, compact, and cost-effective dual-output port widely tunable SOA-based fiber compound-ring laser structure is demonstrated. Such a unique nested ring cavity enables the splitting of optical power into various branches where amplification and wavelength selection for each branch are achieved utilizing low-power SOAs and a tunable filter, respectively. Furthermore, splicing Sagnac Loop Mirrors at each end of the bidirectional fiber compound-ring cavity not only allows them to serve as variable reflectors but also enables them to channel the optical energy back to the same port thus omitting the need for high optical power combiners. Further discussed is how the said bidirectional fiber compound-ring laser structure can be extended to achieve a high-power fiber laser source by exclusively using low power optical components such as N × N couplers and (N > 1) number of SOAs. More than 98% coherent beam combining efficiency of two parallel nested fiber ring resonators is achieved over the C-band tuning range of 30 nm. Optical signal-to-noise ratio (OSNR) of +45 dB, and optical power fluctuation of less than ±0.02 dB are measured over 3 h at room temperature.

M. A. Ummy (✉)
New York City College of Technology, City University of New York, New York, USA
e-mail: maummy@citytech.cuny.edu

S. Bikorimana · A. Hossain · R. Dorsinville
The City College of New York, City University of New York, New York, USA
e-mail: sbikori00@citymail.cuny.edu

A. Hossain
e-mail: ahossai12@citymail.cuny.edu

R. Dorsinville
e-mail: rdorsinville@ccny.cuny.edu

© Springer Nature Switzerland AG 2019
P. Ribeiro et al. (eds.), *Optics, Photonics and Laser Technology 2017*,
Springer Series in Optical Sciences 222,
https://doi.org/10.1007/978-3-030-12692-6_10

10.1 Introduction

Extensive research has been conducted on single-mode fiber resonators of different architectures: linear [1], Fox-Smith [2], ring [3–6], and compound fiber ring [7–10] cavities. Such architectures have been established for designing and building various types of fiber laser sources with single-longitudinal mode operation for both low and high optical power applications, such as optical communication systems, scientific, medical, material processing and military purposes [11] thus underscoring the great interest in output optical power adjustability, scalability and wavelength tunability.

Wavelength selection in a fiber laser is typically achieved with different types of fiber-based optical filters. Most research works have achieved wavelength selection in a linear cavity erbium doped fiber (EDF) lasers using Fabry-Perot (FP) filters as well as Fiber Bragg Grating (FBG) based-optical filters. In addition, single longitudinal mode operation has been attained by the saturable absorption of un-pumped EDF with narrow band FBG [1]. Other architectures, such as the optical-fiber analog of Fox-Smith resonators utilize directional-coupling technologies and optical fibers to couple multiple cavities together to form compounded optical resonators resulting in several periods at the output with various modes of intensity. The highlight of this configuration is that such linked resonators can be designed to allow or suppress spectral modes, which is paramount for laser line narrowing, filtering, and spectral analysis [2].

Furthermore, equivalent phase shift (EPS) FBG filters have been used in fiber ring lasers in order to achieve ultra-narrow single transmission band selection. Typically, sampled fiber gratings (SFBGs) exhibit multiple reflection peaks when periodically changing the refractive index modulation. As previous works have demonstrated, EPS can be introduced into SFBGs by manipulating a single period out of the multiple sampling periods. This phenomenon is exploited to create an ultra-narrow single transmission band FBG filter alongside with semiconductor optical amplifiers (SOAs) as gain media (as opposed to EDFs, which suffer from wavelengths' competition and homogenous line broadening) [5]. EPS FBGs have been successfully used in several fiber-optic systems, such as distributed feedback lasers [12], optical CDMA coding [13], and single longitudinal mode fiber ring lasers [14].

Further works have improved on the ring structured EDF lasers using FBG, FP etalons and Sagnac loops to select longitudinal modes more efficiently. Single frequency and narrow line-width EDF ring lasers have been illustrated utilizing laser diodes as pumps and EDFs as gain media where fiber Faraday Rotator are introduced into the system to eliminate spatial hole burning effect [6], which can also be eliminated by the use of SOAs, for instance, in a bidirectional fiber compound-ring resonator herein discussed.

It is usually desirable to adjust and control output power of fiber lasers. Typically, complex and expensive in-line variable optical attenuators (VOA) with adjustable insertion losses are used to control the output optical power level of laser sources. Mechanical, micro-electromechanical [15–17], acousto-optic [18], electro-wetting [19], optical fiber tapers [20], and hybrid microstructure fiber-based techniques [21] are widely used to adjust the insertion losses of the in-line fiber-based VOA. However, all-fiber based low power variable reflectors or mirrors such as SLM can also be used to adjust the optical power from both low and high-power fiber laser sources. All-single-mode fiber-based SLMs have been widely utilized in highly sensitive temperature [22], strain [23], pressure [24] sensors and wavelength optical switches [25]. Moreover, the SLMs with adjustable reflectivity have been used to form FP linear resonators [26–29] where the output optical power is dependent on the SLMs' reflectivity [30, 31]. In this work, two SLMs were utilized to regulate the optical power delivered from the two output ports of the proposed fiber compound-ring laser. Furthermore, two inexpensive low power SOAs were placed in two parallel nested ring cavities to demonstrate the possibility of achieving a highly power scalable, adjustable and switchable fiber laser structure based on multiple nested compound-ring cavities formed by N × N fiber couplers with two SLM-output couplers.

Various approaches have been employed to scale up optical power of laser sources where beam combining has shown to be a promising alternative technique of achieving high power by scaling up multiple combined laser elements. Several works have demonstrated the coupling of several anti-reflection (AR) coated laser diodes (LDs) into external cavities for operation as coherent ensembles. Placing spatial filters at the cavities' Fourier planes to serve as coherence feedback for each of the individual laser diode beams results in narrow linewidth single mode output. Contemporary research works have investigated high quality active and passive phase locking of multiple LDs using master oscillator power amplifier (MOPA) designs and optical coupling in external resonators, respectively, specifically Talbot cavities, which force collective coupled mode oscillation. The combined output power is naturally limited by the number of phase-locked LDs and their individual power. Consequently, high power laser sources with high beam quality have been demonstrated by using complex coherent and spectral beam combining techniques in external cavities [32–39]. Previous works using coherent combination techniques have illustrated the combination of several MOPAs fiber laser systems and use of single polarization low-nonlinear photonic crystal fibers to achieve relatively high combining efficiencies of up to 95% without beam quality degradation. Such works have paved the way for future works improving on brightness enhancements and increased power scalability. For instance, the use of external output coupling mirrors for coherent beam combining has been used to achieve nearly diffraction limited beam operation with significant improvement in brightness.

In addition, incoherent beam combining method [40, 41] has been used to scale up the optical power by combining individual laser elements as well. Promising results were delivered through spectral beam combination techniques by a three-channel 1-μm fiber laser with a combining efficiency of 93%. Such a system, possessing power amplifier fiber channels with a narrowband, polarized, near diffraction limited output tunable over nearly the entire 1 μm Yb^{3+} gain bandwidth, served to further heighten the prospects of increased combined beam quality, power scalability, and average power. Even improved results were obtained when four narrow linewidth Yb^{3+} doped photonic crystal fiber amplifier chains, each being of a different wavelength were combined by a polarization independent reflective diffraction grating. Michelson and

Mach-Zehnder resonators were used mostly in coherent beam combining to achieve high combining efficiency and nearly diffraction-limited beam quality [41–44]. More significantly, ring resonators have demonstrated high reliability, efficiency and stability [45] for passive coherent beam combining methods. However, achieving high power laser sources with high combining efficiency through the above approaches is not plausible without incorporating sophisticated high-power external optical components such as micro-lenses, isolators, circulators, photonic crystal fibers and MOPAs. As such, they do not serve in the interest of simplicity and cost-effectiveness. Furthermore, rare-earth, ytterbium doped fiber amplifiers (YDFAs) [45] and erbium doped fiber amplifiers (EDFAs) [46] that are usually used as gain media for beam combining to achieve high power laser systems with different types of cavity structures require external laser pumps rendering them bulky and quite inefficient. However, using nested compound-ring cavities to equally split circulating beams into N-number of low power beams for amplification by N-number of low power SOAs, one can achieve a highly efficient and high-power laser system that eliminates extra pump lasers, MOPAs, and other expensive external high power optical components.

SOAs [47], stimulated Raman scattering (SRS) amplifiers [48, 49] and stimulated Brillouin scattering (SBS) amplifiers [50, 51] have also been used as gain media in different fiber laser systems. Of all the mentioned amplifiers, SOAs stand out the most because they are more compact, light, cost effective, efficient, available for different operational regions from a wide range of wavelength spectrum, and can be incorporated into other indium phosphide (InP) based optical components. Extensive theoretical studies and analyses of SOA-based optical systems have shown that they can be used for a wide range of systems, such as, compact SOA-based laser rangefinder [52], optical pulse delay discriminator [53, 54], optical and logic gate [55] for high-speed optical communications by exploiting four-wave mixing [56, 57] and photonic integrated circuits (PIC), which provides functions for information signals imposed on optical wavelengths [58]. It must be noted that utilizing SOAs in the bidirectional fiber ring resonator structure as proposed herein eliminates the need for extra optical components such as optical isolators and circulators; thus, facilitating integration with other optical components for compact fiber laser systems. SOAs have also been used along with EDFAs to suppress optical power fluctuation in pulsed light wave frequency sweepers where the suppression of power fluctuation is attributed to the gain saturation and fast response of SOAs [59, 60].

In this work, a novel technique is proposed and validated for coherent beam combining method based on the passive phase-locking mechanism [61–63] of two C-band low power SOA-based all-single-mode fiber compound-ring resonators at 3 dB fiber couplers connecting two parallel merged ring cavities. Unlike in previous work [64], the non-adjustable multimode fiber output coupler formed by a high power and expensive power combiner with a multimode output fiber (i.e., low brightness) has been replaced by two low power SLMs to create a dual-output port all-single-mode fiber laser structure with switchable and adjustable output power. Additionally, single-mode performance is sustained to improve the brightness at the proposed fiber compound-ring laser output port. The output power of the proposed configuration was almost twice as large as the output power obtained from a single SOA-based fiber ring or FP linear resonator [29]. More than 98% beam combining efficiency of two parallel nested fiber ring resonators is demonstrated over the C-band tuning range of 30 nm. Optical signal-to-noise ratio (OSNR) +45 dB, and optical power fluctuation of less than ±0.02 dB are measured over three hours at room temperature.

The main characteristics of the proposed fiber compound-ring laser, power tunability, and switchable output port can find applications in long-distance remote sensing [65]. Moreover, its wide range wavelength tunability can benefit various applications in fiber sensors [66–70], wavelength division multiplexer (WDM) optical communications [71], and biomedical imaging systems [72] working in the third near infrared biological window [73].

Lastly, a method to realize a highly power-scalable fiber compound-ring laser system via low power and low-cost optical components, such as tunable filters, couplers, and low power SOAs is discussed.

10.2   Experimental Setup

Figure 10.1 illustrates the experimental setup of the C-band SOA-based tunable fiber laser with two nested ring cavities (i.e., compound-ring resonator) and two broadband SLMs that can serve as either dual-output ports or a single output port depending on the reflectivity of each SLM. Each ring cavity is comprised of two branches: I-II and I-III, for the inner and the outer ring cavity, respectively. Both ring cavities share a common branch, I, which contains SOA_1 (Kamelian, OPA-20-N-C-SU), a tunable optical filter (TF-11-11-1520/1570), and a polarization controller, PC_1. Branch II contains SOA_2 (Thorlabs, S1013S), and a polarization controller, PC_2. Due to the lack of availability of a third SOA, branch III only contains a polarization controller, PC_3. As Fig. 10.1 portrays, all branches are connected by two 3 dB fiber couplers, C_1 and C_2, which are connected to SLM_1 and SLM_2, correspondingly. Each SLM (SLM_1 and SLM_2) in conjunction with a PC (PC_4 and PC_5, respectively) acts as a variable reflector. By adjusting PC_4 or PC_5, one can manipulate the reflectivity of SLM_1 and SLM_2, respectively, and switch between single and dual-output port configurations [27, 28]. Inserting the low power tunable optical filter (TF) in the common branch, I, allows wavelength selection and tuning between 1520 and 1570 nm. The three PCs (PC_1, PC_2 and PC_3) control the state of polarization of the light circulating within the compound ring cavity.

10.3   Principal of Operation

When the pump level (i.e., bias current threshold level) of either SOA is larger than the total losses of the fiber compound-ring cavity, amplified spontaneous emission (ASE) emitted from the SOAs propagates in either the forward or backward direction. For example, when a bias current, IB, of about 75 mA is injected into SOA_1 (branch I),

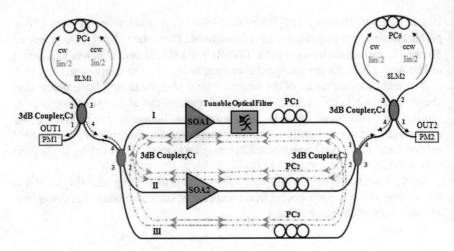

Fig. 10.1 Experimental setup of the dual SLM SOA-based TF compound-ring laser [74]

the ASE emitted by SOA_1 circulates in the clockwise (CW) direction propagating through the tunable optical filter (TF) allowing specific wavelengths to pass through polarization controller PC1 to enter port 1 of the 3 dB fiber coupler C_2 splitting the light beam equally between branches II and III at ports 2 and 3, respectively. The light beam that circulates into branch II propagates through polarization controller, PC_2, and then is amplified by SOA_2 when its bias current level, IB, is around 180 mA. This amplified light beam then reaches port 2 of 3 dB fiber coupler C_1 where it is equally divided between ports 1 and 4. Likewise, the light beam from branch III reaches port 3 after passing through polarization controller, PC_3. A half of the light beam coupled into port 1 of 3 dB fiber coupler C_1 is further amplified by SOA_1. In this fashion, a round-trip is completed in the fiber compound-ring structure, which ensures lasing. The remaining half of the light beam is coupled into output port 4 of 3 dB fiber coupler C_1 and is injected into input port 4 (i.e., I_{in}) of the SLM_1. Polarization controller, PC_4, controls the reflectivity of SLM_1 by adjusting the polarization of the light beams propagating through SLM_1. For a single output port configuration, the polarization controller, PC_4, of SLM_1 is adjusted for minimum power at output port 1 (OUT1). The counter-clockwise (CCW) and CW light beams interfere destructively and constructively at output ports 1 and 4, correspondingly of 3 dB fiber coupler, C_3, which channels the power wholly back to the compound-ring cavity.

Because the fiber compound ring resonator lacks any optical isolators within its branches, the two CCW-propagating light beams circulate in the nested ring cavities as shown in Fig. 10.1. The CCW beam from SOA_1 reaches port 1 of 3 dB coupler C_1 and splits into two equivalent light beams (i.e., 50%), which are transmitted into both ports 2 and 3. The light beam that travels into branch II is amplified by SOA_2, and passes through polarization controller PC_2 before it reaches port 2 of the 3 dB fiber coupler C_2. Similarly, the light beam that propagates through branch III passes through polarization controller PC_3 before it reaches port 3 of 3 dB fiber coupler C_2.

Half of the light beam at 3 dB fiber coupler C_2 is coupled into port 1 and propagates back into branch I to complete one round trip, while the other half of the beam is channeled into SLM_2, which then exits at output port 1 of 3 dB fiber coupler C_4 (OUT2). The output power can be controlled by polarization controller, PC_5. An optical spectrum analyzer (OSA), VOA and optical power meter (PM) were used to characterize the proposed fiber compound-ring laser. Note that the path lengths of both loops are nearly identical since all the branches have identical length and all fiber connections are done by using FC/APC connectors.

10.4   Characterization of the Fiber Compound-Ring Lasers

10.4.1   Gain Medium

The ASEs of SOA_1 and SOA_2 were characterized by using an OSA where both SOAs were set at the same bias current (I_B) of 200 mA. The ASE of both SOA_1 (green solid line) and SOA_2 (blue broken line) are shown in Fig. 10.2. Despite both SOAs being biased identically at 200 mA, they exhibited different ASE spectra. The ASE data in Fig. 10.3 indicates that SOA_1 has higher gain than SOA_2 for the same bias current level suggesting that dissimilar bias current levels are required to obtain the same output power when the SOAs are individually used in the proposed fiber compound-ring resonator.

10.4.2   Lasing Threshold Level

To determine the lasing threshold levels of the fiber compound-ring laser, the TF was set to 1550 nm and each SOA was placed individually in the proposed fiber

Fig. 10.2 ASE spectra of SOA_1 (green solid line) and SOA_2 (blue broken line) of the fiber compound-ring laser with both SOAs set at bias current I_B of 200 mA [74]

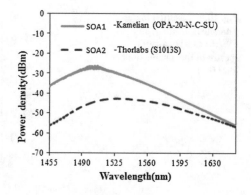

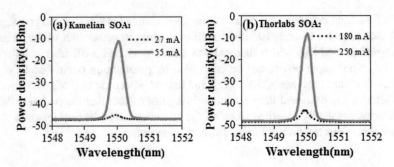

Fig. 10.3 Illustrates the output spectra of the fiber laser threshold for each individual SOA gain medium. **a** Shows the lasing threshold of SOA$_1$; **b** shows the lasing threshold of SOA$_2$ gain medium

compound-ring cavity. As shown in Fig. 10.3a and b, lasing occurs when the bias current levels of SOA$_1$ and SOA$_2$ are above the minimum threshold currents 27 mA and 180 mA respectively.

10.4.3 Tunable Optical Filter

The insertion losses (ILs) and corresponding full-width-half maximum (FWHM) linewidths within the entire tuning range (1520–1570 nm) of the tunable filter is shown in Fig. 10.4. The insertion losses decrease as the wavelength increases. The maximum and minimum ILs of 5.5 dB and 2.2 dB were measured at 1520 nm and 1570 nm, respectively. A similar downward trend was also observed when the FWHM linewidths were plotted against the wavelengths, as shown in Fig. 10.4. The linewidth varies from 0.4 to 0.32 nm at 1520 and 1570 nm, respectively.

This downward trend suggests an opposite upward trend for the output power of the proposed fiber compound-ring laser for a constant gain setting of the SOAs. It

Fig. 10.4 Insertion losses (triangles), IL (dB), and FWHM (nm) (squares) spectra of the tunable optical filter [74]

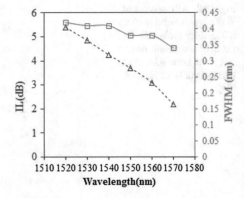

Table 10.1 3 dB-Bandwidth (BW) at different bias current I_B (mA) levels and 1550 nm center wavelength with both SLMs' reflectivity set at <0.1% [74]

SOA$_1$ I_{B1} (mA)	SOA$_2$ I_{B2} (mA)	P_{OUT1} (dBm)	P_{OUT2} (dBm)	3 dB-BW (nm)
75	250	3.40	3.40	0.1985
100	300	5.80	5.70	0.2075
125	350	6.73	6.75	0.2122
150	400	7.65	7.68	0.2131
175	450	8.35	8.35	0.2157
200	500	8.94	8.95	0.2182

follows that a constant output power is achievable over the entire wavelength tuning range by adjusting the bias current (I_B) of the SOAs but at the expense of signal broadening of the fiber laser source. The reflectivity of both SLMs was set to less than 0.1% so that both output ports of the fiber compound-ring lasers (i.e., OUT1 and OUT2) have the same output power. Then, by collecting both CW and CCW light beams through SLM$_1$ and SLM$_2$, respectively, the 3 dB bandwidth was measured at different bias current levels at 1550 nm wavelength by using an OSA. The 3 dB-bandwidth increased from 0.1985 to 0.2182 nm as the bias current was increased to the standard bias current of each of the SOAs (see Table 10.1).

10.4.4 Coherent Beam Combining Efficiency

The principle of the proposed passive coherent beam combining technique of dual compound-ring based fiber lasers with two adjustable output couplers (i.e., SLMs) is based on the passive phase-locking mechanism caused by spontaneous self-organization operation [58, 60]. The wide bandwidth of the SOAs facilitates the passive phase-locking mechanism, thus allowing the fields' self-adjustment to select common oscillating modes or resonant frequencies of the counter-propagating (i.e., clockwise and counter-clockwise) light beams in the two merged ring cavities and optimize their in-phase locking state conditions without any active phase-locking systems.

To assess the beam combining efficiency of the proposed fiber laser structure, utilized each individual SOA was utilized as a gain medium in the common branch, I, of the compound-ring cavity and measured the output power at both its output couplers: OUT1 and OUT2. Following that, both SOAs were placed simultaneously within the compound-ring cavities (branches I and II, respectively) and measured the output power at both output couplers. Note that the SLMs' reflectivity was adjusted to the maximum (i.e. >99.9%) and minimum (i.e., <0.1%), respectively. The tunable filter was manually adjusted from 1535 to 1565 nm and each SOA was driven and kept constant at its standard bias current, 200 and 500 mA, respectively. Figure 10.5 illustrates the passive coherent beam combining efficiency spectrum (right vertical

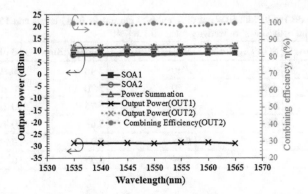

Fig. 10.5 Output power and combining efficiency spectra of the proposed dual-SLM SOA-based tunable fiber compound-ring laser system using two SOAs as gain media. Individual SOA output power spectrum: SOA_1 (filled squares), SOA_2 (unfilled circles), output power summation spectrum of both SOAs (unfilled triangles), and actual measured output power (crosses) at the output port, OUT2 with SOA_1 and SOA_2 driven at 200 mA and 500 mA constant bias current. The PC_1 and PC_2 were maximized for each wavelength [71]

axis) and the output power spectrum (left vertical axis) from the proposed fiber compound-ring laser operating with the individual SOAs as well as both SOAs over the C-band tuning range of 30 nm.

The beam combining efficiency (filled circles) was obtained by dividing the optical power measured at the output port (OUT2) under dual SOA fiber laser operation by the power summation (unfilled triangles) of the same port under individual SOA operation: SOA_1 (filled squares) and SOA_2 (unfilled circles) (i.e., $\eta = [P_{measured}/(P_{SOA1} + P_{SOA2})]$. The leakage optical power spectrum (unfilled squares) at the other output port (OUT$_1$) re-mained below −28.5 dBm. The maximum output power delivered with SOA_1 (Kamelian model) and SOA_2 (Thorlabs model) separately was +8.91 and +8.90 dBm at 1565 nm. On the other hand, when both SOAs were placed in the compound-ring cavities, the maximum measured output power obtained at the output port, OUT2, was +11.9 dBm at 1565 nm, which was double of that obtained with just either SOA. The maximum output power obtained by adding the optical power from a single SOA versus dual SOA fiber laser operation at the output port, OUT2, both at 1565 nm, was +11.91 dBm and +11.9 dBm, respectively. This is where the insertion losses of the tunable filter were the lowest.

The maximum and minimum obtained combining efficiencies (filled circles) were 99.76% and 98.06% at 1565 nm and 1555 nm, respectively, as shown in Fig. 10.5 (right vertical axis).

10.4.5   *Fiber Laser Power Tunability and Its Switchable Dual-Output Port Operation*

The proposed fiber compound-ring laser can operate with two adjustable and switchable output ports (i.e., OUT1 and OUT2). The output power from either output port can be tuned by simply controlling the bias current levels and thus gain levels of the SOAs or by adjusting the reflectivity of the SLMs while keeping the former constant.

The SOA gain was adjusted by setting the tunable filter at 1550 nm wavelength and adjusting the bias current levels of the SOAs. Table 10.2 shows the output power evolution at both output ports, OUT1 and OUT2 as a function of the bias current levels, I_{B1} and I_{B2}. The reflectivity of SLM_1 and SLM_2 was set to $\geq 99.9\%$ and $\leq 0.1\%$, respectively. The achieved maximum dynamic range was 40.75 dB at 1550 nm wavelength and standard bias current levels of 200 and 500 mA for SOA_1 and SOA_2, respectively.

The second approach involves adjusting the reflectivity of both SLM_1 and SLM_2 while keeping the gain of the SOAs constant (i.e., I_{B1} and I_{B2} is fixed at 200 and 500 mA, respectively). Depending on the reflectivity of SLM1 and SLM_2, the proposed fiber compound-ring laser can be operated as either a single or a dual-output configuration.

In a single output configuration, one of the SLMs, either SLM_1 or SLM_2, should be kept at the highest reflectivity (i.e., $\geq 99.9\%$) while the other should be set to its lowest reflectivity of (i.e., $\geq 0.1\%$). For us to characterize the power tunability performance of both output ports of the fiber laser, the tunable filter was to 1550 nm, and initialized the reflectivity settings of SLM_1 and SLM_2 to $\leq 0.1\%$ and $\geq 99.9\%$, respectively as explained earlier. The initial measured output power from both output ports, OUT1 and OUT2 was +11.85 dBm and −28.9 dBm, respectively. The reflectivity of SLM_1 was gradually changed by slowly adjusting polarization controller PC_4, while recording the power meter readings and the output signal spectrum at both output ports, OUT1 and OUT2, and the FWHM at output port, OUT1. The output power from output port OUT1, was thus changed from +11.85 to −28.5 dBm while maintaining the output power at output port OUT2 at −28.9 dBm by adjusting polarization controller PC_5 for SLM_2. The above process was then reversed by

Table 10.2 Optical power from the fiber laser output-port, OUT1 AND OUT2, at different bias current I_B (mA) levels and 1550 nm center wavelength

SOA_1 I_{B1} (mA)	SOA_2 I_{B2} (mA)	P_{OUT1} (dBm)	P_{OUT2} (dBm)
26	180	−36	−1.5
50	200	−32	5
75	250	−29.5	7.8
100	300	−28.9	9.3
150	400	−28.6	11.1
200	500	−28.9	11.85

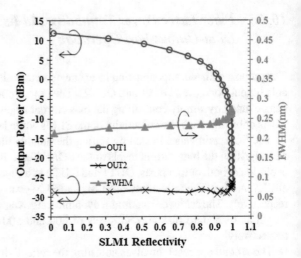

Fig. 10.6 Shows the output power of output port, OUT1 (circles) and OUT2 (crosses), respectively, and the 3 dB-bandwidth of output port, OUT1 (tringles) as a function of different reflectivity values of the Sagnac loop mirror, SML$_1$ for single output port operation

switching the reflectivity of SLM$_1$ and SLM$_2$ to $\geq$99.9% and $\leq$0.1%, respectively and examining the ports' performances as done above. In this case, the measured output power from output port, OUT2 was adjusted from +11.87 to $-$28.9 dBm while maintaining the output port, OUT1, at $-$28.9 dBm. Figure 10.6 illustrates the output from both output ports, OUT1 (circles) and OUT2 (crosses) as a function of the reflectivity of SLM$_1$. Note that both output ports behave similarly and that the 3 dB-bandwidth of the light beam from OUT1 (triangles) increased as the reflectivity of the SLM$_1$ increased while the output power decreased due to strong feedback (i.e., reflected light beam) from each of the SLMs.

In the dual-output port configuration, both output ports can be fixed and adjusted to any output power level between +11.9 and $-$28.9 dBm. As portrayed in Fig. 10.7, both output ports, OUT1 and OUT2, were set to +8.94 and +8.95 dBm, respectively by adjusting the reflectivity of both SLM$_1$ and SLM$_2$ to $\leq$0.1%. Then, the output power from output port OUT1, was gradually tweaked from +8.94 to $-$28.9 dBm by adjusting the reflectivity of the SLM$_1$ from 0.1% to more than 99.9% while simultaneously optimizing the reflectivity of SLM$_2$ to keep the output power at OUT2 constant at +8.95 dBm. The reflectivity of SLM$_2$ was around 50% while that of SLM$_1$ was around 99.9% for the output power at OUT2 constant at +8.95 dBm.

10.4.6 Wavelength Tunability

The wavelength tuning width of the optical filter is 50 nm (see Fig. 10.4); its maximum IL, 5.5 dB, occurs at 1520 nm while its minimum IL, 2.2 dB, occurs at 1570 nm. The bias currents for SOA$_1$ and SOA$_2$ was set at 200 and 500 mA, respectively. Then the reflectivity of SLM$_1$ and SLM$_2$ was adjusted and set constant at $\leq$0.1% and $\geq$99.9%, respectively. Then by optimizing the polarization controllers, PC$_1$, PC$_2$ and PC$_3$ as

Fig. 10.7 Illustrates the output power from both output ports, OUT1 (filled circles) and OUT2 (unfilled squares) for different reflectivity values of the Sagnac loop mirror, SLM$_1$ for dual-output port operation while maintaining constant output power at OUT2 constant at +8.95 dBm [74]

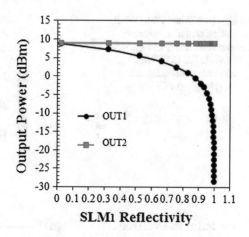

Fig. 10.8 Shows the wavelength spectrum of the fiber compound-ring laser where PC$_1$, PC$_2$ and PC$_3$, were optimized at each wavelength [74]

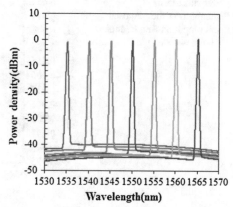

the TF was set to 1535, 1540, 1545, 1550, 1555, 1560 and 1565 nm, the output light beam was measured as expressed in Fig. 10.8.

10.4.7 Optical Signal-to-Noise Ratio

The peak signals measured by the OSA within the wavelength spectrum in question (see Fig. 10.8) were used to calculate the optical signal-to-noise ratio (OSNR) of system by subtracting the peak power value at each center wavelength (i.e., 1535, 1540, 1545, 1550, 1555, 1560 and 1565 nm) from the background noise level of each wavelength spectrum (i.e., OSNR(dB) = $P_{Signal} - P_{Noise}$, where both P_{Signal} and P_{Noise} level are expressed in dBm), as demonstrated in Fig. 10.9. The OSNR remained well above +39 dB over the whole wavelength tuning range in which the maximum OSNR of +44.6 dB was found to be at 1565 nm.

Fig. 10.9 Shows the OSNR at each center wavelength spectrum, 1535, 1540, 1545, 1550, 1550, 1555, 1560 and 1565 nm with optimized PC_1, PC_2 and PC_3

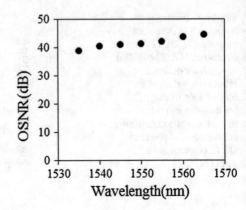

Fig. 10.10 Shows the output power short-term fluctuations of the proposed fiber compound-ring laser at 1550 nm wavelength

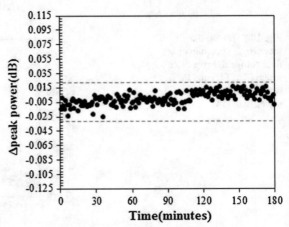

The obtained OSNR spectrum of our fiber compound ring laser is comparable to that of those used in remote sensing [75, 76], fiber sensors that utilize wavelength-peak measurement method [67] and optical communication systems [77].

10.4.8 Short-Term Optical Power

Lastly, using the OSA to monitor results and acquire data, the short-term optical power stability test was conducted at room temperature with the SOAs set to their standard bias current levels and the tunable filter fixed at 1550 nm central wavelength. The optical stability test was carried out over a total duration of 180 min with 1 min intervals and an OSA resolution bandwidth of 0.01 nm without additional data averaging. Figure 10.10 demonstrates that the proposed fiber compound-ring laser, whose power fluctuations were within ±0.02 dB and could have been further reduced by proper packaging of the system, is very stable.

10.5   (N ≫ 1) SOAs Based Fiber Compound-Ring Laser Structure

Thus far, the high passive coherent beam combining efficiency of two SOAs based fiber compound-ring laser without the use of active phase modulators has been successfully demonstrated. The proposed fiber compound-ring cavity has high prospects to realize an efficient, stable, simple, low-cost, compact, and highly power-scalable fiber system by passively combining [53, 54] low power (N ≫ 1) number of SOAs fiber compound-ring lasers by using N × N fiber couplers with high beam combining efficiency. Estimation studies have shown that the beam combining efficiency worsens with a higher number (i.e. N > 8) [78] of individual combined Y-shaped linear cavity based fiber lasers. Yet as demonstrated in other works [79], several individual fiber lasers have been combined in Y-shape coupled arrays with high efficient coherent beam combining from 16 channels designed by cascading two fiber laser arrays using EDFAs as gain media. Thus, in addition to cascading multiple arrays of compound-ring laser cavities for the proposed (N ≫ 1) fiber laser compound-ring laser, one can also utilize SOAs because of their wide bandwidths, which enable sufficient lasing modes within the actual gain bandwidth. Yet, it is expected that the changes in the behavior of mode competition among the oscillating modes will potentially diminish the likelihood of obtaining the same spectral lasing mode [80, 81] thus degrading the beam combining efficiency when many arrays of merged compound-ring resonators are combined. In addition, the nonlinearities, optical damages and nonlinear effects (SRS, SBS, and four-wave mixing in all-single-mode fiber output couplers) that manifest in a SOA [82] at very high optical power will restrict the maximum amount of achievable optical power from the proposed fiber compound-ring laser.

This work has enhanced and improved on previous works [55] by omitting the expensive high-power combiner with a multimode fiber output port. The proposed SOA-based fiber compound-ring laser structure is a dual-output port all-single-mode-fiber laser structure that can be used in single or dual-output operation with adjustable output power. Figure 10.11 illustrates the proposed fiber laser structure with NxN fiber couplers and two SLMs, SLM_1 and SLM_2, which form the output couplers, OUT1 and OUT2.

10.6   Conclusion

Various work and advancements have been realized leading up to and paving the way the for further research on SOA based-fiber laser sources, which can increasingly be established in current optical applications due to their compact, cost effective, and user-friendly nature as opposed to their counterparts that utilize rare-earth YDFAs, EDFAs, SRS amplifiers, SBS amplifiers as well as free-space solid state laser sources. Newer designs aim to eliminate the bulkiness of conventional fiber laser systems

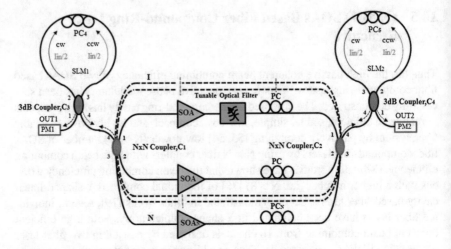

Fig. 10.11 (N ≫ 1) number of SOAs based dual-output port fiber compound-ring laser structure

as discussed earlier by obsoleting optical isolators and circulators thus, paving the avenue for the fabrication of a simple, on-chip scalable, power scalable, adjustable and switchable laser source.

To this end, a dual-SOA based all-single-mode dual parallel nested fiber ring resonator with two SLMs (to enable operational mode switching and output power adjustability thus bypassing the need for VOAs) was designed, which resulted in an impressive coherent beam combining efficiency of 98% over the C-band tuning range of 30 nm, optical signal-to -noise ratio (OSNR) of +45 dB, and optical power fluctuation within 0.02 dB over the course of 3 h at room temperature.

The chapter then concluded by discussing using N × N fused fiber couplers to achieve an all-single-mode-fiber compound-ring laser structure with (N ≫ 1) number of SOAs for power scaling to design a high-power laser source using low power optical components exclusively. Due to the advanced technologies of SOAs and tunable filters covering a wide range of wavelength spectrum in different electromagnetic spectrum bands, the proposed fiber compound-ring laser can be used to build compact laser systems covering different optical wavelength-bands as well.

References

1. X. He, X. Fang, C. Liao, D.N. Wang, J. Sun, A tunable and switchable single-longitudinal-mode dual-wavelength fiber laser with a simple linear cavity. Opt. Express **17**, 21773–21781 (2009)
2. P. Barnsley, P. Urquhart, C. Millar, M. Brierley, Fiber Fox-Smith resonators: application to single-longitudinal-mode operation of fiber lasers. J. Opt. Soc. Am. A **5**, 1339–1346 (1988)
3. K. Iwatsuki, A. Takada, K. Hagimoto, M. Saruwatari, Y. Kimura, M. Shimizu, Er3+-doped fiber-ring-laser with less than 10 kHz linewidth, in *Optical Fiber Communication Conference*,

Vol. 5 of 1989 OSA Technical Digest Series (Optical Society of America, 1989), paper PD5

4. J. Zhang, C.Y. Yue, G.W. Schinn, W.R.L. Clements, J.W.Y. Lit, Stable single-mode compound-ring erbium-doped fiber laser. J. Lightwave Technol. **14**, 104–109 (1996)

5. X. Chen, J. Yao, Z. Deng, Ultranarrow dual-transmission-band fiber Bragg grating filter and its application in a dual-wavelength single-longitudinal-mode fiber ring laser. Opt. Lett. **30**, 2068–2070 (2005)

6. Z. Ou, Z. Dai, B. Wu, L. Zhang, Z. Peng, Y. Liu, Research of narrow line-width Er3+-doped fiber ring laser with FBG F-P etalon and FBG Sagnac loop, in *Microelectronic and Optoelectronic Devices and Integration, 2009 International Conference on Optical Instruments and Technology 2008*

7. J. Zhang, J.W.Y. Lit, Compound fiber ring resonator: theory. J. Opt. Soc. Am. A **11**, 1867–1873 (1994)

8. P. Urquhart, Compound optical-fiber-based resonators. J. Opt. Soc. Am. A **5**, 803–812 (1988)

9. J. Zhang, J.W.Y. Lit, All-fiber compound ring resonator with a ring filter. J. Lightwave Technol. **12**, 1256–1262 (1994)

10. J. Capmany, M.A. Muriel, A new transfer matrix formalism for the analysis of fiber ring resonators: compound coupled structures for FDMA demultiplexing. J. Lightwave Technol. **8**, 1904–1919 (1990)

11. W. Shi, Q. Fang, X. Zhu, R.A. Norwood, N. Peyghambarian, Fiber lasers and their applications [Invited]. Appl. Opt. **53**, 6554–6568 (2014)

12. D.J. Jiang, X.F. Chen, Y.T. Dai, H.T. Liu, S.Z. Xie, A novel distributed feedback fiber laser based on equivalent phase shift. IEEE Photon. Technol. Lett. **16**, 2598 (2004)

13. Y.T. Dai, X.F. Chen, D.J. Jiang, S.Z. Xie, C.C. Fan, Equivalent phase shift in a fiber Bragg grating achieved by changing the sampling period. IEEE Photon. Technol. Lett. **16**, 2284 (2004)

14. X.F. Chen, J.P. Yao, F. Zeng, Z.C. Deng, Single-longitudinal-mode fiber ring laser employing an equivalent phase-shifted fiber Bragg grating. IEEE Photon. Technol. Lett. **17**, 1390 (2004)

15. R.R.A. Syms, H. Zou, J. Stagg, H. Veladi, Sliding-blade MEMS iris and variable optical attenuator. J. Micromech. Microeng. **14**, 1700 (2004)

16. A. Unamuno, D. Uttamchandani, MEMS variable optical attenuator with Vernier latching mechanism. IEEE Photon. Technol. Lett. **18**, 88–90 (2006)

17. C. Marxer, P. Griss, N.F. de Rooij, A variable optical attenuator based on silicon micromechanics. IEEE Photon. Technol. Lett. **11**, 233–235 (1999)

18. Q. Li, A.A. Au, C.-H. Lin, E.R. Lyons, H.P. Lee, An efficient all-fiber variable optical attenuator via acoustooptic mode coupling. IEEE Photon. Technol. Lett. **14**, 1563–1565 (2002)

19. A. Duduś, R. Blue, M. Zagnoni, G. Stewart, D. Uttamchandani, Modeling and characterization of an electrowetting-based single-mode fiber variable optical attenuator. IEEE J. Sel. Topics Quantum Electron. **21**, 253–261 (2015)

20. A. Benner, H.M. Presby, N. Amitay, Low-reflectivity in-line variable attenuator utilizing optical fiber tapers. J. Lightwave Technol. **8**, 7–10 (1990)

21. C. Kerbage, A. Hale, A. Yablon, R.S. Windeler, B.J. Eggleton, Integrated all-fiber variable attenuator based on hybrid microstructure fiber. Appl. Phys. Lett. **79**, 3191–3193 (2001)

22. K.S. Lim, C.H. Pua, S.W. Harun, H. Ahmad, Temperature-sensitive dual-segment polarization maintaining fiber Sagnac loop mirror. Opt. Laser Technol. **42**, 377–381 (2010)

23. G. Sun, D.S. Moon, Y. Chung, Simultaneous temperature and strain measurement using two types of high-birefringence fibers in Sagnac loop mirror. IEEE Photon. Technol. Lett. **19**, 2027–2029 (2007)

24. H.Y. Fu, H.Y. Tam, L.-Y. Shao, X. Dong, P.K.A. Wai, C. Lu, S.K. Khijwania, Pressure sensor realized with polarization-maintaining photonic crystal fiber-based Sagnac interferometer. Appl. Opt. **47**, 2835–2839 (2008)

25. G. Das, J.W.Y. Lit, Wavelength switching of a fiber laser with a Sagnac loop reflector. IEEE Photon. Technol. Lett. **16**, 60–62 (2004)

26. D.S. Lim, H.K. Lee, K.H. Kim, S.B. Kang, J.T. Ahn, M.Y. Jeon, Generation of multiorder Stokes and anti-Stokes lines in a Brillouin erbium-fiber laser with a Sagnac loop mirror. Opt. Lett. **23**, 1671–1673 (1998)

27. S.S. Wang, Z.F. Hu, Y.H. Li, L.M. Tong, All-fiber Fabry-Perot resonators based on microfiber Sagnac loop mirrors. Opt. Lett. **34**, 253–255 (2009)
28. M.A. Ummy, N. Madamopoulos, P. Lama, R. Dorsinville, Dual Sagnac loop mirror SOA-based widely tunable dual-output port fiber laser. Opt. Express **17**, 14495–14501 (2009)
29. M.A. Ummy, N. Madamopoulos, A. Joyo, M. Kouar, R. Dorsinville, Tunable multi-wavelength SOA based linear cavity dual-output port fiber laser using Lyot-Sagnac loop mirror. Opt. Express **19**, 3202–3211 (2011)
30. D.B. Mortimore, Fiber loop reflectors. J. of Lightwave Technol. **6**, 1217–1224 (1988)
31. S. Feng, Q. Mao, L. Shang, J.W. Lit, Reflectivity characteristics of the fiber loop mirror with a polarization controller. Opt. Commun. **277**, 322–328 (2007)
32. S. Klingebiel, F. Röser, B. Ortaç, J. Limpert, A. Tünnermann, Spectral beam combining of Yb-doped fiber lasers with high efficiency. J. Opt. Soc. Am. B **24**, 1716–1720 (2007)
33. V. Raab, R. Menzel, External resonator design for high-power laser diodes that yields 400 mW of TEM$_{00}$ power. Opt. Lett. **27**, 167–169 (2002)
34. C.J. Corcoran, R.H. Rediker, Operation of five individual diode lasers as a coherent ensemble by fiber coupling into an external cavity. Appl. Phys. Lett. **59**, 759–761 (1991)
35. B. Liu, Y. Braiman, Coherent beam combining of high power broad-area laser diode array with near diffraction limited beam quality and high power conversion efficiency. Opt. Express **21**, 31218–31228 (2013)
36. V. Daneu, A. Sanchez, T.Y. Fan, H.K. Choi, G.W. Turner, C.C. Cook, Spectral beam combining of a broad-stripe diode laser array in an external cavity. Opt. Lett. **25**, 405–407 (2000)
37. T.H. Loftus, A. Liu, P.R. Hoffman, A.M. Thomas, M. Norsen, R. Royse, E. Honea, 522 W average power, spectrally beam-combined fiber laser with near-diffraction-limited beam quality. Opt. Lett. **32**, 349–351 (2007)
38. T.Y. Fan, Laser beam combining for high-power, high-radiance sources. IEEE J. Sel. Topics Quantum Electron. **11**, 567–577 (2005)
39. S.J. Augst, A.K. Goyal, R.L. Aggarwal, T.Y. Fan, A. Sanchez, Wavelength beam combining of ytterbium fiber lasers. Opt. Lett. **28**, 331–333 (2003)
40. C. Wirth, O. Schmidt, I. Tsybin, T. Schreiber, T. Peschel, F. Brückner, T. Clausnitzer, J. Limpert, R. Eberhardt, A. Tünnermann, M. Gowin, E.T. Have, K. Ludewigt, M. Jung, 2 kW incoherent beam combining of four narrow-linewidth photonic crystal fiber amplifiers. Opt. Express **17**, 1178–1183 (2009)
41. P. Sprangle, A. Ting, J. Penano, R. Fischer, B. Hafizi, Incoherent combining and atmospheric propagation of high-power fiber lasers for directed-energy applications. IEEE J. Quantum Electron. **45**, 138–148 (2009)
42. D. Sabourdy, V. Kermène, A. Desfarges-Berthelemot, M. Vampouille, A. Barthélémy, Coherent combining of two Nd: YAG lasers in a Vernier–Michelson-type cavity. Appl. Phys. B **75**, 503–507 (2002)
43. G. Bloom, C. Larat, E. Lallier, M. Carras, X. Marcadet, Coherent combining of two quantum-cascade lasers in a Michelson cavity. Opt. Lett. **35**, 1917–1919 (2010)
44. D. Sabourdy, V. Kermène, A. Desfarges-Berthelemot, L. Lefort, A. Barthélémy, P. Even, D. Pureur, Efficient coherent combining of widely tunable fiber lasers. Opt. Express **11**, 87–97 (2003)
45. F. Jeux, A. Desfarges-Berthelemot, V. Kermène, A. Barthelemy, Experimental demonstration of passive coherent combining of fiber lasers by phase contrast filtering. Opt. Express **20**, 28941–28946 (2012)
46. V.A. Kozlov, J. Hernández-Cordero, T.F. Morse, All-fiber coherent beam combining of fiber lasers. Opt. Lett. **24**, 1814–1816 (1999)
47. D.S. Moon, B.H. Kim, A. Lin, G. Sun, W.T. Han, Y.G. Han, Y. Chung, Tunable multi-wavelength SOA fiber laser based on a Sagnac loop mirror using an elliptical core side-hole fiber. Opt. Express **15**, 8371–8376 (2007)
48. C.S. Kim, R.M. Sova, J.U. Kang, Tunable multi-wavelength all-fiber Raman source using fiber Sagnac loop filter. Opt. Commun. **218**, 291–295 (2003)

49. P.G. Zverev, T.T. Basiev, A.M. Prokhorov, Stimulated Raman scattering of laser radiation in Raman crystals. Opt. Mater. **11**, 335–352 (1999)
50. S.P. Smith, F. Zarinetchi, S. Ezekiel, Narrow-linewidth stimulated Brillouin fiber laser and applications. Opt. Lett. **16**, 393–395 (1991)
51. J.C. Yong, L. Thévenaz, B. Yoon Kim, Brillouin fiber laser pumped by a DFB laser diode. J. Lightwave Technol. **21**, 546 (2003)
52. A.J. Lowery, M. Premaratne, Design and simulation of a simple laser rangefinder using a semiconductor optical amplifier-detector. Opt. Express **13**, 3647–3652 (2005)
53. A.J. Lowery, M. Premaratne, Reduced component count optical delay discriminator using a semiconductor optical amplifier-detector. Opt. Express **13**, 290–295 (2005)
54. M. Premaratne, A.J. Lowery, Analytical characterization of SOA-based optical pulse delay discriminator. J. Lightwave Technol. **23**, 2778–2787 (2005)
55. N. Arez, M. Razaghi, Optical and logic gate implementation using four wave mixing in semiconductor optical amplifier for high speed optical communication systems, in *International Conference Network and Electronics Engineering (IPCSIT), 2011*, vol. 11
56. N. Das, M. Razaghi, R. Hosseini, Four-wave mixing in semiconductor optical amplifiers for high-speed communications, in *2012 5th International Conference on Computers and Devices for Communication (CODEC)*, IEEE (2012)
57. N.K. Das, Y. Yamayoshi, H. Kawaguchi, Analysis of basic four-wave mixing characteristics in a semiconductor optical amplifier by the finite-difference beam propagation method. IEEE J. Quantum Electron. **36**, 1184–1192 (2000)
58. H. Heidrich, Multifunctional photonic integrated circuits based on SOA and ring resonators, in *Optical Fiber Communication Conference, Vol. 1 of 2003 OSA Technical Digest Series (Optical Society of America, 2003), paper TuG3*
59. K. Sato, H. Toba, Reduction of mode partition noise by using semiconductor optical amplifiers. IEEE J. Sel. Topics Quantum Electron. **7**, 328–333 (2001)
60. K. Takano, K. Nakagawa, Y. Takahashi, H. Ito, Reduction of power fluctuation in pulsed lightwave frequency sweepers with SOA following EDFA. IEEE Photon. Technol. Lett. **19**, 525–527 (2007)
61. H. Bruesselbach, D.C. Jones, M.S. Mangir, M. Minden, J.L. Rogers, Self-organized coherence in fiber laser arrays. Opt. Lett. **30**, 1339–1341 (2005)
62. J. Lhermite, A. Desfarges-Berthelemot, V. Kermene, A. Barthelemy, Passive phase locking of an array of four fiber amplifiers by an all-optical feedback loop. Opt. Lett. **32**, 1842–1844 (2007)
63. B. Lei, Y. Feng, Phase locking of an array of three fiber lasers by an all-fiber coupling loop. Opt. Express **15**, 17114–17119 (2007)
64. M.A. Ummy, S. Bikorimana, N. Madamopoulos, R. Dorsinville, Beam combining of SOA-based bidirectional tunable fiber nested ring lasers with continuous tunability over the C-band at room temperature. J. Lightwave Technol. **34**, 3703–3710 (2016)
65. J.P. Cariou, B. Augere, M. Valla, Laser source requirements for coherent lidars based on fiber technology. ComptesRendus Phys. **7**, 213–223 (2006)
66. A. Martinez-Ríos, G. Anzueto-Sanchez, R. Selvas-Aguilar, A.A.C. Guzman, D. Toral-Acosta, V. Guzman-Ramos, V.M. Duran-Ramirez, J.A. Guerrero-Viramontes, C.A. Calles-Arriaga, High sensitivity fiber laser temperature sensor. IEEE Sensors J. **5**, 2399–2402 (2015)
67. T.B. Pham, H. Bui, H.T. Le, V.H. Pham, Characteristics of the fiber laser sensor system based on etched-Bragg grating sensing probe for determination of the low nitrate concentration in water. Sensors **17**, 7 (2016)
68. H. Fu, D. Chen, Z. Cai, Fiber sensor systems based on fiber laser and microwave photonic technologies. Sensors **12**, 5395–5419 (2012)
69. N.S. Park, S.K. Chun, G.H. Han, C.S. Kim, Acousto-optic-based wavelength-comb-swept laser for extended displacement measurements. Sensors **17**, 740 (2017)
70. P.C. Peng, J.H. Lin, H.Y. Tseng, S. Chi, Intensity and wavelength-division multiplexing FBG sensor system using a tunable multiport fiber ring laser. IEEE Photon. Technol. Lett. **16**, 230–232 (2004)

71. F. Delorme, Widely tunable 1.55 μm lasers for wavelength-division-multiplexed optical fiber communications. IEEE J. Sel. Topics Quantum Electron. **34**, 1706–1716 (1998)
72. S.H. Yun, G.J. Tearney, J.F. de Boer, N. Iftimia, B.E. Bouma, High-speed optical frequency-domain imaging. Opt. Express **11**, 2953–2963 (2003)
73. E. Hemmer, A. Benayas, F. Légaré, F. Vetrone, Exploiting the biological windows: current perspectives on fluorescent bioprobes emitting above 1000 nm. Nanoscale Horiz. **1**, 168–184 (2016)
74. M.A. Ummy, S. Bikorimana, R. Dorsinville, Beam combining of SOA-based bidirectional tunable fiber compound-ring lasers with external reflectors, in *Optics and Lasers Technology, 2017 5th International Conference on Photonics. PHOTOPTICS, 2017*
75. G. Wang, L. Zhan, J. Liu, T. Zhang, J. Li, L. Zhang, J. Peng, L. Yi, Watt-level ultrahigh-optical signal-to-noise ratio single-longitudinal-mode tunable Brillouin fiber laser. Opt. Lett. **38**, 19–21 (2013)
76. Y. Luo, Y. Tang, J. Yang, Y. Wang, S. Wang, K. Tao, L. Zhan, J. Xu, High signal-to-noise ratio, single-frequency 2 μm Brillouin fiber laser. Opt. Lett. **39**, 2626–2628 (2014)
77. L.S. Yan, X.S. Yao, Y. Shi, A.E. Willner, Simultaneous monitoring of both optical signal-to-noise ratio and polarization-mode dispersion using polarization scrambling and polarization beam splitting. J. Lightwave Technol. **23**, 3290 (2005)
78. D. Kouznetsov, J. Bisson, A. Shirakawa, K. Ueda, Limits of coherent addition of lasers: simple estimate. Opt. Rev. **12**, 445–447 (2005)
79. W. Chang, T. Wu, H. Winful, A. Galvanauskas, Array size scalability of passively coherently phased fiber laser arrays. Opt. Exp. **18**, 9634–9642 (2010)
80. S. Sivaramakrishnan, W. Chang, A. Galvanauskas, H.G. Winful, Dynamics of passively phased ring oscillator fiber laser arrays. IEEE J. Quant. Electron. **51**, 1–9 (2015)
81. E.J. Bochove, S.A. Shakir, Analysis of a spatial-filtering passive fiber laser beam combining system. IEEE J. Sel. Top. Quant. Electron. **15**(320–327), 78 (2009)
82. S. Barua, N. Das, S. Nordholm, M. Razaghi, Comparison of pulse propagation and gain saturation characteristics among different input pulse shapes in semiconductor optical amplifiers. Opt. Commun. **359**, 73–78 (2016)

Chapter 11
Active Fibre Mode-locked Lasers in Synchronization for STED Microscopy

Shree Krishnamoorthy, S. Thiruthakkathevan and Anil Prabhakar

Abstract Mode-locked fibre ring lasers can generate picosecond optical pulse widths with MHz repetition rates. Applications in optical imaging, or in experiments with pump-probe lasers, benefit from being able synchronize two lasers at high repetition rates, while retaining the narrow optical pulse widths. We investigate the characteristics of an actively mode-locked fibre ring laser, designed as a slave laser and driven by a commercial Ti:Sapphire laser acting as a master. The master-slave synchronization was stabilized for frequency detuning by matching the cavity lengths, and the dependence of the output pulse width of the slave laser was studied as its cavity was detuned. The increase in pulse width was asymmetric about the ring cavity resonance frequency, a phenomenon that we were able to establish as a consequence of an asymmetry in the detuning range of the higher order cavity modes. We observed that the detuning range decreased linearly with the mode number, an observation that was supported by a theoretical perturbative analysis of cavity locking.

11.1 Introduction-Nanoscopy and Lasers

Microscopy has seamlessly adopted laser sources to improve imaging. They have been challenging the theoretical boundaries of traditional microscopy to improve the image resolution to nanometers. With continued advancements, microscopic imaging has transitioned to nanoscale imaging, and the field is now aptly being referred to as nanoscopy. To image finer structures, scanning tunneling microscopes (STM) [1] and UV based microscopes have moved to wavelengths below the visible spectrum. However, such techniques are not compatible with live cell imaging as the absorption

S. Krishnamoorthy (✉) · A. Prabhakar
Indian Institute of Technology, Chennai, India
e-mail: shree.krishnamoorthy@gmail.com

S. Thiruthakkathevan
Valarkathir Creatronics Private Limited, Chennai, India

© Springer Nature Switzerland AG 2019
P. Ribeiro et al. (eds.), *Optics, Photonics and Laser Technology 2017*,
Springer Series in Optical Sciences 222,
https://doi.org/10.1007/978-3-030-12692-6_11

and scattering is very high [2]. Consequently, advanced microscopy using superlenses and other artificially created materials [3, 4] are being pursued.

Fluorescent microscopy is widely used, and preferred, as a method to label the proteins of interest. This allows us to study specific phenomena dealing with interactions of molecules in live cells and tissues. However, fluorescent microscopy is limited in spatial resolution by Abbe's limit. Techniques such as structured illumination microscopy rely on the mathematical nature of the imaging instruments and improve the resolution in the post-processing stage [5] e.g. we can use a Moiré pattern to preserve the phase information in the widefield image. Some techniques rely on intensity dependent non-linear interactions like two-photon (2p) and multi-photon [6] absorption within the sample. Similarly, total internal reflection (TIRF) microscopes uses evanescent optical fields to achieve finer resolutions [7]. Near field optics is also used in near-field scanning optical microscopy (NSOM) for nanoscopy. Other techniques, such as photoactivated localization microscopy (PALM) and stochastic optical reconstruction microscopy (STORM), allow a very small fraction of the illuminated molecules to fluoresce, while keeping the a spatially selected illuminated population in dark state using a patterned laser like RESOLFT, GSD and STED [8–10].

11.1.1 STED with Pulsed Lasers

STED is a fast, all optical technique for nanoscopy imaging [11], wherein the fluorescent molecules are de-excited using a depletion laser [8, 12]. The depletion photons interact with the excited molecules only at the periphery of the excitation spot, suppressing the fluorescence there. As a result, only the molecules at the center will fluoresce, cause the imaged spot size to reduce. The suppression of fluorescence is intensity dependent and the effective spot size becomes

$$d \approx \left(\frac{\lambda}{2n \sin \alpha} \right) \frac{1}{\sqrt{1 + \frac{I_d}{I_S}}}. \tag{11.1}$$

This is merely Abbe's resolution, further reduced by depletion laser intensity I_d when compared to the saturation intensity I_S [13]. The mechanism of STED is explained in Fig. 11.1.

For a typical fluorophore (fluorescein here), there are two absorption regions, the first is the single photon excitation, or just excitation and another is the 2p excitation. The emission region is shown for the fluorophore. The depletion laser is shaped into a donut and superposed on the sample with the excitation 2p laser spot, as shown in the spatial domain in Fig. 11.1. 2p excitation occurs as usual and the fluorescence intensity decreases typically as shown in the temporal panel. However, in the yellow regions with strong overlap between the depletion and the excitation, the fluorescence

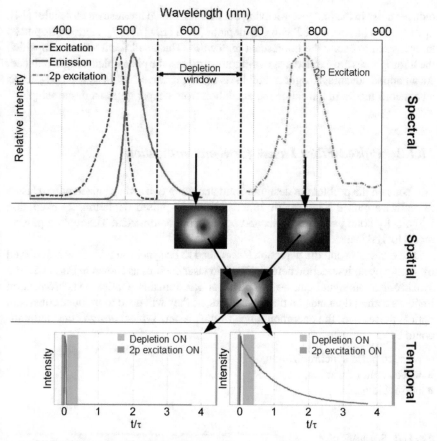

Fig. 11.1 Summary of 2-photon (2p) STED process. In the spectral domain, we see the excitation and 2p excitation spectra and the emission spectrum. Spatially the donut STED beam of wavelength in the depletion window superposes the 2p excitation spot. Yellow shows significant overlap regions. Temporally the fluorescence trace for one excitation cycle is shown. Red pulses are excitation, green are depletion pulses, and the dark green time trace is the resulting fluorescence. The center of the excitation is unaffected, whereas the peripheral regions experience quenching by the depletion laser

is suppressed and the excited molecules are stimulated emission depleted by the depletion laser as shown to the left of the temporal traces.[1] Although the resolution is enhanced with higher depletion laser intensities, we cannot arbitrarily increase the power of the depletion laser. With increased photon energies on the biological sample, there is an increase in photo-bleaching and thermal damage. One way to minimize these effects is to use pulsed lasers which can deliver high energies for a short duration, typically hundreds of picosecond to nanosecond pulses, reducing the average energy incident on the sample. The pulsed excitation also allows for one to use gated detection, by operating the detector at high gains over short durations. This

---

[1]The depletion trace in Fig. 11.1 is a simulation using actual fluorescence traces, not actual data.

reduces noise in the images by leaving out the undepleted fluorescent molecules [14]. 2p excitation has been used with CW depletion in STED [15–17]. Here, an IR pulsed laser is used to excite the fluorescent molecules. The small focal volumes in which the fluorophores are excited in a two-photon process, the photo-bleaching is reduced. As an added advantage, light scatters less at longer wavelengths as it propagates in an aqueous medium, thus allowing excitation from deeper within a tissue sample.

11.1.2 Pulsed Fibre Lasers for Synchronization

We focus on the problem of designing and operating two synchronized pulsed lasers to realize a pulsed STED with 2p excitation and pulsed depletion. 2p excitation is typically done using femto-second pulses from commercial Ti:Sapphire passive mode-locked lasers.

The excitation and the depletion lasers need to be synchronized. This is achieved by setting up an interaction between the two laser cavities as shown in Fig. 11.2. The synchronization signal can be in optical or RF domains. Optical synchronization imposes certain demands on the slave lasers, as they will need to produce consistent optical pulses over the operation range of the master. Typical design considerations could be

- pulse to pulse synchronization,
- wavelength restriction,
- pulse widths,

Fig. 11.2 Schematic of master and slave lasers in synchronization. Synchronization signal is typically around 80 MHz. The master laser is a commercial Ti:Sapphire laser [18] and the slave is a Yb:doped actively mode-locked laser using RF signal derived from master laser optical pulse train [19]

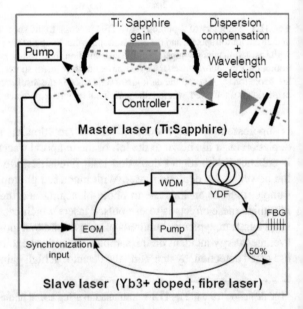

- pulse energies and
- spectral linewidth.

Of all the items, the key factor required in synchronization is the pulse to pulse correspondence between the master and the slave lasers.

There are different implementations of slave lasers in literature. As a depletion laser, an additional pulse stretching module is adopted that converts a part of the master's femto-second laser output into a picosecond laser [17, 20]. The additional module achieves synchronization, but enforces a restriction on the wavelength of the slave laser as the optics is directly coupled to the optical pulses of the master [21]. In other implementations a separate laser is electronically controlled to achieve optical pulses and synchronization [17, 22].

11.2 Actively Mode-locked Fibre Lasers

One way to set up optical synchronization, with a Ti:Sapphire laser as a master and a fibre laser as the slave, is to use a part of the optical pulse energy from the master to injection lock the fibre laser. Fraction spectrum amplification (FSA) extracts and amplifies the spectral energy available above 1 μm from a Rainbow laser [23] and amplifies it using a ytterbium fibre amplifier. In another approach, we could seed the slave laser with sufficient optical energy from the master laser to induce cross phase modulation, as was demonstrated by synchronizing a ytterbium doped fiber laser to a ns pulsed diode laser [24]. Actively mode-locked lasers are capable of high fidelity synchronization using electronic signals. The generation of the electronic signal uses a negligible fraction of the master's optical signal. To obtain repetition rates in the MHz range, actively fibre mode-locked ring lasers are easily constructed with high stability, and easy thermal management. They are self-starting and can be operated at multiple frequencies through harmonic locking. These features makes them attractive for use as pulsed depletion lasers.

11.2.1 History of Actively Fibre Mode-locked Lasers

The evolution of all fibre actively mode-locked lasers is shown pictorially in Fig. 11.3. The fibres were available for the infra-red regime at 1000–1500 nm based on silica. The ability to use silica graded index fibres to form laser cavity was first experimentally demonstrated by Nakazawa et al. [25]. To obtain an all fibre mode-locked laser, the field had to wait for two decades for development of fibre coupled linear components, integrated optical devices and fibre coupled laser diodes. Firstly, with the advances in semiconductor technology, an integrated phase modulator was used to make an actively mode-locked laser at 1 μm which gave 90 ps pulse width [26]. Soliton pulse compression gave 4 ps at 1.5 μm in an active cavity that also used fibre couplers and Erbium doped gain fibre [27]. In the following decade, the figure of 8

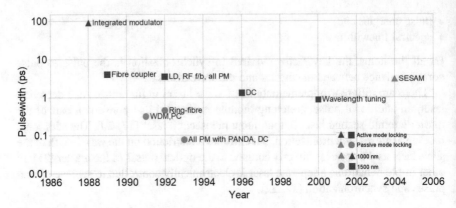

Fig. 11.3 All fibre actively mode-locked laser's development over the years. The major changes are shown in the label. The triangles correspond to 1 μm and circles and squares 1.5 μm. Active mode-locking is shown in blue and passive mode-locking in red

laser (F8L) became popular, and wavelength division multiplexers were available, using this and fibre based polarization control 314 fs passively mode-locked laser was constructed [28]. The ring configuration used by Tamura et al. in the passive mode-locking gave 452 fs pulses and the laser was self starting [29]. In the same year, actively mode-locked lasers were demonstrated with forward, and backward diode laser pumping with fibre pigtails [30]. RF regenerative feedback control provided the stability in the pulse train. It also was a polarization maintaining cavity. However, the pulse widths in actively mode-locked lasers remained high at 3.5 ps compared to the passive ones. Dispersion compensation using large lengths of positive dispersion fibre compressed the gain stretched pulses. This led to shorter pulse widths in the soliton regime at 77 fs [31]. In actively mode-locked lasers, dispersion compensation was employed [32]. This led to reduction in pulse widths to 1.3 ps. In the next decade, researchers attempted to obtain wavelength tuning in the mode-locked lasers [33], which also set the lower limit on pulse width of mode-locked lasers to about 800 fs. The field of ultrafast lasers saw another entrant with the invention of semiconductor saturable absorber mirrors (SESAM). The rapidly evolving field is described very well by U. Keller in her review paper in 2010 [34]. Both Kerr-lens mode-locking and SESAMs have led to ultrafast lasers in passive mode-locking systems. An early adoption of SESAM based passive mode-locking was demonstrated at 1 μm by [35]. With the advances in as listed in the previous paragraph, a typical actively mode-locked laser configuration can now be depicted as shown in Fig. 11.2. Another method of all fibre active mode-locking could be considered to be the pulses generated by cross-phase modulation (XPM). Pulses generated by this method rely on the nonlinear interaction between injected source pulse and the pulses in the nonlinear cavity through XPM process [36].

11.2.2 Active Fibre Mode-locked Laser as Slave

Pulsed electrical signals derived from the Ti:Sapphire were used to injection lock a fibre ring laser (FRL), shown schematically in Fig. 11.2. The FRL was designed to have a cavity resonance close to the repetition rate of the Ti:Sapphire, allowing the FRL to also achieve mode-locking. The slave laser operates at 1064 nm and uses an Yb^{3+} doped fiber with an electro-optic modulator (EOM) that acts as the mode-locker [37, 38]. Typically, the Ti:Sapphire wavelength (λ_i) is tuned between wavelength ranges $690-1020$ nm to obtain the best two-photon excitation of the fluorophore. The electrical synchronization signal derived from the master optical signal varies in amplitude A_i and frequency f_i as shown in Fig. 11.4. As λ_i is changed, f_i also changes between 78.8 and 80.7 MHz, with A_i reaching a maximum for $\lambda_i \sim 800$ nm [37].

We observed that the pulse width of the slave FRL varied over the wavelength range of the master, due to a changing repetition rate of the master laser pulses. This could result in a loss of mode-lock in the FRL. We also observed that the pulse width at the output of the slave FRL varied as the wavelength of the master laser was tuned. To look at the pulse width variation, we took advantage of the sensitivity of nonlinear processes to the peak pulse intensity, and documented the average power at the output of a second harmonic generation (SHG) stage as shown in Fig. 11.5 [37]. The output of the FRL was amplified by 21 dB using two amplifier stages consisting of a Yb:fiber amplifier followed by a master oscillator power amplifier (MOPA). The amplified output was then fed to a SHG stage to produce pulses at 532 nm. The

Fig. 11.4 Synchronization signal voltage (A_i) and frequency (f_i) with changing operating wavelength (λ_i) of the master laser [37, 39].

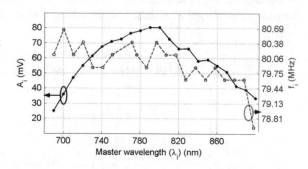

Fig. 11.5 Schematic to monitor the output from a slave laser. Second harmonic generation is inherently sensitive to the intensity of the pulses [39]

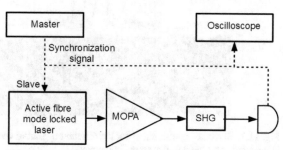

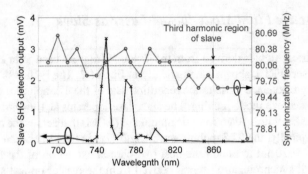

Fig. 11.6 Dependence of Ti:S pulse repetition rate and its effect on SHG power. A peak in SHG occurs when the Ti:S repetition frequency, f_i, matches the third harmonic of the slave laser [37, 39]

recorded peak pulse amplitude at the detector A_{SHG} corresponding to the SHG optical power P_{SHG} for different master laser wavelengths is shown in Fig. 11.6. As expected, P_{SHG} depends on the peak power of the input pulse. Consequently, for a fixed average power in the 1064 nm pulse train, P_{SHG} will depend nonlinearly on the input pulse width. Hence, the SHG process amplifies the effect of high intensity in the amplified 1064 nm pulses. We observe, in Fig. 11.6, that P_{SHG} peaks at 750 nm, 770 nm and 805 nm. Near these peaks, the repetition rate, $f_i \sim 80.07$ MHz, corresponds to the third harmonic frequency ($f_{3^{\mathrm{rd}}{}_s}$) of the slave fiber MLL as indicated by the green lines in the plot.

The peak pulse power and pulse width are not constant in the pulse train produced by the slave laser, as seen in Fig. 11.7. In fact, every third pulse produced by the slave laser has a larger amplitude than the others. A slightly detuned input frequency

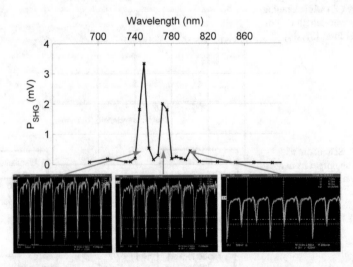

Fig. 11.7 Slave laser pulse traces for the different operating wavelengths. Every third pulse being intense indicates the dominance of fundamental [37, 39]

f_i in the synchronization signal from the master laser causes a modulation in the slave fiber laser. This can be understood on the basis of effects detuning in a fibre mode-locked laser.

11.3 Detuning of Fibre Mode-locked Lasers

Mode-locked lasers produce Fourier limited pulses at resonance [40–42]. The pulse width can be further reduced and soliton regimes are achieved in fibre lasers with addition of nonlinearity [43–45]. However, upon detuning from the mode-locking frequency, the behavior of the laser departs from the ideal mode-locked state.

11.3.1 Deterioration of Mode-locking with Detuning

Deterioration of mode-locking is best understood in actively mode-locked lasers, as they can be driven with sinusoid generators. In a typical case of use as slave laser, the repetition rate of the synchronization signal f_i varies over a range of $\sim 100\,\text{kHz}$. The slave fiber laser is expected to operate over this range. The effect of changing repetition rate f_i is studied by changing the frequency of the input signal to the EOM of the fiber laser using a signal generator [37].

Behavior of a typical actively mode-locked laser upon detuning is shown in Fig. 11.8. Fourier limited pulses are observed at the mode-locking frequency, this yields the shortest pulses and highest energies. Detuning on both sides of the mode-locking frequency causes increase in pulse width and a decrease in pulse energy, this is due to systematic lock-loss in the modes of the laser. Upon further detuning, locking in any mode is completely lost, and the laser is just modulated. In this case, the pulse width is the same as the modulation signal coming from synchronization pulses. In summary, the ring cavity modes that combine to form the optical pulses fall into a mode-locked regime, in a narrow frequency range about the cavity resonance,

Fig. 11.8 Average power and pulse width over the expected operating range of slave laser [37, 39]

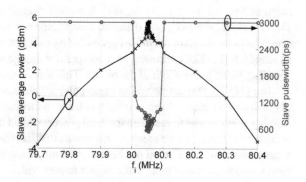

Fig. 11.9 Closer look at effect of detuning on pulse width around f_{3rd_s} [37, 39]

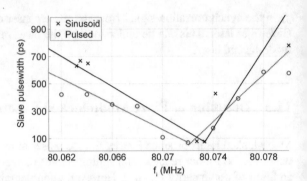

and a modulated regime where broader optical pulses are observed. It has to be noted that the operating regimes of the fiber laser depend on the repetition rate of the master laser's synchronization signal. For wavelength tunable cavities which are optimized for maximizing signal power, the repetition rate varies over a wide range, which will affect the performance of the slave laser. A wavelength independent synchronization mechanism that relies on the active mode-locking phenomenon offers us the ability to synchronize two pulsed lasers over a wide range of repetition rates. One way to ensure that locking is possible is to use optical delays in the laser cavity [46]. If the cavity length of the fiber cavity is maintained, then the operating frequency of the master laser needs to be within the detuning range of all the slave cavity. The narrowest pulses have a pulse width (τ) corresponding to the perfectly mode-locked laser at the mode-locking frequency f_0. For frequency bandwidth f_B [42], τ is given by

$$\tau = \tau_0 \cdot \left(\frac{1}{f_0 f_B} \right)^{1/2}. \tag{11.2}$$

Here, the τ_0 is determined by the gain and modulation depth. The pulse width deterioration due to detuning can be seen for both pulsed and sinusoidal signals in Fig. 11.9. The increase in pulse width is not symmetrical about f_0 for either sinusoidal or pulsed RF inputs. Slopes for the pulse width change for frequencies $f_i > f_0$ is found to be twice as that for $f_i < f_0$. In another demonstration of self-mode locking laser by Thiruthakkathevan [47], two different measurement schemes were used. One measured the pulse width of the pulses generated by using an optical detector followed by an amplifier (LNTIA) followed by a comparator that generated square pulses corresponding to the pulse width of the pulses. This was followed by a low pass filter with a buffer (MVE), that extracted the mean pulse width over many pulses. This scheme of pulse width measurement is incorporated as a feedback component in Fig. 11.10. Another method was to measure the total power content of the higher harmonics. To achieve this, the lower frequency components that include the primary mode and the second harmonic are filtered out before obtaining the average powers. The resultant power measured is due to only the higher modes which is a more reliable indicators

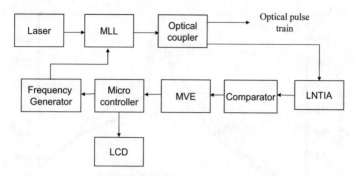

Fig. 11.10 Block diagram of the SMLL, for the pulse width measurement method [47]

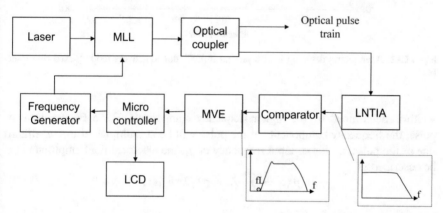

Fig. 11.11 Schematic circuit used to lock the laser cavity by measuring the RF power in the higher harmonics

of mode-locking. This scheme is shown in Fig. 11.11. The asymmetry in pulse width is also found for this Er^{3+} doped mode-locked laser in both methods, as shown in Fig. 11.12. Since, the asymmetry is present for both Yb doped, with FBG limited spectrum, and Er doped lasers, we can assume that it is not an intrinsic property of the rare-earth doped fibre. Rather, we believe that the asymmetry is intrinsic to the method of active mode-locking, and is analogous to similar effects reported in electrical circuits [48]. For harmonic locking at nf_0, the fundamental mode dominates and, and manifests itself as an amplitude modulation on the optical pulse train as shown in Fig. 11.7 (Table 11.1).

11.3.2 Loss of locking in Fourier Space

From Fourier theory, we know that while the lowest frequency components in a spectrum contribute to power and slow varying attributes of the pulses, the pulse

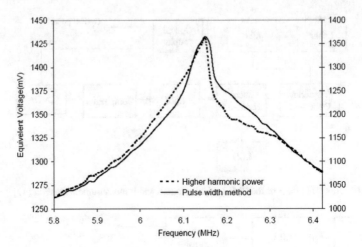

Fig. 11.12 Asymmetric detuning is also present in Er^{3+} doped actively mode-locked ring laser [47]

width is determined by the higher frequency components. For a Fourier limited pulse, the frequency components of the pulse will be at multiples of the repetition rate of the pulse, i.e for an input frequency of f_i, the electrical field amplitude can be described as

$$p(t) = \sum_n A_n \exp\left(j2\pi n f_i t\right), \tag{11.3}$$

where for the nth harmonic A_n is the complex amplitude at frequency nf_i. Thus, the frequency location $\tilde{f}_n$ of each harmonic n would show if the pulse is mode-locked or not. For a harmonic mode n, if the observed frequency location $\tilde{f}_n$ is same as the expected harmonic frequency of nf_i, then we can say that the harmonic mode is locked to the input synchronization signal at the frequency f_i. If the observed frequency $\tilde{f}_n$ and the expected frequency nf_i do not match, i.e,

$$\Delta f_n \triangleq \tilde{f}_n - n f_i \tag{11.4}$$

is nonzero, then the mode is not locked. To quantify this, we must look at the electrical spectrum of the pulses and capture the behavior of the harmonics that are first to get unlocked with detuning.

The experimental scheme to observe the Fourier mode is as given in Fig. 11.13 [39]. In this scheme, the location of the peak frequency $\tilde{f}_n$ and the corresponding amplitude A_n are recorded for each harmonic (or mode number) n for the input synchronization signal frequency f_i. One can observe that at the resonant frequency $f_0 = 26.69$ MHz the power in all the modes is highest as shown in Fig. 11.14. As the input frequency f_i is detuned from the resonant frequency, the power in all the modes declines.

Table 11.1 SMLL output for different RF synchronization inputs

Driving signal type	SMLL output and RF signal	RF in Fourier
Sinusoidal		$\left(\dfrac{\pi}{j}\right)\delta\left(f-f_0\right)-\delta\left(f+f_0\right)$
Square		$\sum_{k=-\infty}^{\infty}\left(-\dfrac{2j}{\pi(2k+1)}\delta\left(f-(2k+1)f_0\right)\right)$
Triangular		$\sum_{k=-\infty}^{\infty}\left(-\dfrac{2}{(\pi(2k+1))^2}\delta\left(f-(2k+1)f_0\right)\right)$

The oscilloscope traces show both the SMLL pulses and the input RF traces. The corresponding Fourier representation of the RF input is also shown [47]

For each input frequency f_i, the peak frequency location, for each mode n, is at $\widetilde{f}_n$. The deviation Δf_n, from the expected frequency location nf_i, was defined in (11.4). Figure 11.15 is obtained by plotting Δf_n for the different harmonics n, as the input frequency f_i is detuned around the resonance f_0. When $f_i = f_0$, all modes follow the expected frequency and $|\Delta f_n| = 0$ for all n. The mode-lock persists for a range of input frequencies around the resonance, in a manner typical of injection locked oscillators [48–51]. It has been shown that mode-locked lasers are also injection locked by the action of the modulation a nonlinear mode-locker [40, 41, 52]. The injection locked laser has all the modes in phase when locked at f_0 and $|\Delta f_n| = 0$ for all the modes n.

Fig. 11.13 Schematic for finding the electrical amplitude spectrum for the modes of the pulses [39]

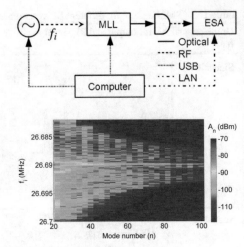

Fig. 11.14 Amplitude A_n for each mode n of the pulse as the input frequency f_i is varied around the resonance $f_0 = 26.69$ MHz [39]

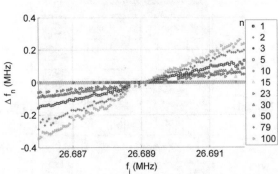

Fig. 11.15 Deviation Δf_n of each mode around the resonance. Lower modes do not deviate in the narrow detuning range and the higher modes deviate faster than the lower modes [39]

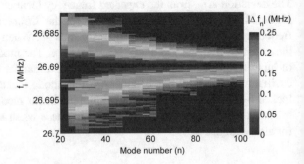

However, when detuned, the higher modes are out of the injection range and are the first to lose lock. In these modes the deviation $|\Delta f_n| > 0$. The deviations in higher modes build up at a faster rate than for lower modes. This implies that the injection locking is dependent on n [39]. This can be observed in Fig. 11.16. Both upper and lower limits decrease as mode number increases, indicating that the injection range

Fig. 11.16 Frequency deviation $|\Delta f_n|$ for each mode n of the pulse as the input frequency f_i is varied around the resonance $f_0 = 26.69$ MHz [39]

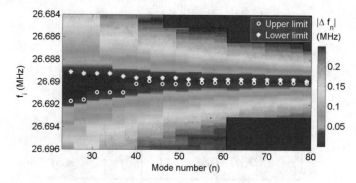

Fig. 11.17 Upper $\left(f_{(U,n)}\right)$ and lower $\left(f_{(L,n)}\right)$ limits of the injection region [39, 53]

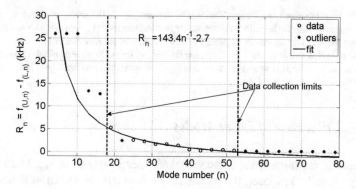

Fig. 11.18 Injection range R_n for each mode n, showing the dependency on the mode number. Modified from [53]

narrows for higher modes. However, the slope of decrease for upper limit is larger by about three times when compared to the lower limit. The region where the deviation in the frequency is nearly zero is the region with mode-locking for any given mode, this is recognized as the the injection range R_n of the mode n [40, 41, 49, 50]. The range R_n is bound by an upper frequency $\left(f_{(U,n)}\right)$ and a lower frequency $\left(f_{(L,n)}\right)$. R_n is defined as

$$R_n \triangleq f_{(U,n)} - f_{(L,n)}. \tag{11.5}$$

As expected, upon further detuning away from R_n, the higher modes start to show non-zero deviation and lose mode-locking. The injection range R_n for each mode n is as shown in Fig. 11.18, and is found to decrease with increasing mode number. The reduction in the injection range is proportional to the mode number as n^{-1} as found by the fit to the ranges. For modes >45, our observations of injection range were limited to $\sim 28\,\text{kHz}$, by our measuring apparatus (Fig. 11.17).

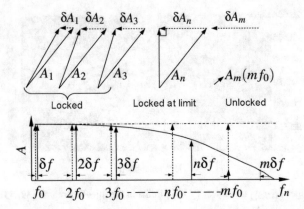

Fig. 11.19 Effect on cavity mode amplitudes with detuning δf. In the mode amplitude spectra, the mode envelope deviates to the solid line envelope from ideal mode-locked envelope shown in dotted-dashed line. The mode locations in frequency follow the Fourier frequencies of $(f_n = nf_0 + n\delta f)$, until mode n, that is still in the locking limit, the further modes $(m > n)$ are unlocked and oscillate at mf_0. The locking limit is where the injection signal is quadrature to the signal in the cavity as shown in the above schematic [39]

11.3.3 Unlocking of Laser Modes

The detuning mechanism for a mode-locked laser is shown in Fig. 11.19. For a perfectly mode-locked laser, the modes are at $f_n = nf_0$, as shown as dotted delta functions. Upon detuning by δf, the modes occur at $f_n = nf_0 + n\delta f$, interpreted as a frequency deviation $n\delta f$ from the locked state of each mode. The detuned state is shown with solid delta functions in frequency spectrum. Once the frequency deviation $n\delta f$ is outside the locking range R_n for the mode, the mode is no longer locked. This is shown for mode m which is at a higher frequency than the modes n that are within the locking range. This mode acts as free running oscillator, and the frequency of oscillation is same as at resonance, i.e. mf_0. However, the phase of the oscillation is no longer related to the rest of the modes [48, 49] and these modes are said to be unlocked.

In the unperturbed, mode-locked state, the amplitudes A_n for all modes are equal. This makes the spectral envelope of the frequency spectrum uniform, as shown by the dotted horizontal line in Fig. 11.19. The mode amplitudes follow the mode-locking equation for the nth mode given by Haus as [40]

$$\left\{ 1 + jb + j\frac{nf_i}{f_L}(\sigma + g) - g\left[1 - \left(\frac{nf_i}{f_L}\right)^2 \right] \right\} A_n = M\left(A_{n-1} - 2A_n + A_{n+1}\right),$$

$$(11.6)$$

where, quality factor is Q, and laser linewidth f_L. The laser gain is g, the empty-cavity resonance is f_C and detuning parameter is given by $\sigma = \dfrac{2Qf_L}{f_s}\left(\dfrac{f_i - f_C}{f_i}\right)$. Any detuning δf_s to the central oscillating frequency f_s is accounted for by $b = \dfrac{2Q}{f_s}\delta f_s$. When mode-locked $b = 0$ and $(\sigma + g) = 0$. The terms of the right constitute the injection signal due to the action of the modulator with strength M.

Upon detuning, the spectral envelope deviates from the mode-locked envelope as shown by the solid curve in the frequency spectrum. δA_n for each mode is the change in strength of injection for the mode. It has been shown that the ratios of injection signal to the mode amplitude does not change with the detuning [53]. A portion of the amplitude is required to compensate the phase deviation that builds up with detuning. When the mode frequency f_n is within the locking range R_n for the mode, the deviation is the difference phasor between the mode amplitude A_n at resonance, and the resulting mode amplitude $A_n + \delta A_n$ after perturbation [48]. The phasors for different modes with respect to the amplitude deviation δA_n are shown at the top of the respective mode at nf_0 in the frequency spectrum. The frequency f_n, in mode is at the edge of the injection range at either $f_{(U,n)}$ or $f_{(L,n)}$. A condition where the deviation and the resulting amplitude are at quadrature to each other is shown for the mode n in Fig. 11.19. Using perturbation analysis on (11.6), Krishnamoorthy et al. [53] have shown that R_n is given by

$$R_n = \left[\frac{f_s f_0}{Qf_C}\left(\frac{1}{n}\frac{A_{i1}}{A_{o1}}\right)\right],\qquad(11.7)$$

where, $\dfrac{A_{i1}}{A_{o1}}$ is the ratio of the mode amplitude to the injection signal for the first mode as determined by laser parameters. Which can be further simplified as

$$R_n = R_1\left(\frac{1}{n}\right).\qquad(11.8)$$

where, $R_1 = \dfrac{f_s f_0}{Qf_C}\left(\dfrac{A_{i1}}{A_{o1}}\right)$ is the range for the fundamental mode $n = 1$. This equation clearly shows that there is an n^{-1} dependence on injection range.

A mode m greater than n is unlocked, the phasor $A_m + \delta A_m$ makes an angle greater than $90°$ with the deviation δA_m. The large phase required to maintain the angle between the injection and the resultant mode amplitude cannot be provided by the laser, thus the mode is not locked and acts as a free running oscillator. The resulting frequency of the unlocked mode is mf_0 and not $mf_0 + m\delta f$ as would be expected for a locked mode. This is shown by the small delta function at mf_0 in the frequency spectrum in Fig. 11.19. To obtain the narrowest pulses from the slave laser, the repetition rate of the synchronizing master laser needs to be at a harmonic of the locking frequency of the slave laser, with the available detuning range decreasing as n as we attempt to lock n cavity modes together.

11.4 Conclusion

Fibre lasers are already the focus for adoption for nonlinear microscopy [54]. Other than 2p STED [15, 17], many techniques in time resolved spectroscopy use optical pump and probe methods that involve multiple lasers working in tandem, e.g. in time resolved CARS [55]. Both microscopy with STED and spectroscopy like time resolved coherent anti-Stokes Raman scattering (CARS), require two pulsed lasers working in synchronization to obtain the desired results. This review shows that active fibre mode-locked lasers are an excellent candidate for synchronization with another master laser in the configuration shown in Fig. 11.2. In this review, the authors have focused on using a Ti:Sapphire laser as a master laser, while driving a ytterbium doped fibre (YDF) ring laser in a master-slave configuration using EOM as the mode-locking element.

Actively mode-locked laser is driven using the electrical signal generated form the master laser, which ensures pulse to pulse synchronization. Wavelength selectivity is easily achieved using FBGs in the cavity that restrict the lasing wavelength and linewidth. There are many techniques, like using PZT, that can be employed to tune the wavelength of operation [33]. Other wavelength conversion schemes like second harmonic generation (SHG) or parametric conversion schemes can be used to achieve even wider wavelength options. The pulse energy is increased using fibre amplifiers and master oscillator optical amplifiers (MOPA) as shown in Fig. 11.5. The fibre mode-locked laser can be tuned in active configuration to a large frequency range of ~500 kHz, while providing pulse widths in the order of few ps to few ns. At resonant frequencies, they provide the sharpest pulses, that can be further compressed using positive index fibre and other techniques as discussed previously. To obtain uniform pulse trains, methods like regenerative feedback, and higher harmonic power optimization are required to provide the necessary stabilization [33, 47].

For the FRLs to act as slave lasers effectively, it is required that they be operable away from their locking frequencies. This requires a renewed understanding of the detuning mechanism in the lasers. The increase in the pulse width and the decrease in power away from the locking frequencies is discussed here. There is an inherent asymmetry in all FRLs with detuning. The asymmetry in pulse width is shown for Yb:doped FRLs and Er:doped FRLs in Figs. 11.9 and 11.12. The power in the higher modes also decrease asymmetrically as shown with higher harmonic power method. The understanding of the mechanism of the unlocking of lasers is understood by looking at the constituent modes n. The tolerance to detuning depends on the injection locking range of each mode, which decreases with mode number n as discussed using experiments and analytically. Further understanding of workings of FRLs is needed to build adaptable fibre lasers for use in biology. While, the lasers find the inroads to biology, the time for bio-inspired lasers has also come [56].

Acknowledgements S. Krishnamoorthy thanks Prof. Satyajit Mayor from NCBS-TIFR for guidance and support, the BioEngineering Research Initiative at NCBS for supporting her research and providing the opportunity to be associated with the project. The authors thank Jayavel D., Yusuf Panbiharwala and Sathish for help with construction of the lasers. The authors thank Central Imaging and Flow Cytometry Facility (CIFF) at NCBS-TIFR, Bangalore and the photonics@IITM group and Jitu-lab for facilities. Jayant lab and Rama Reddy for support with fluorescence experiments. The authors wish to thank the anonymous reviewers for their valuable suggestions.

References

1. G. Binnig, H. Rohrer, C. Gerber, E. Weibel, Surface studies by scanning tunneling microscopy. Phys. Rev. Lett. **49**, 57–61 (1982)
2. S.L. Jacques, Optical properties of biological tissues: a review. Phys. Med. Biol. **58**, R37 (2013)
3. X. Zhang, Z. Liu, Superlenses to overcome the diffraction limit. Nat. Mater. **7**, 435 (2008)
4. J.B. Pendry, Negative refraction makes a perfect lens. Phys. Rev. Lett. **85**, 3966–3969 (2000)
5. M.G.L. Gustafsson, Surpassing the lateral resolution limit by a factor of two using structured illumination microscopy. J. Microsc. **198**, 82–87 (2000)
6. W. Denk, J. Strickler, W. Webb, Two-photon laser scanning fluorescence microscopy. Science **248**, 73–76 (1990)
7. D. Axelrod, Cell-substrate contacts illuminated by total internal reflection fluorescence. J. Cell Biol. **89**, 141–145 (1981)
8. S.W. Hell, Nobel lecture: nanoscopy with freely propagating light. Rev. Mod. Phys. **87**, 1169 (2015)
9. E. Betzig, Single molecules, cells, and super-resolution optics (nobel lecture). Angew. Chem. Int. Ed. **54**, 8034–8053 (2015)
10. W.E. Moerner, Single-molecule spectroscopy, imaging, and photocontrol: Foundations for super-resolution microscopy (nobel lecture). Angew. Chem. Int. Ed. **54**, 8067–8093 (2015)
11. K. Nienhaus, G.U. Nienhaus, Where do we stand with super-resolution optical microscopy? J. Mol. Bio. **428**, 308–322 (2016). Study of biomolecules and biological systems: Proteins
12. S.W. Hell, J. Wichmann, Breaking the diffraction resolution limit by stimulated emission: stimulated-emission-depletion fluorescence microscopy. Opt. Lett. **19**, 780–782 (1994)
13. S.W. Hell, Far-field optical nanoscopy. Science **316**, 1153–1158 (2007)
14. G. Vicidomini, A. Schönle, H. Ta, K.Y. Han, G. Moneron, C. Eggeling, S.W. Hell, STED nanoscopy with time-gated detection: theoretical and experimental aspects. PLOS ONE **8**, 1–12 (2013)
15. J.N. Farahani, M.J. Schibler, L.A. Bentolila, A. Mendez-Vilas, J. Diaz, Stimulated emission depletion (STED) microscopy: from theory to practice. Microsc. Sci. Technol. Appl. Educ. **2**, 1539 (2010)
16. T.A. Klar, S. Jakobs, M. Dyba, A. Egner, S.W. Hell, Fluorescence microscopy with diffraction resolution barrier broken by stimulated emission. Proc. Natl. Acad. Sci. **97**, 8206–8210 (2000)
17. K.T. Takasaki, J.B. Ding, B.L. Sabatini, Live-cell superresolution imaging by pulsed STED two-photon excitation microscopy. Biophys. J. **104**, 770–777 (2013)
18. M.T. Asaki, C.P. Huang, D. Garvey, J. Zhou, H.C. Kapteyn, M.M. Murnane, Generation of 11-fs pulses from a self-mode-locked Ti:sapphire laser. Opt. Lett. **18**, 977–979 (1993)
19. A. Prabhakar, S. Mayor, S. Krishnamoorthy, Mode locked laser for generating a wavelength stabilized depletion pulse and method thereof (2014)
20. Y. Wu, X. Wu, L. Toro, E. Stefani, Resonant-scanning dual-color STED microscopy with ultrafast photon counting: a concise guide. Methods **88**, 48–56 (2015)
21. M.A. Lauterbach, M. Guillon, A. Soltani, V. Emiliani, STED microscope with spiral phase contrast. Sci. Rep. **3**, 2050 (2013)
22. A. Honigmann, S. Sadeghi, J. Keller, S.W. Hell, C. Eggeling, R. Vink, A lipid bound actin meshwork organizes liquid phase separation in model membranes. eLife **3**, e01671 (2014)
23. W. Li, Q. Hao, Y. Li, M. Yan, H. Zhou, H. Zeng, Ultrafast laser pulse synchronization, in *Coherence and Ultrashort Pulse Laser Emission* ed. by F.J. Duarte (InTech, 2010)

24. M. Rusu, R. Herda, O.G. Okhotnikov, 1.05-μm mode-locked ytterbium fiber laser stabilized with the pulse train from a 1.54-μm laser diode: errata. Opt. Express **12**, 5577–5578 (2004)
25. M. Nakazawa, M. Tokuda, N. Uchida, Continuous-wave laser oscillation with an ultralong optical-fiber resonator. J. Opt. Soc. Am. **72**, 1338–1344 (1982)
26. G. Geister, R. Ulrich, Neodymium-fibre laser with integrated-optic mode locker. Opt. Commun. **68**, 187–189 (1988)
27. J.D. Kafka, D.W. Hall, T. Baer, Mode-locked erbium-doped fiber laser with soliton pulse shaping. Opt. Lett. **14**, 1269–1271 (1989)
28. I.N. Duling, Subpicosecond all-fibre erbium laser. Electron. Lett. **27**, 544–545 (1991)
29. K. Tamura, H. Haus, E. Ippen, Self-starting additive pulse mode-locked erbium fibre ring laser. Electron. Lett. **28**, 2226–2228 (1992)
30. H. Takara, S. Kawanishi, M. Saruwatari, K. Noguchi, Generation of highly stable 20 GHz transform-limited optical pulses from actively mode-locked Er3+-doped fibre lasers with an all-polarisation maintaining ring cavity. Electron. Lett. **28**, 2095–2096 (1992)
31. K. Tamura, E. Ippen, H. Haus, L. Nelson, 77-fs pulse generation from a stretched-pulse mode-locked all-fiber ring laser. Opt. Lett. **18**, 1080–1082 (1993). cited By 654
32. T.F. Carruthers, I.N. Duling, 10-GHz, 1.3-ps erbium fiber laser employing soliton pulse shortening. Opt. Lett. **21**, 1927–1929 (1996)
33. M. Nakazawa, E. Yoshida, A 40-GHz 850-fs regeneratively fm mode-locked polarization-maintaining erbium fiber ring laser. IEEE Photonics Technol. Lett. **12**, 1613–1615 (2000)
34. U. Keller, Ultrafast solid-state laser oscillators: a success story for the last 20 years with no end in sight. Appl. Phys. B **100**, 15–28 (2010)
35. R. Herda, O.G. Okhotnikov, Dispersion compensation-free fiber laser mode-locked and stabilized by high-contrast saturable absorber mirror. IEEE J. Quantum Electron. **40**, 893–899 (2004)
36. M. Rusu, R. Herda, O.G. Okhotnikov, 1.05-μm mode-locked ytterbium fiber laser stabilized with the pulse train from a 1.54-μm laser diode. Opt. Express **12**, 5258–5262 (2004)
37. S. Krishnamoorthy, M. Mathew, S. Mayor, A. Prabhakar, Actively mode locked fiber laser for synchronized pulsed depletion in STED, (ThP-T1-P-17) in *6th EPS-QEOD Europhoton Conference, Neuchatel, Switzerland* (2014)
38. S. Krishnamoorthy, D. Jayavel, M. Mathew, S. Mayor, A. Prabhakar, Depletion laser for pulsed sted using wavelength stabilized actively mode locked lasers, in *ICOL, Dehradun, India* (2014)
39. S. Krishnamoorthy, S. Mayor, A. Prabhakar, Synchronization between two fixed cavity mode locked lasers, in *Proceedings of the 5th International Conference on Photonics, Optics and Laser Technology - Volume 1: PHOTOPTICS, INSTICC* (SciTePress, 2017), pp. 273–282
40. H.A. Haus, A theory of forced mode locking. IEEE J. Quantum Electron. **11**, 323–330 (1975)
41. C.J. Buczek, R.J. Freiberg, M. Skolnick, Laser injection locking. Proc. IEEE **61**, 1411–1431 (1973)
42. D.J. Kuizenga, A. Siegman, FM and AM mode locking of the homogeneous laser-Part I: theory. IEEE J. Quantum Electron. **6**, 694–708 (1970)
43. F.X. Kärtner, D. Kopf, U. Keller, Solitary-pulse stabilization and shortening in actively mode-locked lasers. J. Opt. Soc. Am. B **12**, 486–496 (1995)
44. L. Nelson, D. Jones, K. Tamura, H. Haus, E. Ippen, Ultrashort-pulse fiber ring lasers. Appl. Phys. B **65**, 277–294 (1997)
45. J. Kim, Y. Song, Ultralow-noise mode-locked fiber lasers and frequency combs: principles, status, and applications. Adv. Opt. Photonics **8**, 465–540 (2016)
46. A. Takada, H. Miyazawa, 30 GHz picosecond pulse generation from actively mode-locked erbium-doped fibre laser. Electron. Lett. **26**, 216–217 (1990)
47. S. Thiruthakkathevan, Scheme for coherent state quantum key distribution. Master's thesis, Indian Institute of Technology Madras, India (2011)
48. B. Razavi, A study of injection pulling and locking in oscillators, in *Proceedings of IEEE Custom Integrated Circuits Conference* (2004), pp. 305–312
49. R. Adler, A study of locking phenomena in oscillators. Proc. IRE **34**, 351–357 (1946)

50. K. Kurokawa, Injection locking of microwave solid-state oscillators. Proc. IEEE **61**, 1386–1410 (1973)
51. S.H. Strogatz, *Nonlinear Dynamics and Chaos: With Applications to Physics, Biology, Chemistry, and Engineering.* (Westview, Boulder, 2014)
52. A. Siegman, *Lasers* (University Science Books, Sausalito, 1986)
53. S. Krishnamoorthy, A. Prabhakar, Mode unlocking characteristics of an RF detuned actively mode-locked fiber ring laser. Opt. Comm. 431,39–44 (2019)
54. C. Xu, F. Wise, Recent advances in fibre lasers for nonlinear microscopy. Nat. Photonics **7**, 875–882 (2013)
55. F. El-Diasty, Coherent anti-Stokes Raman scattering: spectroscopy and microscopy. Vib. Spectrosc. **55**, 1–37 (2011)
56. A. Jonáš, D. McGloin, A. Kiraz, Droplet lasers. Opt. Photonics News **26**, 36–43 (2015)

Chapter 12
Physics-Based Feature Engineering

Bahram Jalali, Madhuri Suthar, Mohammad Asghari and Ata Mahjoubfar

Abstract We describe a new paradigm in computational imaging for performing edge and texture recognition with a superior dynamic range compared to other methods. The algorithm has its origin in Photonic Time Stretch, a realtime instrumentation technology that has enabled observation of ultrafast, non-repetitive dynamics and discovery of new scientific phenomena. In this chapter, we introduce the mathematical foundation of this new transform and review its intrinsic properties including the built-in equalization property that leads to high performance in visually impaired images. The algorithm is spearheading the development of new methods for feature engineering from visually impaired images with unique and superior properties compared to conventional techniques. It also provides a new approach to the computation of mathematical derivatives via optical dispersion and diffraction. The algorithm is a reconfigurable mathematical operator for hyper-dimensional feature detection and signal classification. It has also shown promising results in super-resolution single molecule imaging.

B. Jalali (✉) · M. Suthar · M. Asghari · A. Mahjoubfar
Department of Electrical and Computer Engineering, University of California Los Angeles, Los Angeles, CA, USA
e-mail: jalali@ucla.edu

M. Suthar
e-mail: madhurisuthar@ucla.edu

B. Jalali · A. Mahjoubfar
California NanoSystems Institute, Los Angeles, CA, USA
e-mail: ata.m@ucla.edu

B. Jalali
Department of Bioengineering, University of California Los Angeles, Los Angeles, CA, USA

Department of Surgery, David Geffen School of Medicine, University of California Los Angeles, Los Angeles, CA, USA

M. Asghari
Department of Electrical Engineering and Computer Science, Loyola Marymount University, Los Angeles, CA, USA
e-mail: asghari@ucla.edu

© Springer Nature Switzerland AG 2019
P. Ribeiro et al. (eds.), *Optics, Photonics and Laser Technology 2017*,
Springer Series in Optical Sciences 222,
https://doi.org/10.1007/978-3-030-12692-6_12

255

12.1 Introduction

"Human subtlety will never devise an invention more beautiful, more simple or more direct than does nature". The elegant quote by Leonardo Da Vinci underscores the important role of nature as a source of inspiration for human ingenuity. Inspirations from nature need not be limited to design of physical machines but should be extended to creation of new computational algorithms. We expect this new paradigm to lead to a new class of algorithms that are direct and energy efficient while providing unprecedented functionality.

Every day, the world creates nearly 2.5 Exabyte (10^{18} bytes) of data. Surprisingly, 90% of the data present in the world today has been created in the last two years alone highlighting the exponential rise in the amount of digital data [1]. To process this enormous data, the field of electronics has experienced a "digital revolution" leading to development of high performance digital signal processors (DSPs). However, a large fraction of this digital data is generated by translating real world analog signals into their digital representations using electronic Analog-to-Digital Converters (ADCs). These electronic ADCs cannot capture fast waveforms due to limited temporal resolution as well as impose restrictions on real time operations owing to their slower speed. Additionally, processing massive data in datacenters accounts for 50–60% of their electricity budget and a rapidly growing fraction of total electricity consumption. This calls for development of new computing technologies that offer speed, energy efficiency and ease of implementation. Fortunately, nature and in particular, photonics can provide a solution to certain class of problems.

12.1.1 Photonic Time Stretch

Photonic time stretch, a temporal signal processing technology, employs group-velocity delay (GVD) dispersion to slow down an analog signal in time and thereby, compresses the bandwidth of an analog signal allowing digital processing of fast waveforms which are otherwise not supported by slower electronic ADCs [2–4]. This method also known as the time-stretch dispersive Fourier transform (TS-DFT) has been the most successful solution to solve the critical problems associated with electronic ADCs in terms of temporal resolution as well as dynamic range and has been instrumental in the development of highest performance ADCs [5]. Time stretch technology has been employed widely for real time acquisition of ultrafast signals but the most remarkable application of this technology has been to study the relativistic electron structure [6].

Time stretch spectrometer is an extension of time stretch technology for high throughput single-shot spectroscopy that has led to observation of non-repetitive ultrafast events in optical systems such as optical rogue waves [7], soliton molecules [8] and mode locking in lasers [9]. Similarly, time stretch camera [10], a MHz-frame-rate bright-field imager, uses amplified dispersive Fourier transform to analyze images and has enabled the detection of rare cancer cells in blood with false positive rate of one cell in a million [11]. By integrating artificial intelligence with time stretch imaging, label-free cancer cell detection with record accuracy has been achieved [12, 13]. These high throughput imaging systems generate a torrent of data which can be compressed by foveated sampling using warped stretch that exploits signal sparsity [14].

12.1.2 From Optical Physics to Digital Algorithms

Photonic time stretch technique can be understood by considering the propagation of an optical pulse through a dispersive optical fiber. The optical pulse propagation in an optical fiber is governed by non-linear Schrodinger equation as discussed in [15]. By disregarding the loss and non-linearity in an optical fiber and considering only the group velocity dispersion, this equation upon integration reduces to

$$E_o(z, t) = \frac{1}{2\pi} \int\limits_{-\infty}^{+\infty} \tilde{E}_i(0, \omega) \cdot \left[e^{\frac{-j\beta_2 z\omega^2}{2}} \right] \cdot e^{j\omega t} d\omega \tag{12.1}$$

where $\tilde{E}_i(0, \omega)$ is the input pulse spectrum, $\beta_2 = $ GVD parameter, z is propagation distance, $E_o(z, t)$ is the reshaped output pulse at distance z and time t. The response of a dispersive element in a time-stretch system can be approximated as a phase propagator as presented in [16, 17]

$$H(\omega) = e^{i\phi(\omega)} = e^{i\sum_{m=0}^{\infty} \phi_m(\omega)} = \prod_{m=0}^{\infty} H_m(\omega) \tag{12.2}$$

Therefore, Equation (12.1) for a pulse that propagates through the time-stretch system and is reshaped into a temporal signal with a complex envelope can be written as follows

$$E_o(t) = \frac{1}{2\pi} \int\limits_{-\infty}^{\infty} \tilde{E}_i(\omega) \cdot H(\omega) \cdot e^{j\omega t} d\omega \tag{12.3}$$

By considering sufficient linear dispersion, the stationary phase approximation can be satisfied resulting in a mapping of spectrum to time. Using this spectro-temporal mapping, we can evaluate the sparsity in the spectrum of a signal [16]. In particular, one can create an information gearbox for matching the time-bandwidth of fast real-time optical data to that of the much slower electronics. These photonic hardware accelerators proposed in [16] act as a means to boost the speed and reduce the power consumption of electronics.

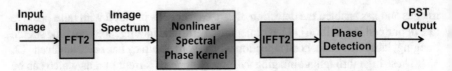

Fig. 12.1 Operation Principle of Phase Stretch Transform (PST). The input image is processed in frequency domain. The nonlinear spectral phase kernel encodes frequency components into the spatial phase of the output image such that high frequency components have higher phase

The time stretch operation $\mathbb{S}\{E_i(t)\}$ on an input pulse $E_i(t)$ defined above in terms of phase can be extended to operation on amplitude of a signal in optical domain as following

$$\mathbb{S}\{E_i(t)\} = \int\limits_{-\infty}^{+\infty} F\{E_i(t)\} \cdot \left[e^{j\phi(\omega)} \cdot \tilde{L}(\omega)\right] \cdot e^{j\omega t} d\omega \qquad (12.4)$$

where $FE_i(t)$ is the fourier spectrum of input pulse $e^{j\phi(\omega)}$ is the phase filter and $\tilde{L}(\omega)$ is the amplitude filter. And for a discrete signal, the stretch operation can be defined as,

$$\mathbb{S}\{E_i[t]\} = IFFT\left\{FFT\{E_i[t]\} \cdot \tilde{K}[\omega] \cdot \tilde{L}[\omega]\right\} \qquad (12.5)$$

where $\tilde{K}[\omega]$ is the phase filter or propagator, $\tilde{L}[\omega]$ is the amplitude filter, IFFT is Inverse Fast Fourier Transform and FFT is Fast Fourier Transform.

The dispersion operation on a 1D temporal signal is equivalent to parallax diffraction on a 2D input space. This led us to analyze the application of above mentioned dispersion-based stretch operations to digital images which resulted in an optics-inspired edge detection algorithm called Phase Stretch Transform (PST) [18, 19]. PST is a qualitatively new method for feature engineering and is discussed at length in the next section.

12.2 Phase Stretch Transform

Phase Stretch Transform (PST) is a recently introduced computational approach for signal and image processing that emerged out of the research on Photonic Time Stretch. This algorithm transforms an image by emulating propagation of electromagnetic waves through a diffractive medium with an engineered 3D dispersive property (refractive index profile) [18, 19].

As shown in Fig. 12.1, by applying a nonlinear spectral phase kernel to the image spectrum (by operating 2D Fast Fourier Transform (FFT2) on the image), the frequency distribution is mapped to the spatial phase of the output image. The nonlinear phase kernel encodes a high phase value to high frequency component. The spatial output after the 2D Inverse Fast Fourier Transform (IFFT2) is no longer a real quantity but instead has a complex phase associated with it. Upon phase detection of the output followed by thresholding operation, high phase values corresponding to high frequency components survive. Hence, edges i.e. high frequency components in an image are detected.

The time stretch stretch operation $\mathbb{S}\{\}$ for an image can be represented as

$$\mathbb{S}\{E_i\,[x,y]\} = IFFT^2 \left\{ FFT^2\,\{E_i\,[x,y]\} \cdot \tilde{K}\,[u,v] \cdot \tilde{L}\,[u,v] \right\} \tag{12.6}$$

In the above equations, $E_i\,[x,y]$ is the input image, x and y are the spatial variables and, u and v are spatial frequency variables. The function $\tilde{K}\,[u,v]$ is called the warped phase kernel and the function $\tilde{L}\,[u,v]$ is a localization kernel implemented in frequency domain for image processing. PST operator [18, 19] is defined as the phase of this Warped Stretch Transform output as follows,

$$PST\{E_i[x,y]\} \triangleq \angle\langle \mathbb{S}\{E_i\,[x,y]\}\rangle \tag{12.7}$$

where $\angle\langle\,\cdot\,\rangle$ is the angle operator.

The transform has been applied for feature extraction in biomedical images to develop diagnostic assistant tools [20, 21] and for edge detection in Synthetic Aperture Radar (SAR) images [22]. It has also been applied for resolution enhancement in super-resolution photon activated localization microscopy (PALM) for imaging of single molecule [23]. The transform drastically improved the localization of point spread function, reduced the computational time by 400% and increased the emitter density by the same amount. The algorithm has been open sourced on GitHub and Matlab Central File Exchange [24] and has received extraordinary endorsements both by the software as well as image processing community. The transform exhibits superior performance over conventional derivative based edge operators in particular for visually impaired images. It is able to reveal features invisible to human eye and to conventional algorithms used today. Because of these unique intrinsic properties offered by PST, it has promising application for feature enhancement in visually impaired images which is discussed in next section.

12.2.1 Feature Enhancement in Visually Impaired Images

In the field of computer vision, feature detection from digital images plays a pivotal role for solving problems associated with image registration, object recognition, content-based image retrieval and deep learning [25–27]. To improve feature detection, prior works such as [25, 28], have exploited grey level statistics of the image and various edge detection methods, respectively. Few efforts have also focused on the use of color distinctiveness and color models [29, 30] and scale selections [31] in images to enhance feature detection.

(a) Original Image   (b) Feature detection   (c) Feature detection
                          using smooth derivative   using PST

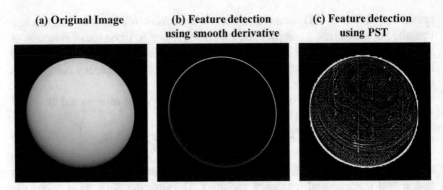

Fig. 12.2 Comparison of feature detection using conventional derivative based edge operator to the case of feature detection using Phase Stretch Transform (PST). The derivative is the fundamental operation used in the popular Canny, Sobel and Prewitt edge detection methods. These derivative based methods are unable to capture the low contrast on the surface of the Uranus planet. On the other hand, PST extracts these surface variations efficiently as shown in [34]

(a) Original Image   (b) Feature detection   (c) Feature detection
                          using smooth derivative   using PST

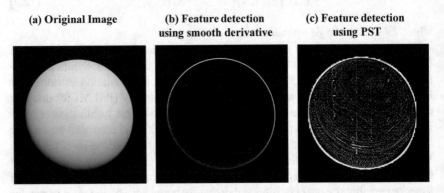

Fig. 12.3 Comparison of feature detection using conventional derivative based edge operator to the case of feature detection using Phase Stretch Operator (PST) on an image of the planet Uranus captured from a different angle of view as compared to the one shown in Fig. 12.2. Original image is shown in (**a**). Results of feature detection using conventional derivative based edge operator and PST operator are shown in (**b**) and (**c**), respectively. PST is able to locate the low contrast on the surface of the planet which are consistent with the edges located in Fig. 12.2 as shown in [34]

The principal aim of feature detection is to classify objects with a higher accuracy and at the same time be robust to changing viewing conditions which include variations in illumination or environmental conditions, object orientations as well as the zoom factor of a camera. Detection and localization of objects in images can be severely impaired by environmental conditions. For instance, images acquired in a foggy weather suffer from various visual impairments such as reduced contrast, blur and noise which leads to low resolution and low contrast [32, 33]. This poses a major problem for many computer vision applications, in particular, for autonomous robots and self-driven vehicles.

(a) Original image   **(b) Feature detection using**   **(c) Feature detection**
                            smooth derivative   **using PST**

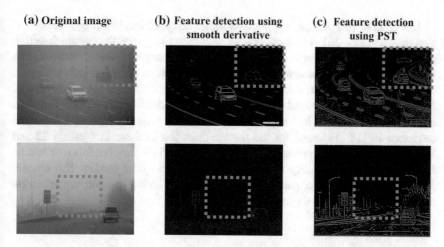

Fig. 12.4 Comparison of feature detection using conventional derivative based edge operator to the case of feature detection using Phase Stretch Transform (PST) in case of visually impaired images. Original traffic images taken in a foggy weather are shown in (**a**). Detected features using conventional derivative based edge operator and PST operator are shown in (**b**) and (**c**), respectively. It can be seen that the conventional derivative based edge detection operator fails to visualize the low contrast details in the visually impaired regions of the images (as shown in green dashed boxes). However, PST captures these low contrast details in the low resolution regions (as shown in green dashed boxes) due to its unique reconfigurable mechanism that detects features over a wide dynamic range. The strength of features detected using PST over both low and high resolution regions of the images is consistent unlike derivative operator

Emerging imaging technologies such as High Dynamic Range (HDR) imaging hold promise to solve these problems in the field of computer vision by improving feature detection. However, the slow frame rate of these technologies restricts their practice in self-driven cars, autonomous robotics and other real-time applications. PST has unique intrinsic properties which are not offered by the state-of-the-art algorithms. To validate this claim, we refer to Fig. 12.2 that shows an image of the planet Uranus processed by the conventional derivative based edge operator and by the PST. The derivative method is the underlying function utilized by the popular Canny, Sobel and Prewitt algorithms. The result clearly shows the dramatic advantage offered by the optics-inspired PST. The inherent equalization ability of PST gives a response ideal for feature detection in low contrast visually impaired images.

Figure 12.3 compares the effect of feature detection using conventional derivative based edge operator with feature detection using PST on another image of planet Uranus captured from a different angle of view. Conventional derivative based edge operator fails to visualize the low contrast in the bright areas of the image over the surface of the planet Uranus. However, PST can clearly show these small intensity changes even in the intensity-saturated areas due to its natural equalization mechanism. These surface variations over the planet are consistent with the edges detected in the Fig. 12.2 highlighting the efficiency of PST.

In Fig. 12.4, PST has been applied on traffic images taken in a foggy weather as also shown in [35, 36]. As we can see, it significantly improves the feature detection and also, outperforms conventional derivative based edge detection methods. The conventional edge detection method based on derivative of the image fails to capture low contrast details in the high intensity but low resolution areas of the image. However, our technique clearly shows the low contrast details in the visually impaired regions of the image. The warp W, and strength S, parameters of PST kernel as described in [18, 19] used for the feature detection in these images are 22 and 500, respectively. This property emerges because PST's transfer function has an inherent equalization ability, derived analytically in the next section. This chapter also discusses the superior performance of PST at low light levels and its application to HDR images towards the end.

12.2.2 Mathematical Foundations of Phase Stretch Transform

The superior performance of Phase Stretch Transform (PST) in the low contrast regime is proved here mathematically by deriving closed-form analytical expressions for its transfer function. The results reveal that the transform has an inherent intensity equalization property leading to high dynamic range performance. Analytical results are also supported by numerical simulations confirming the dynamic range enhancement. The stretch operator $\mathbb{S}\{\}$ on an input image act as follows

$$E_o[x, y] = S\{E_i[x, y]\} \triangleq IFFT2\left\{\tilde{K}[u, v] \cdot \tilde{L}[u, v] \cdot FFT2\{E_i[x, y]\}\right\}$$
(12.8)

and $E_o[x, y]$ is a complex quantity defined as,

$$E_o[x, y] = |E_o[x, y]| \, e^{j\theta[x,y]}$$
(12.9)

Here for simplicity, we assume the localization kernel $\tilde{L}[u, v] = 1$. Now, without the loss of generality and in order to keep the notations manageable in what follows, we consider operation of PST on 1D data, i.e.,

$$PST\{E_i[x]\} \triangleq \angle\{S\{E_i[x, y]\}\} = \angle\{E_o[x]\} = \angle\left\langle IFFT\left\{\tilde{K}[u] \cdot FFT\{E_i[x]\}\right\}\right\rangle$$
(12.10)

The warped phase kernel $\tilde{K}[u]$ is defined by a phase function that has a nonlinear dependence on frequency, u,

$$\tilde{K}[u] = e^{j \cdot \varphi[u]}$$
(12.11)

By expanding the phase term in the warped phase kernel $\tilde{K}[u]$ using Taylor series we have,

$$\tilde{K}[u] = e^{\left(j \sum_{m=2}^{M} \frac{\varphi^{(m)}}{m!} u^m\right)} \tag{12.12}$$

where $\varphi^{(m)}$ is the mth-order discrete derivative of the phase $\varphi[u]$ evaluated for $u = 0$ and values of m are even numbers. PST phase term $\varphi[u]$ only contains even-order terms in its Taylor expansion due to the even symmetry of the phase term $\varphi[u]$ as first considered in [18, 19]. By using the expansion of warped phase kernel as described in (12.12), output complex-field, $E_o[x]$, can be calculated as follows,

$$E_o[x] = IFFT\left\{\tilde{E}_i[u] \times \tilde{K}[u]\right\}$$

$$= IFFT\left\{\tilde{E}_i[u] \times e^{\left(j \sum_{m=2}^{M} \frac{\phi^{(m)}}{m!} u^m\right)}\right\} \tag{12.13}$$

where $\tilde{E}_i[u]$ is the spectrum of the input computed using Fast Fourier transform (FFT). Simulation show that when the applied phase is small, PST works best. Under these conditions, we can use small value approximation for the applied phase kernel. Therefore, the phase term in (12.13) can be simplified to,

$$E_o[x] = IFFT\left\{\tilde{E}_i[u] \times \left[1 + j\left(\sum_{m=2}^{M} \frac{\phi^{(m)}}{m!} u^m\right)\right]\right\} \tag{12.14}$$

$$\rightarrow E_o[x] \approx \left[1 \times E_i[x] + j\sum_{m=2}^{M} \frac{(-1)^{m/2} \phi^{(m)}}{m! (2\pi)^m} E_i[x]^{(m)}\right] \tag{12.15}$$

where $E_i[x]^{(m)}$ is the mth-order discrete mathematical derivative of the input $E_i[x]$. As the input is a real quantity, we can calculate the output phase as,

$$PST\{E_i[x]\} = \angle\{E_o[x]\} \approx \tan^{-1}\left\{\frac{\sum_{m=2}^{M} \frac{(-1)^{m/2} \phi^{(m)}}{m! (2\pi)^m} E_i[x]^{(m)}}{E_i[x]}\right\} \tag{12.16}$$

As can be seen in the (12.16), the transfer function of PST is consisting of a summation of even order mathematical derivatives of the input in the numerator divided by the input amplitude (brightness) in the denominator. A hyper dimensional feature set corresponding to different measures of the curvature of the edge is computed in the numerator while the denominator renders the response nonlinear in such a way that low-light-levels in the input are enhanced. Above analytical results are derived by considering a general expression for the phase kernel and only when the applied phase of the PST kernel is small. We now expand our analytical findings by considering different scenarios that reveal further insights into the unique properties of PST.

Case 1: Let's consider the Phase Kernel $\tilde{K}[u]$ as a quadratic function of frequency variable u. Under this condition $\tilde{K}[u] = u^2$ and by using small phase approximation, the phase term in (12.14) can be simplified to,

$$E_o[x] = IFFT \left\{ \tilde{E}_i[u] \times \left[1 + j \left(u^2 \right) \right] \right\} \tag{12.17}$$

$$\rightarrow E_o[x] \approx \left[1 \times E_i[x] - j \frac{1}{(2\pi)^2} * \frac{d^2 E_i[x]}{dx^2} \right] \tag{12.18}$$

We also assume that the phase of the complex output is restricted to small values. Therefore, phase of the output in (12.18) can be simplified to,

$$PST\{E_i[x]\} = \angle E_o[x] \approx \frac{\frac{-1}{(2\pi)^2} * \frac{d^2 E_i[x]}{dx^2}}{E_i[x]} \tag{12.19}$$

Case 2: We consider here the same Phase Kernel (as a quadratic function of frequency variable u), $\tilde{K}[u] = u^2$ as discussed in Case 1. However, we do not restrict to small phase approximation. The exponential term in (12.13) can now be represented as,

$$E_o[x] = IFFT \left\{ \tilde{E}_i[u] \times \left[\cos\left(u^2 \right) + j \sin\left(u^2 \right) \right] \right\} \tag{12.20}$$

We expand the sine and cosine terms using Euler expansion up to third order and then by applying small value approximation to the complex output of (12.16), the PST output can be computed as shown below

$$PST\{E_i[x]\} = \angle E_o[x] \approx \frac{\frac{-1}{(2\pi)^2} * \frac{d^2 E_i[x]}{dx^2} + \frac{1}{3!(2\pi)^6} * \frac{d^6 E_i[x]}{dx^6} - \frac{1}{5!(2\pi)^{10}} * \frac{d^{10} E_i[x]}{dx^{10}}}{E_i[x] - \frac{1}{2!(2\pi)^4} * \frac{d^4 E_i[x]}{dx^4} + \frac{1}{4!(2\pi)^8} * \frac{d^8 E_i[x]}{dx^8}} \tag{12.21}$$

The closed-form expression of the transfer function of PST shown in (12.16) relates the output to the input in the case of an arbitrary phase kernel valid under small phase approximation. For certain scenarios, the core functionality of PST as a feature detector can be established by the closed-form expression presented in (12.16). As we can see, the output of the PST operator is directly proportional to the even-order derivatives of the input with weighting factors of $\frac{(-1)^{m/2}}{m!} \frac{\phi^{(m)}}{(2\pi)^m}$. Each computed mathematical derivative highlights a different feature of the input. The weighting factors can be modified to selectively enhance features of interest. Hence, our transform acts as a reconfigurable operator that can be tuned to emphasize different features in an input image.

The intrinsic equalization ability of PST can be seen in (12.16). The PST output is inversely related to the input brightness level valid under small phase approximation. Therefore, for a same contrast level, the PST output is higher in dark low-light-level areas of an image. This important observation from (12.16) confirms the fact that PST has an inherent property to equalize the detected output with the input brightness

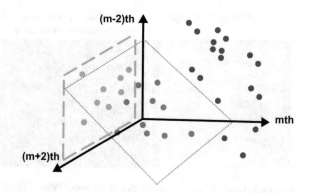

Fig. 12.5 Phase Stretch Transform (PST) as a hyper dimensional classifier. PST operator can act as a reconfigurable operator to compute mth-order derivative. Here the dimensions are the order (even) of derivatives and hyper planes are shown as green and red boxes

level and therefore, allows for a more sensitive feature detection as also reported in [35, 36].

Brightness level equalization is a well-studied technique to improve feature detection algorithms in High Dynamic Range (HDR) images (see [37] for example). One approach to achieve brightness level equalization in images is by applying a logarithmic (log) function to the input before application of feature detection algorithms. By applying a log function, we can achieve high gain for low brightness input. This brightness equalization results in a more sensitive feature detection. Fortunately, the PST operator has a built-in logarithmic behavior which gives it excellent dynamic range. However, this property does not completely describe the functionality of our transform. As observed in (12.16), our transform computes a hyper-dimensional feature set for signal classification (shown in Fig. 12.5). These results also demonstrate a method for computation of mathematical derivatives via group delay dispersion operations.

12.2.3 Clinical Decision Support Systems Using PST

Medical images act as a very important source of information in order to understand the anatomy and organ function as well as aid in diagnosis of diseases. Lately, machine learning tools are unified with image processing techniques for application to medical imaging leading to production of computer-aided diagnostics (CAD) and decision making tools. As discussed previously, PST has superior performance over conventional edge detectors, therefore, its application to medical images for feature detection is promising for accurate segmentation and development of CAD tools.

One such application of PST is to develop a diagnostic assistant tool for pneumothorax [20, 21]. Pneumothorax is a medical situation in which air leaks into the space between the lungs and the chest wall due to a chest injury or a lung disease or even sometimes due to certain medical procedures. The major risk factor that increases the mortality caused by pneumothorax is the failure to identify it at any early

(a) Original Image   (b) Feature detection   (c) Feature detection
                            using smooth derivative   using PST

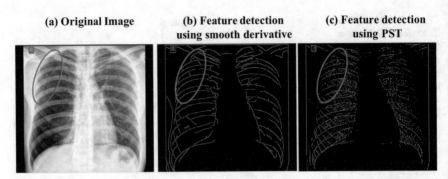

Fig. 12.6 Comparison of feature detection using Phase Stretch Transform (PST) algorithm with the features detected using a conventional derivative based edge operator for a lung X-ray image. In the figure, **a** X-ray of lung of a patient suffering from pneumothorax. **b** Edge detection using conventional derivative based edge operator. **c** Edge detection using PST. The red oval indicates the region of collapsed lung. As shown, PST traces the low contrast lung edge with an equalized response due to intrinsic nonlinear behavior. Conventional derivative based edge operator work well only for high contrast regions and therefore, fails to locate the collapsed lung boundary as described in [20, 21]

examination by a radiologist. By the application of PST to the chest X-rays' of patient suffering from pneumothorax, the boundary of collapsed lung is easily located otherwise difficult to be identified during an initial visual inspection by a radiologist as shown in Fig. 12.6. This tool, which is first of its kind, traces the collapsed lung and aids the radiologist to take correct decision in this life-critical examination. With pneumothorax being very common in ventilated critically ill patients, it becomes important to develop tools for accurate diagnosis as failures in diagnosis can cause life threatening complications [38].

12.3 Simulation Results

In this section, we present simulation results that validate the closed-form analytical expression of the transfer function of PST derived in the previous section. In order to reinforce the new theory developed above, we also demonstrate several visual examples of operation of PST on HDR images. In our first simulation result, we show the PST output for a given 1D data and compare it to the output computed by the analytical expression derived in (12.16). Figure 12.7a shows the phase kernel $\varphi[u]$ of the PST operator designed for this simulation. The warp W, and strength S, parameters of the phase kernel as described in [18, 19] are 12.5 and 4000, respectively. The derivative of the PST phase kernel is shown in Fig. 12.7b. The simulated 1D input data is shown in Fig. 12.7c. In Fig. 12.7d, numerically simulated PST output is compared to the output estimated by the analytical expression of (12.16) using red-solid and blue-dotted lines, respectively. Evidently, the simulation result confirms

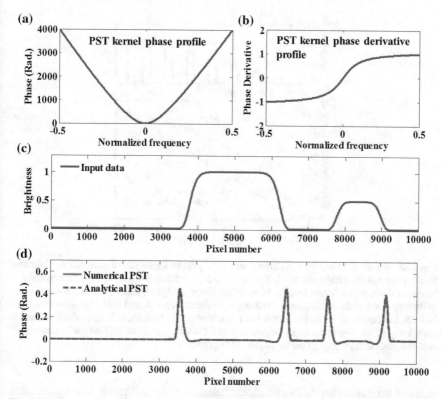

Fig. 12.7 Comparison of numerically simulated output of Phase Stretch Transform (PST) algorithm with the output given by the closed-form analytical expression derived in (12.16). The phase kernel of the PST operator and the corresponding derivative profile of the phase kernel are shown in (**a**) and (**b**), respectively. The input data is shown in (**c**). Numerically calculated PST output data and the output data estimated by the closed-form expression of the PST transfer function, derived in (12.16), is shown in (**d**) using red-solid and blue-dotted lines, respectively. The above simulation result validates the accuracy of the closed-form analytical expression of PST transfer function derived in (12.16)

the accuracy of the closed-form analytical expression of the transfer function of our algorithm as derived in (12.16).

In the next simulation example, we examine the effect of PST operation for feature detection on a signal with varying contrast levels at a constant brightness level and compare it to the case of using the conventional technique of differentiation to detect features in the same input. The warp W, and strength S, parameters of the PST operator are 12.15 and 0.48, respectively. The input was designed to have varying contrast levels at a constant brightness level, as shown in Fig. 12.8a. We compare numerically simulated PST output to the output using differentiation in Fig. 12.8b. As expected, the output of the differentiator is linearly proportional to the contrast level and is insensitive to the input brightness level. On the other hand, PST output is directly dependent on the input contrast level but inversely dependent on the

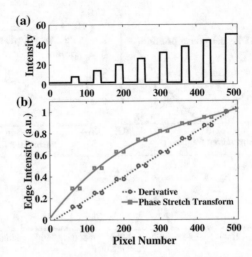

Fig. 12.8 Effect of Phase Stretch Transform (PST) on an input signal with various contrast levels at a constant brightness level. The input 1D data which is designed to have various contrast levels at a constant brightness level is shown in (**a**). Numerically calculated PST output is compared to the output using differentiation in (**b**) using red-solid and blue-dotted lines, respectively. As can be seen, the output of the differentiator has a linear response to contrast level in the input and is completely insensitive to the input brightness level. On the other hand, PST output is nonlinearly related to the contrast level in the input at fixed brightness

constant brightness level. This nonlinear behavior is due to the inherent equalization mechanism of PST as described in (12.16).

The behavior of PST operation for feature detection on a signal with a constant contrast level and various brightness levels is evaluated in Fig. 12.9. The input data, designed to have a constant contrast level and various brightness levels, is shown in Fig. 12.9b. The warp W, and strength S, factors used for the PST operator are 12.15 and 0.48, respectively. The red solid line representing the output data confirms that the relation of PST to same contrast level at various brightness levels, is logarithmic as estimated in (12.16). Therefore, the simulation result presented in Fig. 12.9 further reinforce the accuracy of the closed-form equation to estimate the output of the PST algorithm [34].

Figure 12.10 shows a visual example of using PST for feature enhancement on a 14 bit HDR image. The image has features of interest in extremely low-light-level regions, as seen in the red boxes. We now compare the performance of the derivative operator with PST for feature detection. The derivative operator was implemented from native smooth derivative function. For a fair comparison, both methods use the same localization kernel (a gaussian function) with sigma factor of 2. The warp W, and strength S, parameters used for the PST operator are 12.15 and 0.48, respectively. Results of feature detection using smooth derivative operator and PST operator are shown in Fig. 12.10b and Fig. 12.10c, respectively. The derivative operator is unable to unveil the small contrast details in the dark areas of the image, as can be seen

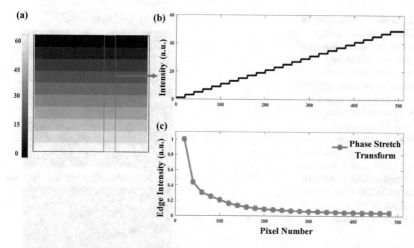

Fig. 12.9   Effect of Phase Stretch Transform (PST) on an input signal with a constant contrast level and various brightness levels. The input data which is designed to have a constant contrast level and various brightness levels is shown in (**b**). Numerically calculated PST output data for feature detection in a signal with a constant contrast level varies with the input brightness level is shown in (**c**). This shows that PST output has an inverse dependence on the input brightness level

in Fig. 12.10b. However, PST extracts these low contrast details in dark areas due to its natural equalization mechanism, see dashed box in Fig. 12.10c. It also can be observed that the intensity of detected edges in the case of smooth derivative is related linearly to the brightness level of the original image, compare solid box areas in Fig. 12.10a, b. In contrast, PST has automatically equalized the brightness level in the solid box region in the image and outputs relatively constant feature intensity for that region, see Fig. 12.10c. We note that PST has failed to visualize features in high contrast areas in the image. This is because of the inverse dependence on brightness level as derived in (12.16). This issue can be mitigated by setting a higher maximum threshold for detected features or by equalizing the image brightness before passing through PST operator.

We can further examine the role of PST for feature detection in low-light-level and high-light-level regions by considering a line scan of a HDR image shown previously in Fig. 12.10. The blue box in the Fig. 12.11 demonstrates the response of PST to low-light-level regions where it outperforms conventional derivative based edge operator. Similarly, for high-light-level regions of the image (shown in green and purple box in the Fig. 12.11), PST outperforms when the contrast is low (shown in green box in the Fig. 12.11). In contrast, the conventional derivative based edge operator response is dominating only in high contrast regions (shown in purple box in the Fig. 12.11).

Fig. 12.10 Comparison of feature detection using smooth derivative operator to the case of feature detection using Phase Stretch Transform (PST). Original image is shown in (**a**). Smooth derivative operator is unable to efficiently visualize the low contrast details in the dark areas of the image. However, PST captures these contrast changes in low-light-level areas due to its intrinsic equalization property

(a) **Original image**

(b) **Feature detection using smooth derivative**

(c) **Feature detection using PST**

To enhance the dynamic range of operation, we introduced a hybrid system in [35, 36] that aggregates edge responses from both the PST and the conventional edge detection filter. As shown in Fig. 12.12, the hybrid system combines the edge response from PST and conventional derivative based edge operator, providing edge detection in both low-light and high-light levels regions by capturing both low and high contrast details.

(a) Input line scan

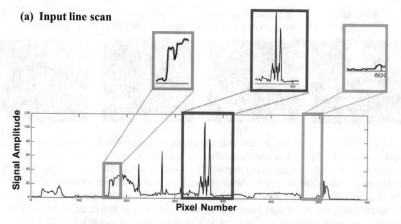

(b) Feature detection using PST and conventional derivative operator

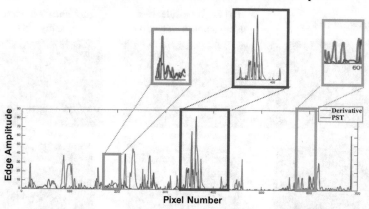

Fig. 12.11 **Comparing feature detection using conventional derivative based edge operator and Phase Stretch Transform (PST) operator under low-light-level and high-light-level conditions. a** Original input line scan corresponding to Row 524 from the image shown in Fig. 12.10. Feature detection of this input line scan using the derivative and the PST operator is shown in (**b**). The blue box demonstrates that the response of PST is higher than the derivative operator under low-light-level conditions. The green and purple box shows the response of PST and derivative operator for feature detection under high-light-level conditions. While PST enhances low contrast features under low-light-level as wells as under high-light-levels (see green box) unlike derivative operator which identifies high contrast features (see blue box)

Finally, Fig. 12.13 shows another example application of PST for Optical Character Recognition (OCR). As shown in Fig. 12.13, in this painting of "Minerva of Peace" there are optical characters in the scroll that need to be recognized (see red solid box in Fig. 12.13a). Results of feature detection using conventional derivative based edge operator and PST operator are shown in Fig. 12.13b and 12.13c respectively. Clearly, conventional operator fails to efficiently visualize the sharp features of the alphabets in the scroll. However, PST traces the edges of alphabets efficiently

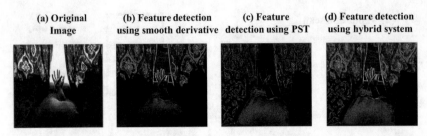

(a) Original Image / (b) Feature detection using smooth derivative / (c) Feature detection using PST / (d) Feature detection using hybrid system

Fig. 12.12 Hybrid system for feature detection using conventional derivative based edge operator and Phase Stretch Transform (PST). Original image is shown in (**a**). Results of feature detection using smooth derivative operator and PST operator are shown in (**b**) and (**c**), respectively. And the output of hybrid system is shown in (**d**). Note that in (**d**), the strength of the detected features in both the high-light-level and low-light-level regions is same. The hybrid system selects the low contrast features detected using PST operator and the high contrast features computed by the smooth derivative operator resulting in a wide dynamic range of operation

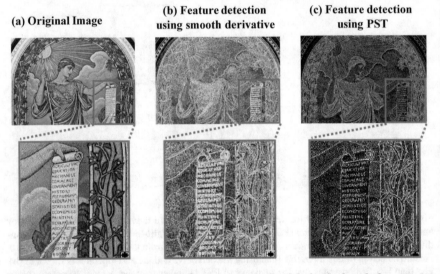

(a) Original Image / (b) Feature detection using smooth derivative / (c) Feature detection using PST

Fig. 12.13 Comparison of feature detection using conventional derivative based edge operator to the case of feature detection using Phase Stretch Transform (PST). Original image is shown in (**a**). Results of feature detection using conventional detectors and PST operator are shown in (**b**) and (**c**), respectively. Enlarged view of the scroll in the painting, shown in the red boxes, establishes the superiority of PST to trace the edges of alphabets in the scroll and thus, provide more information on the contrast changes in dark areas due to its natural equalization mechanism, see Fig. 12.13c. Conventional derivative based edge operator was implemented from find edge function in ImageJ software. The warp W, and strength S, parameters used for the PST operator are 13 and 0.4, respectively.

12.4   Conclusions

In this chapter, we presented that the physics of light propagation in a dispersive or a diffractive media has natural properties that can be exploited for various applications. Photonic time stretch technology utilizes dispersion to slow down an analog signal in time. Phase Stretch Transform employs dispersion to extract features from the data. We showed via analytical derivations as well as numerical simulations that this physics-inspired transform has an intrinsic equalization property. This inherent ability of PST significantly improves feature detection in visually impaired images and thereby, results in a high dynamic range of operation for feature extraction.

The phase kernel of PST can be tuned to compute different orders of mathematical derivative via group delay dispersion operations. This inbuilt reconfigurability of our transform can be used to generate a hyper-dimensional feature set consisting of different orders of derivative of input for signal classification.

Acknowledgements This research work presented in this chapter was partially supported by the National Institutes of Health (NIH) grant no. 5R21 GM107924-03 and the Office of Naval Research (ONR) Multidisciplinary University Research Initiatives (MURI) program on Optical Computing.

References

1. IBM, Bringing big data to the enterprise (2016), https://www-01.ibm.com/software/in/data/bigdata/
2. A.S. Bhushan, F. Coppinger, B. Jalali, Time-stretched analogue-to-digital conversion. Electron. Lett. **34**(9), 839–841 (1998)
3. F. Coppinger, A.S. Bhushan, B. Jalali, Photonic time stretch and its application to analog-to-digital conversion. IEEE Trans. Microw. Theory Tech. **47**(7), 1309–1314 (1999)
4. Y. Han, B. Jalali, Photonic time-stretched analog-to-digital converter: fundamental concepts and practical considerations. J. Lightwave Technol. **21**(12), 3085 (2003)
5. W. Ng, T. Rockwood, A. Reamon, Demonstration of channel-stitched photonic time-stretch analog-to-digital converter with ENOB 8 for a 10 GHz signal bandwidth, in *Proceedings of the Government Microcircuit Applications & Critical Technology Conference (GOMACTech14)* (2014)
6. C. Szwaj, C. Evain, M. Le Parquier, S. Bielawski, E. Roussel, L. Manceron, M. Labat, et al. Unveiling the complex shapes of relativistic electrons bunches, using photonic time-stretch electro-optic sampling, in *Photonics Society Summer Topical Meeting Series (SUM)* (IEEE, 2016), pp. 136–137
7. D.R. Solli, C. Ropers, P. Koonath, B. Jalali, Optical rogue waves. Nature **450**(7172), 1054 (2007)

8. G. Herink, F. Kurtz, B. Jalali, D.R. Solli, C. Ropers, Real-time spectral interferometry probes the internal dynamics of femtosecond soliton molecules. Science **356**(6333), 50–54 (2017)
9. G. Herink, B. Jalali, C. Ropers, D.R. Solli, Resolving the build-up of femtosecond mode-locking with single-shot spectroscopy at 90 MHz frame rate. Nat. Photonics **10**(5), 321–326 (2016)
10. K. Goda, K.K. Tsia, B. Jalali, Serial time-encoded amplified imaging for real-time observation of fast dynamic phenomena. Nature **458**(7242), 1145 (2009)
11. K. Goda, A. Ayazi, D.R. Gossett, J. Sadasivam, C.K. Lonappan, E. Sollier, C. Photonic, High-throughput single-microparticle imaging flow analyzer. Proc. Natl. Acad. Sci. **109**(29), 11630–11635 (2012)
12. C.L. Chen, A. Mahjoubfar, L.C. Tai, I.K. Blaby, A. Huang, K.R. Niazi, B. Jalali, Deep learning in label-free cell classification. Sci. Rep. **6**, 21471 (2016)
13. A. Mahjoubfar, C.L. Chen, B. Jalali, *Artificial Intelligence in Label-free Microscopy* (Springer, 2017)
14. C.L. Chen, A. Mahjoubfar, B. Jalali, Optical data compression in time stretch imaging. PLoS ONE **10**(4), e0125106 (2015)
15. G.P. Agrawal, *Nonlinear Fiber Optics* (Academic press, Chicago, 2007)
16. B. Jalali, A. Mahjoubfar, Tailoring wideband signals with a photonic hardware accelerator. Proc. IEEE **103**(7), 1071–1086 (2015)
17. A. Mahjoubfar, D.V. Churkin, S. Barland, N. Broderick, S.K. Turitsyn, B. Jalali, Time stretch and its applications. Nat. Photonics **11**(6), 341–351 (2017)
18. M.H. Asghari, B. Jalali, Physics-inspired image edge detection, in *2014 IEEE Global Conference on Signal and Information Processing (GlobalSIP)* (IEEE, 2014), pp. 293–296
19. M.H. Asghari, B. Jalali, Edge detection in digital images using dispersive phase stretch transform. J. Biomed. Imag. **2015**, 6 (2015)
20. M. Suthar, A. Mahjoubfar, K. Seals, E.W. Lee, B. Jalali, Diagnostic tool for pneumothorax, in *2016 IEEE Photonics Society Summer Topical Meeting Series (SUM)* (IEEE, 2016), pp. 218–219
21. M. Suthar, Decision Support Systems for Radiologists based on Phase Stretch Transform (2016), http://escholarship.org/uc/item/39p0h9jp
22. C.V. Ilioudis, C. Clemente, M.H. Asghari, B. Jalali, J.J. Soraghan, Edge detection in SAR images using phase stretch transform, in *2nd IET International Conference on Intelligent Signal Processing 2015 (ISP)* (IET, 2015), pp. 1–5
23. T. Ilovitsh, B. Jalali, M.H. Asghari, Z. Zalevsky, Phase stretch transform for super-resolution localization microscopy. Biomed. Optics Express **7**(10), 4198–4209 (2016)
24. JalaliLabUCLA/Image-feature-detection-using-Phase-Stretch-Transform, https://github.com/JalaliLabUCLA/Image-feature-detection-using-Phase-Stretch-Transform/
25. B.S. Manjunath, C. Shekhar, R. Chellappa, A new approach to image feature detection with applications. Pattern Recognit. **29**(4), 627–640 (1996)
26. B. Zitova, J. Flusser, Image registration methods: a survey. Image Vision Comput. **21**(11), 977–1000 (2003)
27. Y. LeCun, Y. Bengio, G. Hinton, Deep learning. Nature **521**(7553), 436–444 (2015)
28. G. Medioni, R. Nevatia, Matching images using linear features. IEEE Trans. Pattern Anal. Mach. Intell. **6**, 675–685 (1984)
29. H. Stokman, T. Gevers, Selection and fusion of color models for image feature detection. IEEE Trans. Pattern Anal. Mach. Intell. **29**(3), 371–381 (2007)
30. J. Van de Weijer, T. Gevers, A.D. Bagdanov, Boosting color saliency in image feature detection. IEEE Trans. Pattern Anal. Mach. Intell. **28**(1), 150–156 (2006)
31. T. Lindeberg, Feature detection with automatic scale selection. Int. J. Comput. Vision **30**(2), 79–116 (1998)
32. M. Pavli, H. Belzner, G. Rigoll, S. Ili, Image based fog detection in vehicles, in *2012 IEEE Intelligent Vehicles Symposium (IV)* (IEEE, 2012), pp. 1132–1137
33. S. Sivaraman, M.M. Trivedi, Looking at vehicles on the road: a survey of vision-based vehicle detection, tracking, and behavior analysis. IEEE Trans. Intell. Trans. Syst. **14**(4), 1773–95 (2013)

34. B. Jalali, M. Suthar, M. Asghari, A. Mahjoubfar, Optics-inspired computing, in *Proceedings of the 5th International Conference on Photonics, Optics and Laser Technology—Volume 1: PHOTOPTICS* (2017), pp. 340–345. ISBN: 978-989-758-223-3. https://doi.org/10.5220/0006271703400345

35. M. Suthar, H. Asghari, B. Jalali, Feature enhancement in visually impaired images. IEEE Access **6**, 1407–1415 (2018)

36. M. Suthar, M. Asghari, B. Jalali, Feature enhancement in visually impaired images (2017). arXiv:1706.04671

37. Z. Hameed, C. Wang, Edge detection using histogram equalization and multi-filtering process, in *2011 IEEE International Symposium on Circuits and Systems (ISCAS)* (IEEE, 2011), pp. 1077–1080

38. J.J. Rankine, A.N. Thomas, D. Fluechter, Diagnosis of pneumothorax in critically ill adults. Postgrad. Med. J. **76**(897), 399–404 (2000)

Author Index

© Springer Nature Switzerland AG 2019
P. Ribeiro et al. (eds.), *Optics, Photonics and Laser Technology 2017*,
Springer Series in Optical Sciences 222,
https://doi.org/10.1007/978-3-030-12692-6

Subject Index

B
Beam combining, 215, 216, 221, 222, 227, 228
Binary optics, 62
Bistability, 99, 101–104, 112, 114–116

C
Chemical etching, 126, 135
Cholesteric liquid crystal, 75
CMOS process, 193
Coated microfiber knot resonator, 150, 152, 154, 155, 158
Collection Gate, 193, 194, 199, 202, 203, 209, 211

D
Diffraction gratings, 31–33, 46, 49, 50, 215

E
Electron transfer, 193, 194, 196, 198, 200, 201, 203, 206–211
Er-doped fibers, 165, 166, 168, 170, 171, 173–175, 178–180, 182–184, 188–190
Er-Yb fibers, 176, 179, 183

F
Fabry-Perot filter, 75, 87, 88, 92
Feature detection and resolution enhancement, 259
Feature extraction, 259, 273
Fiber lasers, 165, 166, 168, 170, 172–174, 176, 178, 179, 181–188, 190
Fibre ring laser, 233, 239
Fluorescence, 99, 105–115, 130–134, 234, 235
Fs-laser writing, 130, 135, 137–139

Functions, 4–6, 9, 10, 12, 21, 23, 26, 32–37, 40, 43–45, 55–61, 66, 67, 69, 75, 83, 92, 99–101, 105, 108, 112, 113, 115, 116, 120, 131, 133, 137, 149, 151, 160, 178, 185–189, 199, 206, 216, 223, 224, 248, 249, 256, 259, 261–268, 273
Fused silica, 46, 123, 124, 126–128, 130–132, 134, 136–139

H
High Dynamic Range (HDR), 261, 262, 265, 273
High-order-diffraction suppression, 31, 33, 37, 55–57, 62
High-power lasers, 216, 228
Hyper-dimensional classification, 255, 265, 273

I
Image analysis, 233, 234, 236, 257
Infrared, 43, 55, 57, 75, 76, 82, 86, 90, 93, 95, 125, 155, 175, 217
Injection-locked, 245

J
42.40.Jv Computer Generated Holograms, 2

K
42.30.Kq Fourier Optics, 2

L
Large pixel, 193
Laser tuning, 217
Liquid crystal, 3, 75, 76, 100, 149

© Springer Nature Switzerland AG 2019
P. Ribeiro et al. (eds.), *Optics, Photonics and Laser Technology 2017*,
Springer Series in Optical Sciences 222,
https://doi.org/10.1007/978-3-030-12692-6

Printed in the United States
By Bookmasters